Drug Delivery across Biological Barriers

Drug Delivery across Biological Barriers

Editors

Tatiana B. Tennikova
Evgenia G. Korzhikova-Vlakh

MDPI

Basel • Beijing • Wuhan • Barcelona • Belgrade • Novi Sad • Cluj • Manchester

Editors

Tatiana B. Tennikova
Institute of Chemistry
Saint-Petersburg State
University
St. Petersburg
Russia

Evgenia G. Korzhikova-Vlakh
Institute of Macromolecular
Compounds
Russian Academy of Sciences
St. Petersburg
Russia

Editorial Office
MDPI
St. Alban-Anlage 66
4052 Basel, Switzerland

This is a reprint of articles from the Special Issue published online in the open access journal *Pharmaceutics* (ISSN 1999-4923) (available at: www.mdpi.com/journal/pharmaceutics/special_issues/biological_barriers).

For citation purposes, cite each article independently as indicated on the article page online and as indicated below:

Lastname, A.A.; Lastname, B.B. Article Title. *Journal Name* **Year**, *Volume Number*, Page Range.

ISBN 978-3-0365-9041-7 (Hbk)
ISBN 978-3-0365-9040-0 (PDF)
doi.org/10.3390/books978-3-0365-9040-0

Contents

About the Editors

Tatiana B. Tennikova

Tatiana B. Tennikova is a Doctor of Chemical Sciences, a Professor, and an Honored Scientist of the Russian Federation. Professor Tennikova is one of the inventors of new-generation sorbents, specifically polymer monoliths. At present, the product under the trade name Convection Interaction Media (CIM) has been manufactured successfully and is distributed across the world by the company BIA Separations, Sartorius division (Slovenia–Germany). She is the Chair of the Department of Medicinal Chemistry at the Institute of chemistry, Saint-Petersburg State University. Professor Tennikova is a well-known expert in different scientific fields, including in the fields of adsorption and chromatography, polymer chemistry, bio-organic chemistry, medicinal chemistry, biomaterials, and analytical biotechnology. At present, her scientific interests focus on polymer systems for drug delivery, smart scaffolds for tissue engineering, and nanochips for medical diagnostics. Prof. Tennikova has co-authored over 200 scientific papers in peer-reviewed international journals (h-index of 30 and more than 3800 total citations in Scopus).

Evgenia G. Korzhikova-Vlakh

Evgenia G. Korzhikova-Vlakh studied chemistry at Saint-Petersburg State University (SPBU, Russia) and obtained her PhD from the Institute of Macromolecular Compounds, Russian Academy of Sciences (IMC RAS). Since 2012, she has been an Associate Professor at SPBU, and since 2016, she has been a leading scientist as the head of the Laboratory of Polymer Materials for Medicine and Biotechnology at IMC RAS. Her research focuses on the development and study of polymeric materials for bioanalysis, biocatalysis, solid-phase extraction, drug delivery, and regenerative medicine. Dr. Korzhikova-Vlakh has published more than 100 scientific papers in peer-reviewed international journals (h-index of 20 and 1400 total citations in Scopus).

Preface

This reprint presents a collection of original papers and reviews in the field of modern drug delivery systems and strategies to ensure efficient transport of therapeutic drugs in the body. Control of drug transport through physiological barriers, such as through the blood–brain barrier, mucosal and endothelial barriers, and cell membranes, is a very urgent task aimed at enhancing therapeutic efficacy. The reprint will be of interest and use to scientists working in the fields of pharmacy, medicine, and biomaterials, as well as students specializing in life sciences.

Tatiana B. Tennikova and Evgenia G. Korzhikova-Vlakh
Editors

pharmaceutics

MDPI

Review

Established and Emerging Strategies for Drug Delivery Across the Blood-Brain Barrier in Brain Cancer

Alessandro Parodi [1,*], Magdalena Rudzińska [1], Andrei A. Deviatkin [1], Surinder M. Soond [1], Alexey V. Baldin [1] and Andrey A. Zamyatnin Jr. [1,2,*]

1 Institute of Molecular Medicine, Sechenov First Moscow State Medical University, 119991 Moscow, Russia; magdda.rudzinska@gmail.com (M.R.); andreideviatkin@gmail.com (A.A.D.); surinder.soond@yandex.ru (S.M.S.); alexeyvbaldin@gmail.com (A.V.B.)

2 Belozersky Institute of Physico-Chemical Biology, Lomonosov Moscow State University, 119992 Moscow, Russia

* Correspondence: aparodi.sechenovuniversity@gmail.com (A.P.); zamyat@belozersky.msu.ru (A.A.Z.J.); Tel.: +7-495-609-1400 (ext. 3277) (A.P.); Tel.: +7-495-609-1400 (ext. 3028) (A.A.Z.J.)

Received: 4 April 2019; Accepted: 20 May 2019; Published: 24 May 2019

check for updates

Abstract: Brain tumors are characterized by very high mortality and, despite the continuous research on new pharmacological interventions, little therapeutic progress has been made. One of the main obstacles to improve current treatments is represented by the impermeability of the blood vessels residing within nervous tissue as well as of the new vascular net generating from the tumor, commonly referred to as blood-brain barrier (BBB) and blood-brain tumor barrier (BBTB), respectively. In this review, we focused on established and emerging strategies to overcome the blood-brain barrier to increase drug delivery for brain cancer. To date, there are three broad strategies being investigated to cross the brain vascular wall and they are conceived to breach, bypass, and negotiate the access to the nervous tissue. In this paper, we summarized these approaches highlighting their working mechanism and their potential impact on the quality of life of the patients as well as their current status of development.

Keywords: brain cancer; blood-brain barrier; drug delivery; FUS; CED; nanomedicine

---

1. Introduction

Tumors of the central nervous system (CNS) account for about 3% [1,2] of the worldwide diagnosed neoplastic diseases and represent one of the most frequent causes of solid tumor-related deaths in childhood [3]. More than 85% of the CNS tumors affect the brain, which is also a primary metastatic site for tumors originating in other organs including the bladder, breast, kidney, and lung [4]. Gliomas are the most common tumors of the brain, and they can originate from different cell phenotypes that constitute the glia (astrocytes, oligodendrocytes, microglia, ependymal cells). Further categorizations are based on cancer aggressiveness which is evaluated on a scale ranging from grade I to IV, with grade IV being the most malignant, challenging to treat and likely to reoccur. In this scenario, treatments vary from simple observation for grade I glioma (with 5–15 years median survival) to surgical resection in combination with radio and chemotherapy for grade IV glioma (with 9–12 months median survival). Resection is by far the most effective treatment at least in terms of mass tumor reduction, but it is limited by the structural complexity and the primary function of the brain. Tumor debulking is usually referred to as "maximal safe resection" [5], implying a high risk of cognitive loss following the surgical procedure and incomplete removal of the tumor. Surgical limitations contribute to the high incidence of brain cancer recurrence, usually detected within 2 cm from the primary tumor [6].

Glioblastoma multiforme (GBM) is the most common tumor of the brain in adults, representing about 50% of all diagnosed primary brain cancers and usually classified as a grade IV glioma [7]. GBM is characterized by cellular and molecular heterogeneity that makes the optimization of the pharmacological interventions very difficult. The Stupp protocol is the gold-standard treatment for GBM [8], and it consists of surgical resection, postoperative radiotherapy, and temozolomide (TMZ), often used in association with adjuvant therapies including carmustine and PCV (procarbazine, lomustine, and vincristine). Despite their significant cytostatic properties in vitro, many Food and Drug Administration approved chemotherapeutics have shown limited curative benefits in the clinic. In the case of brain tumors, the development of more effective treatments is hampered by the specialized barrier function that characterizes the blood vessels residing in the central nervous system and usually referred to as the blood-brain barrier (BBB). In its physiological function, the BBB thoroughly selects and controls the mass transport occurring in and out the brain, limiting the healthy (and tumor) tissue diffusion of the administered pharmaceuticals while increasing the therapeutic doses in the patients that do not respond to the treatments is rarely a viable option. Also, the new blood vessels originating from the neoplastic lesions and often referred to as blood-brain tumor barrier (BBTB) are significantly less permeable than the neovasculature of the tumors developing in other organs being that their development is driven by the nervous system microenvironment. Herein, we describe new clinical and experimental approaches that aim to disrupt, bypass and negotiate these vascular barriers to favor the accumulation of therapeutics in brain cancer tissue.

1.1. Anatomy of the BBB: Tight Junctions

The very first researcher that introduced the concept of BBB was Lena Stern [9], a pioneer in the neuroscience field that coined the term *hematoencephalic* barrier to describe the BBB. Other scientists worthy of mention for their contribution to the discovery of the BBB's functional and anatomical organization are Ehrlich, Lewandowsky, and Goldmann [10]. According to Sweeney et al. [11], the BBB is defined as "a continuous endothelial membrane within brain microvessels that has sealed cell-to-cell contacts and is sheathed by mural vascular cells and perivascular astrocyte end-feet." In the human, the BBB characterizes over 100 billion capillaries that cover a total length of around 400 miles and a surface area of 20 M^2 [12]. BBB vessels control the exchange of circulating molecules, nutrients and gas between the blood and the nervous tissue. In its physiological function, the BBB protects the brain from larger particles, proteins and hydrophilic molecules including potential neurotoxins and bacteria. It is believed that only 2% of small molecules and 0% of the large molecules can cross the BBB. Theoretically, only highly hydrophobic molecules with a molecular mass not higher than 400–500 Da can diffuse through this barrier [13]. BBB properties are due to many factors including (but not limited to) highly selective cellular sorting mechanisms regulating the transcellular traffic and the expression of tight junctions (TJs) between adjacent endothelial cells, limiting the paracellular transport.

TJs are composed of different transmembrane proteins including (but not limited to) the family of claudins, occludin, and junctional adhesion molecules (JAM-A, -B, and -C) and they interact with the cell cytoskeleton through membrane-associated guanylate kinases called zonula occludens proteins (ZO-1, ZO-2, and ZO-3). It is believed that all these proteins have a pivotal role in determining BBB function and a specific work performed on claudin-5 demonstrated that inhibiting its expression increased BBB permeability for molecules as large as 800 kDa [14]. This demonstration highlights the fine regulation that stands at the basis of BBB permeability, suggesting that TJ targeting could be a viable strategy to increase it. The efficiency of these proteins in closing the gaps between endothelial cells can be experimentally evaluated in vitro by measuring transendothelial electric resistance (TEER) that determines the resistance associated with ionic transport via the transcellular and the paracellular route. In the case of proper BBB reconstruction, TEER needs to be significantly higher (at least above 900 $\Omega \times cm^2$) than in other endothelial settings (2–20 $\Omega \times cm^2$). This value is considered the cut-off for the permeability of IgG, considering this under physiological conditions, TEER values range from 1500 to 8000 $\Omega \times cm^2$ [15,16]. However, these values can vary as a function of the animal origin and the quality of the endothelial cells (primary or immortalized cell lines) [16]. Usually, immortalized cell lines do not provide TEER values higher than 200 $\Omega \times cm^2$ while endothelial cells derived from inducible pluripotent stem cells can provide TEER values higher than 1500 $\Omega \times cm^2$. Recent discoveries highlighted the possibility that, despite their sealing action, these proteins could determine two distinct mechanisms of BBB crossing. The first is known as "charge pore pathway' in which the claudins form a molecular channel permeable only to small ions. The second is known as "size selective pathway" in which the passage to larger molecules occurs via a transient dissociation of TJ complexes [17]. A deeper understanding of these protein organizations could open new avenues of drug delivery as described later in the text.

1.2. Cellular and Enzymatic Elements of the Neurovascular Unit

The barrier function of the CNS endothelium is also determined by other cell phenotypes and biological structures including astrocytes, pericytes, microglia cells, neurons, and basement membranes which when taken with the endothelial cells, constitute what is commonly known as the neurovascular unit (Figure 1). Astrocytes are glial cells that interact with the endothelial cells through their polarized end-feet formations and control the BBB blood flow, development, and functions likely by enhancing the TJ expression in the mature BBB, even though they do not participate in its embryonic development [18,19]. In this context, some authors believe that astrocytes are not crucial for TJ expression, while others indicate that they can control TJ expression via Src-suppressed C-kinase substrates [20]. The modulation of BBB permeability occurs via secretion of important protein factors like the glial-derived neurotrophic factor, transforming growth factor-β1, basic fibroblast growth factor, interleukin 6, angiopoietin 1, retinoic acid, and Wnt [21,22]. Astrocytes also control the water exchange between intracellular, interstitial, vascular, and ventricular compartments by inducing the expression of the potassium channel kir4.1 and the water channel aquaporin-4. Pericytes have structural functions stabilizing the small BBB vessels and modulating the process of neovascularization and angiogenesis [23]. They are believed to significantly contribute to induce BBB gene expression as well as astrocyte end feet polarization, even though more investigations are needed to reveal the complete spectrum of their activities in determining BBB and BBTB characteristics [24]. They control endothelial cell proliferation, survival, differentiation [18], and induce TJ mRNA expression in the embryonic formation of the BBB [25]. Microglia cells are the resident macrophages of the brain and contribute to the barrier function by modulating the innate immunity in the perivascular regions of the brain [22] and participating in the regulation of the expression of the TJ components [26]. Finally, neurons can induce the expression of TJ proteins like occludin and this phenomenon occurs synergistically with astrocytes [27]. BBB permeability also depends on enzymatic and immunological barriers limiting the molecular diffusion of blood solutes in the brain parenchyma. The endothelial cells composing the BBB express efflux transporters that are very efficient in transporting back to the luminal side the small hydrophobic molecules that crossed the BBB [28]. Efflux carriers are mostly adenosine triphosphate-binding cassette (ABC) transporters [27], and they are fundamental in clearing brain tissue from small lipophilic molecules. Between them, the P-glycoprotein (P-gp) and breast cancer resistance protein (ABCG2) were shown to have a significant role in the efflux of xenobiotics that penetrated the endothelial cell membrane, limiting the diffusion of chemotherapeutics in the brain parenchyma. P-gp is the most investigated pump, and its impact on brain transport was shown in knockout mice, where brain delivery increased up to 10–100 times [29]. This efflux pump is responsible for hampering the diffusion of many chemotherapeutics including doxorubicin (DOX), daunorubicin, vinblastine, vincristine, etoposide, and teniposide [30]. Also, together with the absence of endothelial fenestration, CNS endothelial cells showed a higher negative surface charge [31] and a lower transport rate through pinocytosis [32]. These parameters are highly considered for the designing and the development of more efficient delivery approaches (see later) since they constitute the physical and biological features of the BBB.

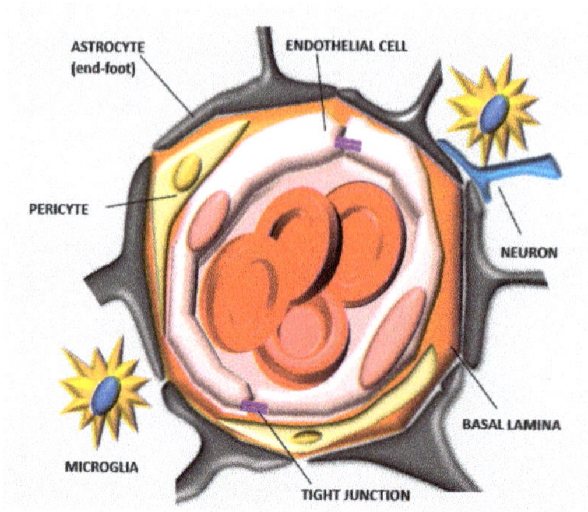

Figure 1. Anatomy of the neurovascular unit: the blood-brain barrier (BBB) structure is determined by different biological components that organize together in forming the neurovascular unit. Endothelial cells form the lumen of the capillary, interact with the basal lamina and the pericytes embedded in this matrix. The astrocytes, neurons, and microglia cells further support this cellular backbone. Other physical agents determining the barrier function of this specialized endothelium are the tight junctions (TJs) that are expressed between adjacent endothelial cells.

2. Models of BBB

One of the major obstacles in developing effective drug delivery across the BBB is the current lack of appropriate experimental in silico, in vitro and in vivo models allowing for cost-effective and high-throughput screening for different therapeutics. In silico models [33,34] of brain cancer are extensively developed for predicting tumor growth and infiltration in response to the treatments, while only a few cases are focused on predicting drug delivery in the brain neoplastic lesions [35,36]. The development of predictive computational models is critical in this field, also considering that mice have a brain structure extremely different from humans, counting for a 1:10 glial cell-to-neuron ratio versus a 1:5 ratio registered in humans [37]. Current in vitro and in vivo models are not reliable in mimicking and measuring BBB permeability respectively, but the research in this area is very active to discover new targets for favoring BBB accumulation as well as to understand the molecular dynamics that control TJ expression in the neurovascular unit. In the next sections, traditional and advanced methods to measure BBB function are described.

2.1. Traditional In Vitro Models of BBB

Three important parameters need to be consistent in establishing in vitro models of BBB: (1) low permeability validated through high TEER values, (2) expression of specific BBB biomarkers (i.e., TJ components and specific transporters and enzymes) [38], and (3) evaluation of barrier integrity through specific size molecular markers (sodium fluorescein, lucifer yellow, fluorescein isothiocyanate (FITC)-inulin, FITC-dextrans, and FITC- bovine serum albumin) [39]. In vitro models vary from simple acellular systems to very complex, multi-phenotype cellular models. Acellular models usually consist of parallel artificial membrane permeability assays (PAMPA) [40] and are based on synthetic lipophilic membranes that can only partially reproduce the physical properties of the BBB in vivo. These membranes are used to predict the passive diffusion of molecules through the barrier as a function of their hydrophobic or hydrophilic character. Few attempts to isolate brain capillaries and test BBB properties ex vivo have been performed, but the complexity of the isolation protocols, low reproducibility, and the difficulties to flow the tested molecules in the lumen of the isolated blood vessels affect their ordinary use [40,41]. On the other hand, new advances in cell isolation allowed for reconstructing the BBB with endothelial cells isolated from the brain, even though non-endothelial surrogate cellular models (i.e., Caco-2, ECV304) [42], that can still express TJs, are used for research purposes [40]. Many attempts at reconstructing the neurovascular unit were performed by co-culturing the endothelium with astrocytes, C6 glioma cells, pericytes, mixed glial cells, and conditioned media. Two-dimensional (2D) in vitro models are generated by seeding the endothelial cells on the apical side of a porous membrane while interacting with another cell phenotype (i.e., astrocyte or pericyte) seeded on the other side of the membrane via cellular protrusions extended through the pores. A third cell phenotype can be included in the system by seeding it on the bottom of the well to generate a conditioned culture environment and allowing for investigating the direct effect of cancer cells on endothelial cells forming the BBB [43]. The system can be further refined by coating the porous membrane with proteins belonging to the basal lamina and by decreasing serum concentration to favor the movement of the TJs from the cytoplasm to the basolateral region of the cells [44]. The serum can contain protein factors (i.e., vascular endothelial growth factor) that increase the permeability of the reconstructed endothelium in vitro, while supplementing the media with hydrocortisone or Adenosine 3′,5′-cyclic monophosphate (cAMP) analogs can increase endothelial barrier function since this second messenger is involved in maintaining the ultrastructure conformation of the TJs [44].

2.2. D Models and In Vivo Methods to Evaluate BBB Permeability

Three-dimensional (3D) models are currently one of the most advanced technologies to reconstitute in vitro the BBB, and are constituted of different cell phenotypes including cancer cells, normal astrocytes, and endothelial cells. The cells can assembly in spheroid units supported by hydrogels, scaffolds, and adhesion molecules. The group of Pasqualini developed 3D spheroids (1 mm in diameter) through magnetic levitation, by seeding glioma cells on a hydrogel composed by gold, magnetic iron oxide nanoparticles, and filamentous bacteriophage targeting cell integrins to favor cell interactions [45]. They showed that the spheroids could resemble in vitro the protein expression of tumor biomarkers (*N*-cadherin) registered in vivo and that multiple cell phenotypes could be mixed in the same spheroid unit to investigate cell interaction, biology, and drug diffusion while providing effective implantable tumors. As it occurs in vivo, a necrotic core characterized the spheroids and, by modulating the external magnetic field, it was possible to control their size and shape. Also known as organ-on-chip, new advances in microfluidic devices were utilisied to better recapitulate the characteristics of the BBB tissue by combining geometrical, physical, and biological features of this tissue [46,47]. These tools can also be implemented with sensors providing real-time and continuous measurements of the changes occuring in BBB permeability under different conditions. These systems usually consist of polydimethylsiloxane that provides optimal integration with microscopy analysis and fine-tuned engineering via soft lithography on the microscale, which supports the organized culturing of cellular layers derived from the nervous tissue (i.e., endothelial cells, neurons, and astrocytes). In addition, they can be integrated with channels in which the media flows and supports the growth of endothelial cells to mimick the characteristics of primary tissue [48,49]. The different compartments allow for intercellular interactions to establish the critical cues of cellular communications for generating a functional BBB in vitro. In this scenario, the generation of refined 3D models can represent a breakthrough in the development of more advanced tools to investigate the biology of the neurovascular unit since they can: (1) include multiple interacting cell phenotypes and (2) evaluate BBB in flow conditions. However, to date, these systems are too complex to be ordinarily used worldwide and drug screening is still mostly performed in traditional transwell systems. For more information about these systems, we suggest the following reviews [16,50].

In vivo pharmacokinetic evaluation in the brain depends on different biological parameters including blood flow in the BBB, the density of influx and efflux transporters as well as the affinity of the drug for these transporters. The goal of these measurements is to quantify the product between the amount of therapeutic that crossed the BBB and the surface area of the BBB [51]. In vitro pharmacokinetics methods are not considered reliable because drug passive diffusion is generally over-estimated, while the active transport is frequently underestimated [52]. Different advanced techniques allow for calculating drug accumulation in the brain parenchyma like ex-vivo equilibrium dialysis performed on brain homogenates or slices or by using dialysis fibers directly implanted in vivo. This second method is generally preferred when possible because it allows for measuring drug concentration in the brain in the presence of normal blood flow. Also known as brain microdialysis, this method consists of implanting a small capillary in the brain parenchyma under continuous perfusion (Figure 2). The tip of the capillary is semipermeable and allows for collecting tissue fluids. However, the insertion of the capillary in the brain parenchyma could damage the BBB continuity with consequent leakage of blood fluid leading to an overestimation of the drug concentration. Overall there are three significant challenges in increasing brain drug delivery: (1) targeting the vasculature of the brain, (2) overcoming the BBB, and (3) favoring drug diffusion in the brain diseased tissue. In the next chapters, available information about current strategies for crossing the BBB will be described with a focus on their working mechanisms as well as the pros and cons of the different methods.

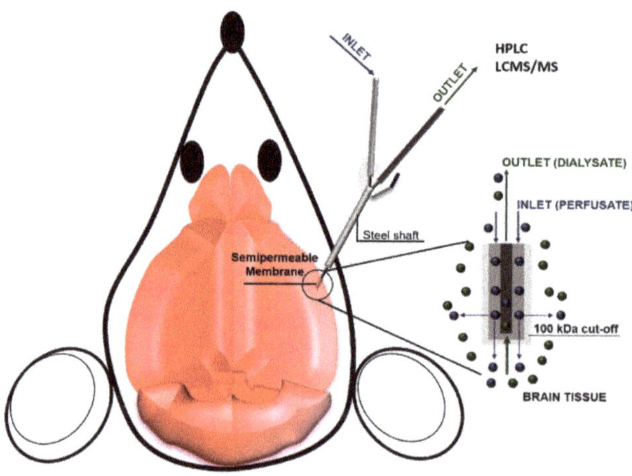

Figure 2. Scheme of brain microdialysis: A catheter is inserted in the brain tissue, while a controlled system (i.e., a syringe pump) injects in the brain a perfusate solution. At the end of the catheter is applied a semi-permeable membrane that allows for the injection of the perfusate, as well as for the collection of the dialysate composed by the perfusate and the brain tissue fluids. The collected dialysate can be eventually analyzed for its molecular content.

3. Breaching the BBB

Considering the importance of the brain and the physiological relevance of the BBB, barrier disruption by affecting TJ integrity and/or endothelial cell continuity has to be fine-tuned and reversible. These properties are fundamental because potential extravasation of circulating factors (i.e., albumin) can be very toxic for the neurons [53]. Traditional approaches to transiently affect BBB integrity are based on the injection of a hyperosmotic solution (usually consisting of a highly concentrated solution of mannitol [54]) just before the administration of the therapeutics. Hyperosmotic solutions can induce endothelial cell shrinking with a consequent increase in vascular leakage in the brain parenchyma. This approach was effective in increasing the overall survival of the patients (from 11 to 17 months), but it requires repeated hospitalization and is also considered very invasive (it needs patient sedation), unspecific, and accompanied by severe systemic toxicity, including neurological deficits, strokes, seizures, and new tumor-nodule formation [55]. Current clinical trials are devoted to optimizing the use of hyperosmotic solution based on mannitol [56] or NaCl [56] to increase chemotherapy and antibody delivery to the brain tumor and decrease intracranial pressure. Recently it was shown in rats that the osmotic disruption of the BBB (achieved via intracarotid injection of a 25%

solution of mannitol) could be exploited to increase the delivery of hydrophobic siRNA, previously modified with phosphocholine (PC)-docosahexanoic acid. The increase in the hydrophobicity of this biological therapeutic was shown to enhance the retention of the siRNA in the brain without affecting its therapeutic action. The group of Chung developed a polymeric carrier of polydixylitol with high osmotic power that showed high efficiency in nucleic acid delivery in vitro and in vivo. More importantly, they showed that the osmotic BBB opening could induce caveolae-mediated transcytosis of the carriers while having a low toxicity profile [57]. More advanced methods to breach the BBB are described in the following sections.

3.1. Focused Ultrasounds

Advanced options available to breach the BBB consist of physical mechanisms that can be remotely applied with low invasiveness. Ultrasounds were shown to be effective in increasing the BBB permeability and implants able to deliver fine-tuned acoustic pressures in the brain are currently under clinical trial evaluation [58]. Ultrasound therapeutic potentialities were known since the 1940s, but it was only recently that technical ameliorations to this technique in avoiding skull overheating and in improving energy transfer have made it extremely non-invasive and transformational [59]. The control of the applied forces is critical considering that current brain cancer treatments (i.e., TMZ) require multiple drug administrations. In this effort, the transfer of the acoustic energy can be compared to the use of a magnifying glass that converges high levels of light to ignite one small area (focal point), while outside this point the ultrasound can penetrate the tissue with no effect (Figure 3). Focused ultrasound (FUS) showed the most promising results when used in combination with microbubbles representing one of the most advanced ways to breach the BBB safely. As it is reversible, the effect lasts for a few hours [60,61] and the treatment can be targeted to some brain regions sparing healthy nervous tissue sites [59]. Microbubbles (5–10 micrometer in diameter) usually consist of lipids, polymers, surfactants, or proteins like albumin, and are loaded with gases like perfluorocarbon. These materials usually have amphipathic characteristics, and their stability depends on the formation of hydrophobic, covalent, and disulfide bonds, respectively [62]. Microbubbles are currently commercially available for diagnostic purposes in echocardiography (Optison® Definity®, and Sonovue® microbubbles) since when exposed to ultrasounds their vibration generates a strong echogenic signal based on the difference in acoustic impedance between the gas and the surrounding tissue. Today there are at least three clinical trials focused at testing FUS to improve the conditions of patients affected by GBM. To evaluate the safety of this approach, the group of Liu exposed rats to repetitive FUS at three different acoustic pressures defined as a function of erythrocyte extravasation and at different microbubble doses. The study concluded that high acoustic pressure and microbubble doses could cause brain hemorrhage, tissue necrosis, cell apoptosis, astroglial activation, and glial scarring [63]. However, moderate acoustic pressures and adequate microbubble administrations allowed for a safe breach of the BBB. This work is fundamental in the field because it demonstrates the need to finely tune the conditions of the system to avoid tissue damage [63]. In support of this data, repetitive FUS applications were applied on primates by using an implantable ultrasound device to evaluate the long term effects of this procedure. The system was used in combination with microbubbles, and the animals were exposed to ultrasounds for a total of seven times in four months. Magnetic Resonance Imaging (MRI) was used to assess the successful breaching of the BBB and positron emission tomography coupled with fluorine-18-labeled fluorodeoxyglucose detection did not show any changes in glucose metabolism. At the end of the experiment, the animals did not show any neurological distress and histology showed limited extravasation of red blood cells [64]. Different hypotheses on the mechanism at the base of BBB breaching were elaborated. High acoustic pressures induce inertial cavitation with the generation of heat and microbubble collapse allowing for a controlled microdamage of the brain vasculature (sonoporation) through the formation of heat, shock waves, and microjets. Experimental data are showing that BBB opening can occur at the level of the TJs and that this procedure likely increases paracellular permeability following endothelial spasm. On the other hand, relatively low acoustic pressures can generate stable cavitation of the microbubbles where the carriers oscillate in a nonlinear fashion without destruction. This phenomenon occurring in the proximity of the endothelial bed increases the flow dynamics of the liquid that surrounds the microparticles increasing shear stress forces (micro-streaming) [65] that can modulate the BBB ion channel and receptor activity [66] and induce caveolae-dependent transport [59,67] in brain parenchyma. FUS showed to increase the delivery of a plethora of anticancer therapeutics, including chemotherapeutics, antibodies, small interfering siRNA, and nanoparticles like superparamagnetic iron and gold nanocarriers as theranostic tools [7]. Ultrasound applications can also favor nanoparticle diffusion in the brain parenchyma by increasing the physiological porosity of the extracellular space, where the tissue architecture depends on electrostatic bonds connecting cells and extracellular matrix. The pioneering work of the group of Frenkel [68] demonstrated on ex-vivo brain slices that pulsed ultrasounds can create pores up to 500 nm in the perivascular space, potentially favoring therapeutics and nanoparticle diffusion after BBB crossing, even though in vivo experimentations are necessary to confirm this data. Also, it was shown that this approach could locally decrease the BBB expression of P-gp following a mechanism probably related to changes in blood flow [30]. Current limitations of FUS are represented by the short half-life of the microbubbles (average circulation time is estimated in 3–15 min) that are readily entrapped in the organs of the mononuclear phagocytic system due to their size in the microscale [69]. Other limitations are related to the need for coupling FUS with constant imaging monitoring to avoid major side effects and better target the tumor tissue.

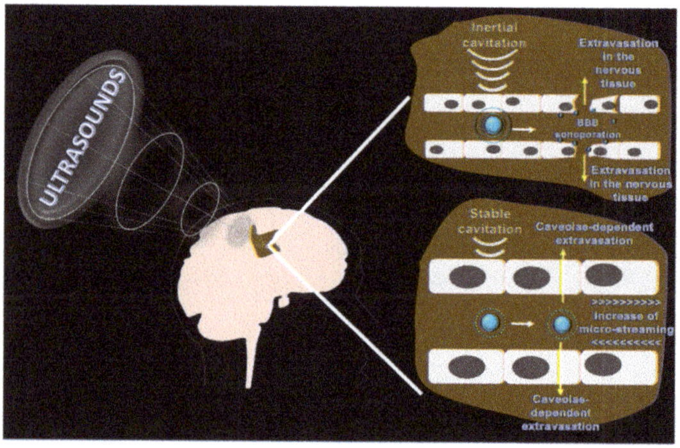

Figure 3. Effects of focused ultrasound on BBB permeability: External FUS is applied in combination with microbubbles injection. As a function of the acoustic pressure applied it can be obtained by: inertial cavitation and sonoporation, where relatively high levels of acoustic pressures induce the microparticles to collapse with a consequent controlled breaching of the BBB; and stable cavitation and micro-streaming where relatively low levels of acoustic pressure induce the microbubbles to vibrate increasing the flow shear stress in the proximity of the vascular wall with a consequent increase of the endothelial transport through caveolae.

3.2. Photodynamic Therapy

The photodynamic effect is another example of locally increased BBB permeability. This approach is based on light irradiation of photosensitive molecules (i.e., 5-aminolevulinic acid) and it was first developed for ameliorating current imaging and surgical techniques through intraoperative irradiation. Also in this case, the treatment can be focused to a minimal area of the brain and, compared to the healthy tissue, it was shown that some photosensitizers have a natural tropism for the neoplastic lesions [70]. Photodynamic irradiation can increase the delivery of large molecules and nanoparticles [33] while its invasiveness can be reduced through the generation of efficient optical clearing windows in the skull tissue to avoid the high light scattering generated by the bone. Tissue optical clearing windows are obtained by immersing the naked skull in hyperosmotic solutions before irradiation [71] or by implanting irradiation devices. A recent work of the group of Zhu demonstrated that this approach increases reactive oxygen species generation that in turn induce tight (CLDN-5 and ZO-1) and adherent (VE-cadherin) junction protein internalization with a consequent decrease of the barrier function. This evidence is fundamental because they confirm the reversibility of the BBB breaching through the rearrangement of the junctions while providing a working mechanism of this procedure.

4. Bypassing the BBB

There are essentially two extensively investigated pharmacological approaches that can be referred as to interstitial treatments for brain cancer: the application of biodegradable wafers and convection-enhanced delivery (CED), and both are designed to bypass the BBB. Generally, they are considered extremely invasive; however, both are already included in the clinical practice even though a lot of research is still dedicated to increasing their therapeutic benefits and their safety.

4.1. Biodegradable Wafers

Biodegradable wafers were designed to exploit the surgery step in which the tumor is removed. As aforementioned in the text, in the case of brain cancer, it is impossible to perform massive tissue debulking and the cavity that results from cancer tissue removal is usually still positive for the presence of cancer cells. In this scenario, therapeutic wafers are implanted in the area of the tumor bed, where they locally release chemotherapeutics killing the residual cancer tissue. Gliadel® wafers represent the gold standard therapeutic of this approach. Gliadel® consists of the copolymer polifeposan (1,3-bis(p-carboxyphenoxy) propane and sebacic acid, 4:1 molar ratio) [72] loaded with the alkylating agent carmustine (3.85% *w*/*w*). The wafers are disk-shaped of 14 mm in diameter and 1 mm in thickness, with an overall weight of 200 mg. A previous investigation in primates (without tumor or tissue removal) receiving the wafers demonstrated that carmustine can diffuse in the brain parenchyma between 2 and 8 mm from the implantation site [73]. These wafers are FDA approved in many countries and they are designed for being completely biodegradable and support a sustained release of at least five days after implantation [74], while the wafer is supposed to be completely degraded in 2–3 weeks. The disks are directly applied on the surface of the brain cavity where the tumor was removed and up to eight disks can be implanted (Figure 4). This number can sometimes offer limitations in the efficacy of Gliadel® wafers, because eight disks may not cover all the area of the tumor cavity. For this reason, the surgery needs the support of intraoperative frozen section diagnosis [75] to apply the disks in the areas of the cavity where more neoplastic tissue is detected. Different studies in the last decades demonstrated that, compared to the patients that received only placebo treatment, Gliadel® was efficient in increasing, by several weeks, the overall survival of the patients. In a typical clinical scenario, Gliadel® application is followed by the Stupp protocol and it is not indicated for patients with not resectable tumors or with tumors infiltrating the ventricular system, which still represent the majority of the cases. Major concerns about this treatment are related to the potential adverse effects that the wafers can generate (i.e., cognitive loss, seizures, poor wound healing, intracranial hypertension, cerebrospinal fluid leakage, and cyst formation) via the formation of carboxylic acids as a byproduct of the polyanhydride polymer. Additionally, the process of drug loading involves the presence of toxic residues (several hundred parts per million) like dichloromethane, acetone, tetrahydrofuran, and ethyl acetate. Edema, in particular, frequently occurs (from 6.8% to 25% of the cases [76]) in patients that received Gliadel® and it was speculated that the toxicity could derive from cell necrosis of the neoplastic tissue sensitive to carmustine. Usually, the edema is refractory to corticosteroids treatments, while a controlled use of bevacizumab (anti-VEGF-A treatment) just after the surgery was shown to reduce the risk of these side effects [77]. Infection is another documented potential adverse effect of the wafers even though a large study performed by the group of Chaichana, demonstrated that the previous presence of diabetes mellitus, multiple resections, and prolonged hospitalization are probably the main causes at the base of this phenomenon [78]. In addition, the group of Carpentier described a clinical case in which they analyzed not fully degraded wafers removed from the brain of a patient that underwent a second surgery due to a tumor recurrence [73]. The analysis demonstrated that incomplete degradation occurred at the level of the sebacic acid units (hydrophilic material), causing the scientists to speculate that a hydrophobic layer of unspecified biological material generated on the surface of the wafers reducing their degradation. In an attempt to define the patient population that more likely could benefit from Gliadel®, the group of Urbschat demonstrated that high expression of miRNA-181d is associated with low overall and progression-free survival of the patients [79], even though more investigations are necessary to understand the mechanism of this phenomenon. Current research focuses on generating implantable systems that can overcome Gliadel® limitations like extending this approach to more therapeutics and prolonging the drug release period. In this scenario, materials like biocompatible silk [80] and poly(lactic-co-glycolic acid) (PLGA) [81] represent optimal candidates to advance this kind of technology since they can be loaded with different therapeutics and their ultrastructure can be tuned to achieve the fine controlled release of different drugs [82].

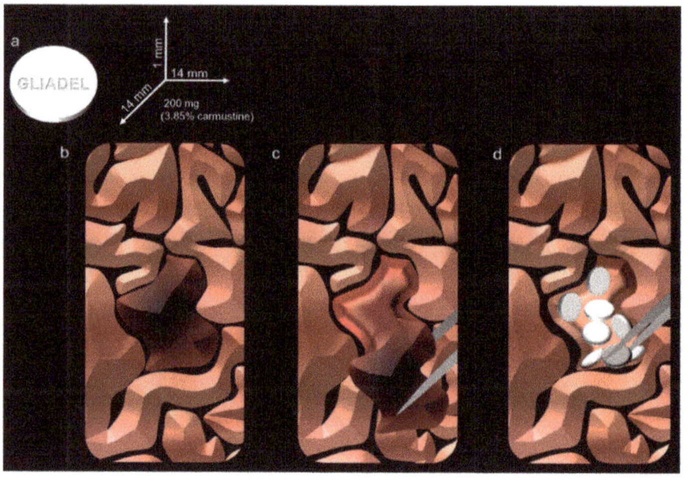

Figure 4. Gliadel wafers treatment: (**a**) physical and pharmaceutical features of Gliadel; (**b**) identification of the brain tumor, (**c**) maximal safe resection of the tumor tissue, (**d**) application of the Gliadel wafers in the tumor bed. Up to eight disks can be applied.

4.2. CED

CED allows for intra-tumoral local drug infusion via a catheter placed directly into the tumor parenchyma. Despite its invasiveness, CED was evaluated in clinical trials for the treatment of GBM and it demonstrated an adequate safety profile for several convection-delivered agents [83]. However, in vitro and in vivo experimentations unveiled that the convective flow can favor glioma and cancer stem cells invasion via the activation of the CXCR4-CXCL12 signaling pathway. The binding of chemokine CXCL12 to its receptor triggers multiple responses including increasing in intracellular calcium flux, gene transcription, chemotaxis, cell survival, and proliferation [84]. To prevent this side effect, it was recently shown that the co-administration of the CXCR4 antagonist AMD3100 can decrease this phenomenon [85]. Another limitation related to CED is the occurrence of backflow during the treatment, described as the fluid discharge around the catheter shaft with consequent leakage of the therapy out of the brain instead of into the nervous tissue. Backflow-free catheters are under evaluation to increase the flow rates at higher values to mitigate this phenomenon [84]. CED was investigated for the delivery of various agents, such as conventional chemotherapy [86], cytotoxin-ligand conjugates targeting cell surface receptors [87], antisense oligonucleotides [86,88], and nanovectors [86,89]. CED showed promising results when used to infuse brain-penetrating nanoparticles (BPNP) that resulted in significant tissue penetration thanks to their small size and their stability as monodispersed agents and allowed for drug release in a controlled fashion. BPNP are PLGA nanoparticles loaded with paclitaxel and modified with nile red to permit their imaging. The presence of polyethylene glycol (PEG) was shown to favor their diffusion into the tumor parenchyma highlighting the need to improve the design of the therapeutics that are locally injected to support their distribution in tumor tissue [90].

5. Negotiation of the BBB

New approaches of drug delivery aimed at negotiating the passage through the BBB have been proposed based on current knowledge of the transport mechanisms used by this specialized endothelium. Some of them exploit the physical properties of the BBB; others are based on the BBB biochemical receptor and transporter profile. To date, three main routes of BBB negotiation have been developed and referred to as adsorptive-mediated transcytosis (AMT), transporter-mediated transcytosis (TMT), and receptor-mediated transcytosis (RMT) (Figure 5). In this effort, the development of rationally designed nanocarrier surface modifications was shown to be useful to exploit these transport routes. Also, nanomedicine provided a mean to protect the encapsulated drug in the blood environment as well as to increase its bioavailability.

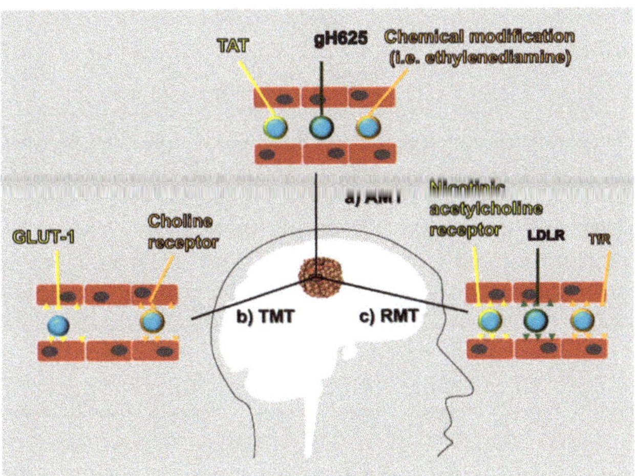

Figure 5. BBB negotiation: Current methods to negotiate BBB are obtained by modifying the therapeutic molecules or the carrier surface to increase their affinity for the BBB. They are generally referred to as: (**a**) adsorptive-mediated transcytosis (AMT) which is based on a positive surface charge of the therapeutics, (**b**) transporter-mediated transcytosis (TMT) which exploits the affinity of the therapeutics for endothelial transporters (i.e., GLUT1 and choline receptor), and (**c**) receptor-mediated transcytosis (RMT) which exploits the affinity of the therapeutics for endothelial receptors (i.e., nicotinic acetylcholine receptor, low-density lipoprotein receptor (LDLR), and transferrin receptor (TfR)).

5.1. AMT

The highly negative surface charge of the BBB surface can be exploited to deliver in the brain parenchyma molecules like cell-penetrating peptides (CPP) that are composed of positively charged and amphipathic amino acids. The transactivator of transcription (TAT) [91] derived from the human immunodeficiency virus and the peptide gH625 [92] derived from the glycoprotein H of herpes simplex virus type are typical examples of CPP used for brain delivery. CPP can be directly conjugated with a therapeutic or applied on the surface of nanocarriers to favor their transport across the BBB. Peptide design is often associated with activable structures that can be sensitive to the acidic pH or to the action of metalloproteases, typical features of tumor environment providing additional cues for targeting [92]. The surface of nanoparticles can also be chemically modified to be positive like in the case of albumin nanoparticles loaded with DOX [93]. Given these properties, AMT allowed for the successful delivery of many biologicals (including nucleic acids) and chemical therapeutics. In the pioneering work of the group of Sabel, three parameters were related to nanoparticle AMT, including size, surface charge, and hydrophobicity (by incorporating surfactants in the structure of the particles) were evaluated [94]. In this study, they used polybutylcyanoacrylate nanoparticles as a carrier model, evaluating their diffusion across an in vivo model of a blood-retina barrier. Contrary to other findings, demonstrating that size (indirectly proportional to nanoparticle ability to overcome BBB) and surface charge (positive particles can better cross BBB) are fundamental in designing nanocarriers for BBB, they showed that hydrophobicity was the key to overcome this endothelium. In particular, they demonstrated that non-ionic surfactants have a higher impact on BBB permeability than anionic ones since they favor the occurrence of an apolipoprotein E corona that was previously shown to increase BBB nanoparticles incorporation via RMT [95,96]. This work is important for the future development of drug delivery systems for the brain because it demonstrated that the surface modifications that occur in the blood milieu are the real key players in determining the brain accumulation of the carriers. However, targeting strategies based on AMT are considered very unspecific since they aim to target some physical features of the BBB that are present throughout all the vascular system, and their internalization can efficiently occur also in off-site organs with a consequent decrease in treatment efficacy and potential occurrence of side effects.

5.2. RMT and TMT

These approaches exploit membrane receptors that are generally over-expressed on the surface of the BBB. However, as highlighted by Warren [97], the BBB is not a static structure, as it can differentially modulate its permeability in physiologic and pathologic conditions. Experimental and clinical data are showing that in different areas of the brain, the BBB can be extremely variable in terms of permeability and expression of transporters, influx, and efflux pumps [98]. TMT and RMT have transformative potential in the generation of new drug delivery strategies for brain cancer because, despite their role in the molecular transport of the brain, they are generally overexpressed on the BBB, representing an optimal target for this tissue. Even though receptors and transporters have very different biological functions, their relevance in drug delivery is similar since they negotiate the internalization of the therapeutic (or the carrier) via interaction with specific ligands and for this reason they are dealt in the same section of this review. TMT is usually referred as the transport route allowing for the passage across BBB of small polar nutrients like sugars, vitamins, hormones, and amino acids [99]. In this effort, nanoparticles can be functionalized with mannose to overcome the BBB via GLUT1 [100] or with quaternary ammonium to exploit the choline transporter [101]. The group of Lu recently developed a method to generate a thin layer of polymeric acetylcholine and choline analog as a coating around individual proteins to favor their delivery in the brain parenchyma by exploiting choline transporters [102]. The coating is generated via a biodegradable crosslinker to facilitate the release of the proteins in the brain parenchyma. With this method, they showed the successful delivery of different proteins including the antineoplastic agent rituximab. On the other hand, RMT is referred to drug delivery strategies exploiting receptors that favor the transport across the BBB of larger proteins. To this category belong the transferrin receptor (TfR) [103,104], low-density lipoprotein receptor (LDLR) [105], diphtheria toxin receptor [106], and nicotinic acetylcholine receptor [107]. TfR has been extensively investigated in the field because it is overexpressed in glioma cells and the BBB, while it is not expressed in the blood vessels of other tissues [103]. In this context, TfR targeting can be functional both at the level of the BBB and at the level of the cancer cells which have undergone further extravasation to the brain parenchyma. The antibody for TfR OX26 was shown to favor BBB transcytosis of therapeutics [108] and nanoparticles [109], but recent evidence contradicted this data, highlighting the need of more research to understand this phenomenon. TfR targeting was recently used also to deliver theranostic agents [110] and to modify liposomal nanoparticles in combination with p-aminophenyl-α-D-manno-pyranoside, targeting GLUT-1 to achieve a double targeting of BBB and cancer cells [111]. RMT also represents a viable strategy to deliver antibodies across the BBB. The antibodies can be conjugated with therapeutics exploiting the RMT trafficking to favor drug delivery across the BBB [112]. To this end, many investigations were performed to understand the optimal antibody affinity towards the targeted receptor and in the case of TfR it was shown that a weaker binding improved antibody delivery by avoiding receptor dimerization and internalization to the endolysosomal compartment [113,114]. In this scenario, recent advances in the field have allowed for the generation of bispecific antibodies capable of recognizing two different targets, usually represented by a targeting receptor and a tumor molecular target [115]. Bispecific antibodies are under consideration also for redirecting T cell specificity towards cancer lesions, including malignant glioma [116]. Compared to TMT, RMT manages the transport of larger molecules, but it is fundamental to state that neither of these transport mechanisms have evolved to negotiate the passage of nanocarriers and there is no evidence to support that these receptors can physically mediate nanoparticle transcytosis. However, it was shown several times that their targeting can also increase nanoparticle trafficking to the abluminal side, therefore more investigation in this area is necessary to dissect the working mechanism of this kind of transport.

5.3. Opportunities in Targeting Endothelial Junctions

Targeting the proteins involved in the formation of adherens junctions can be useful to increase the transport across the BBB. These approaches are usually described as methods of BBB disruption, but we believe that they better fit in the BBB negotiation section because (1) the opening effect is relatively shorter and tunable compared to other methods of BBB disruption [117] and (2) they rely on specific biochemical interactions targeting junction stability like HAV6 and ADTC5 peptides designed to interact with cadherins [118,119]. The group of Siahaan demonstrated that it is necessary to take into consideration peptide stability in the biological serum as well as their structural rigidity to enhance the junction-peptide interaction. In particular, they showed that the use of cyclic versions of a peptide could provide better results than the linear amino acid sequences [117]. In the last two decades, peptidomics studies allowed for generating peptides interacting with claudin-1 (i.e., C1C2 [120]) or occludin [121] (i.e., OCC1 and OCC2). Interestingly it was shown that the destabilization of these proteins could favor their internalization and cellular recycling in combination with a parallel decrease of their mRNA expression [120]. Claudin and cadherin regulation can also be achieved via RhoA signaling activated by the A2A adenosine receptor [122]. Angubindin-1 is a peptide (200 amino acids) derived from the iota toxin of *Clostridium perfringens* and able to bind angulin-1 and -3, known to destabilize the proteins of the tight junctions [123]. After intravenous injection, it was shown to increase BBB permeability probably by increasing the size-selective pathway, and enhancing the delivery of antisense oligonucleotides with no toxic effects [17,123]. Recent evidence showed that the family of the lysophosphatidic acid receptors (LPARs) is overexpressed in the CNS microvasculature and more specifically LPAR1 is overexpressed in the brain vasculature [124]. Upon interaction with lysophosphatidic acid, this receptor can increase the permeability of the BBB possibly via RhoA activation. Intravenous treatments with lysophosphatidic acid can transiently (20 min post administration) increase BBB permeability for small and large molecules including Gd-DTPA, the infrared dye 800cwPEG, and Rhodamine 800 (often used to measure the activity of the efflux pump P-gp) [125,126]. More importantly lysophosphatidic acid was efficient in triplicate negative charged iron oxide nanoparticle (estimated size 30 nm) brain deposition favoring their potential use as theranostics [127]. Histamine [128] and bradykinin (BK) [129] and its more stable compounds [130] can modulate TJ expression via the beta receptor expressed on the endothelial cells while controlling the intracellular concentration of Ca^{2+}, in particular in the cells composing BBTB [131]. This ion is at the base of the TJ ultrastructure and when released in the cell cytoplasm can affect their tightness probably by increasing the expression of e-NOS and n-NOS with the parallel decrease in nuclear expression of the transcription factor ZONAB and decrease of the mRNA and protein expression of claudin-5 and occludin [131] (Figure 6). On the other hand, the zonula occludens toxin (ZOT) [132] was shown to affect TJ stability through a mechanism dependent on phospholipase C and protein kinase C. These enzymes can directly affect actin cytoskeleton reorganization and in turn, increase BBB permeability to chemotherapeutics like paclitaxel.

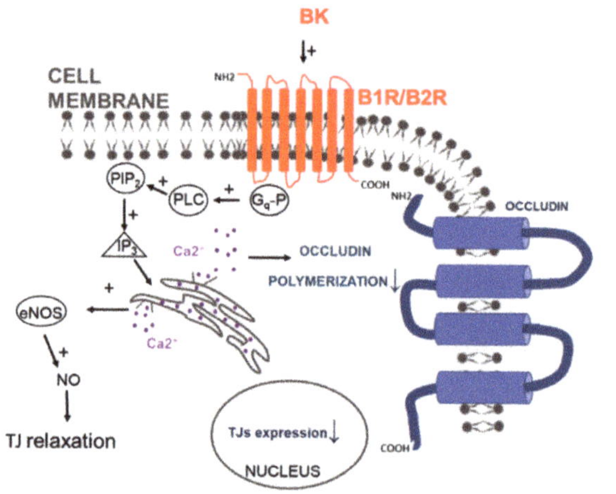

Figure 6. Bradykinin effect on TJs: bradykinin (BK), interacting with its receptors B1R and B2R, induces the intracellular release of Ca^{2+} that can affect occludin polymerization as well as increase the production of nitric oxide that relaxes the junctions. The activation of this pathway is also associated with a decrease in TJ mRNA expression.

6. Crossing Blood-Brain Tumor Barrier

Unlike the BBB, the BBTB has to be considered a pathological tissue since it is the product of the neoplastic lesion. Compared to regular BBB, BBTB is generally considered more permeable, even though as aforementioned, its barrier function (estimated cut-off of around 12 nm) [133] is significantly higher than what usually registered for the neo-vasculature generated from tumors in other organs. Even though the leaky behavior of BBTB can be appreciated also through regular MRI via brain edema detection, its dysfunction is not homogenous in the tumor tissue [28], and high functional variability was also appreciated between different patients. In the case of BBTB, the investigation of peculiar surface markers overexpressed in this tissue represent the best strategy to design carrier targeting, because it provides the opportunity to target the pathological tissue specifically. Despite the traditional targets described for BBB, BBTB can theoretically be targeted exploiting the typical surface biomarkers of growing blood vessels. For example, it was shown that targeting integrin $\alpha v\beta 3$ through the cyclic RGD peptide applied on the surface of polymeric polylactic acid and polyethylenimine particles [134] increased the brain delivery of encapsulated nucleic acids and paclitaxel, respectively, when compared to non-functionalized carriers. Recent findings also demonstrated that brain drug delivery could benefit from strategies aimed at normalizing pathological vasculature like administration of the Ang2-binding and Tie2-activating antibody [135]. More importantly, the group of Koh demonstrated that this approach could eventually enhance brain drug delivery by decreasing the interstitial pressure while increasing blood vessels perfusion and tissue oxygen levels modulating immune cell infiltration [135]. The group of Moses analyzed commercial GBM cell lines and 70 tumor samples from patients affected by GBM and identified the integrin $\alpha 2$ (ITAG2) as a novel surface biomarker for this BBTB. This integrin is involved in cell migration and the surface functionalization of DOX-loaded liposomes (with an antibody specific for this protein) showed cytostatic effects in vitro and in vivo, highlighting the importance of more research in the discovery of novel endothelial surface biomarkers for the treatment of brain tumors [136]. Compared to the surrounding healthy tissue, a brain tumor is characterized by significant changes in cell metabolism tissue that can represent an important targeting cue. Albumin, for example, is normally excluded from the brain parenchyma by the presence of the BBB, but it was shown that neoplastic lesions can increase its uptake likely to exploit this circulating protein as a source of amino acids. The group of Huang demonstrated that brain cancer overexpressed secreted protein acidic and rich in cysteine (SPARC) and GP-60, increasing the albumin endothelial transcytosis and cancer uptake, respectively (Figure 7). To target these receptors, they generated albumin nanoparticles (100 nm) encapsulated with paclitaxel and fenretinide and modified their surface with a CPP to favor particle diffusion in the brain parenchyma [3]. Finally, it is worth mentioning that brain tumors can generate new blood vessels via vascular mimicry, a phenomenon that can occur as a drug resistance mechanism upon the use of anti-angiogenic adjuvant therapies [137]. Both in human and in pre-clinical models, it was shown that the presence of red blood cells within vessel walls lined up with cancer cells and basal lamina. These cells were positive to periodic acid-Schiff but negative to CD34 immune staining, excluding their endothelial nature. In this case, further investigation is necessary to understand the advantages of targeting vascular mimicry and potential therapeutic effects of this approach.

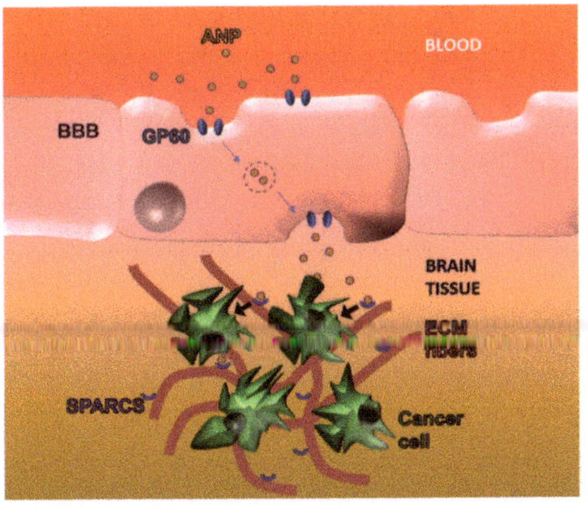

Figure 7. Exploiting tumor metabolic changes to overcome the BBB: Brain cancer lesions overexpress GP60 and secreted protein acidic and rich in cysteine (SPARC) that favor albumin nanoparticle trafficking over the BBB as well as cancer cell internalization in the abluminal side, respectively.

7. Cell and Gene Therapy

Approaches based on local delivery were used to inject healthy neural stem cells [138,139] to exploit their ability to infiltrate neoplastic lesions in the CNS. Genetically modified neural stem cells can be manipulated to generate and release cytotoxic molecules including prodrug-activating enzymes, apoptosis-inducing agents, antibodies [140], and oncolytic viruses [141]. The group of Portnow used neural stem cells modified for expressing cytosine deaminase to convert the prodrug 5-fluorocytosine (that can cross the BBB) to 5-fluorouracil. They directly injected the cells close to an established glioma or in the opposite hemisphere and they showed successful infiltration of the stem cells in the tumor parenchyma as well as higher cytostatic properties upon treatment with the prodrug [142]. Unfortunately, this procedure is affected by low efficiency in implanting viable cells. A way to avoid this issue is to seed the cells in vitro on a biocompatible scaffold (i.e., fibrin) and, like in the case of Gliadel, to insert the scaffold in the cavity obtained after brain tumor removal [143]. In this scenario, HEK 293 EBNA modified to release endostatin were encapsulated in an alginate scaffold prior to brain implantation, inhibiting in vivo GBM-induced angiogenesis process [144], while polymeric biodegradable scaffolds seeded with stem cells overexpressing secretable tumor necrosis factor apoptosis-inducing ligand were implanted to inhibit brain tumor growth [145]. Recent advances in biological drug delivery systems demonstrated that neutrophils could be exploited to overcome the BBB and increase drug delivery for brain cancer. The group Zhang loaded neutrophils in vitro with cationic liposomes and, after systemic administration, they infiltrated the neoplastic lesion guided by inflammatory cytokines and chemokines. The authors loaded the carriers with paclitaxel (which compared to other chemotherapeutics showed a minor impact on neutrophils biology) and exploited the cytokine gradient induced by the surgical removal of the tumor, exactly reproducing the clinical scenario [146]. Other biological agents used to treat brain cancer are adeno-associated viruses (AAV) since they are safe, effective, and one of the most promising methods to enhance gene delivery through the BBB [147,148]. The ability of different serotypes to effectively overcome the BBB is well known [149] even though the mechanism used to overcome BBB has still to be elucidated [150]. Engineering efforts have yielded several AAV variants that can efficiently transduce the CNS via systemic delivery in adult mice [151]. The group of Gao [149] tested nine different AAV vectors encoding green fluorescent protein (GFP) (injected into the superficial temporal vein of the mice) showing that they could increase GFP intensity in different brain compartments. Recently AAV targeted evolution technique revealed a novel recombinant AAV-PHP.B that transfers genes throughout the CNS with an efficiency that is at least 40-fold greater than that of the natural viruses [152]. Despite the optimistic perspectives regarding AAV-based delivery, some drawbacks need to be considered. The insert capacity of the vector is limited by 4 kb due to the AAV nature limiting some possible implementations of this method. In addition, the immune response to viruses can dramatically decrease the efficiency of gene transfer by systemic delivery. On the other hand, virus-induced expression of transgenes in the central nervous system can last for years [153], while the ectopic expression of the transgene can cause side effects. This limitation can be overridden by using cell type-specific promoters [151].

8. Summary

Brain cancer is characterized by an extremely fast and lethal outcome. While radiotherapy and surgery represent viable options to treat this condition, current pharmacological interventions are inadequate to increase patient survival as well as to improve their quality of life. Preclinical experience suggests that current chemotherapeutics are very effective on brain cancer growth, but unfortunately their delivery is affected by the BBB and the BTBB. Finding new pharmaceutical strategies to increase drug delivery in the brain is of primary importance considering that this condition is more frequent in the elderly and that by 2050, more than 1.7 billion people will be over 60 years old [2,154]. Current trends in drug delivery aim to bypass, breach, and negotiate the BBB. Between them, the use of hyperosmotic solutions [56], FUS [155], biodegradable implants [156], and CED [157] are extensively used and/or investigated in the clinic for their safety and efficacy. On the other hand, to our knowledge, only one model of targeted nanoparticle is under investigation in humans for the treatment of GBM [158]. While these strategies were shown to be effective, currently they can increase overall patient survival only by a few weeks. Future perspective needs to take in consideration the complexity of the BBB by developing improved models to study the biology of this tissue as well as the variability within brain cancer tissues [97], including the regional and country variation registered in the incidence of this disease [159]. These data induce to believe that the development of more effective therapeutic strategies should take into consideration the combination of different delivery approaches, possibly with the support of improved imaging techniques, whose role is fundamental to target the cancer tissue.

Funding: This work was supported by the Russian Academic Excellence Project 5-100.

Conflicts of Interest: The authors declare no conflicts of interest.

References

1. Mortazavi, S.; Mortazavi, S.; Paknahad, M. Cancers of the brain and CNS: Global patterns and trends in incidence. *J. Biomed. Phys. Eng.* **2018**, *8*, 151–152.

2. Organization, W.H. *Cancer Incidence and Mortality Worldwide: IARC*; World Health Organization (WHO): Geneva, Switzerland, 2015.
3. Lin, T.; Zhao, P.; Jiang, Y.; Tang, Y.; Jin, H.; Pan, Z.; He, H.; Yang, V.C.; Huang, Y. Blood–brain-barrier-penetrating albumin nanoparticles for biomimetic drug delivery via albumin-binding protein pathways for antiglioma therapy. *ACS Nano* **2016**, *10*, 9999–10012. [CrossRef]
4. Lin, X.; DeAngelis, L.M. Treatment of brain metastases. *J. Clin. Oncol.* **2015**, *33*, 3475–3484. [CrossRef] [PubMed]
5. Li, Y.M.; Suki, D.; Hess, K.; Sawaya, R. The influence of maximum safe resection of glioblastoma on survival in 1229 patients: Can we do better than gross-total resection? *J. Neurosurg.* **2016**, *124*, 977–988. [CrossRef] [PubMed]
6. Leuthardt, E.C.; Duan, C.; Kim, M.J.; Campian, J.L.; Kim, A.H.; Miller-Thomas, M.M.; Shimony, J.S.; Tran, D.D. Hyperthermic laser ablation of recurrent glioblastoma leads to temporary disruption of the peritumoral blood brain barrier. *PLoS ONE* **2016**, *11*, e0148613. [CrossRef]
7. Liu, H.-L.; Fan, C.-H.; Ting, C.-Y.; Yeh, C.-K. Combining microbubbles and ultrasound for drug delivery to brain tumors: Current progress and overview. *Theranostics* **2014**, *4*, 432–444. [CrossRef] [PubMed]
8. Stupp, R.; Mason, W.P.; Van Den Bent, M.J.; Weller, M.; Fisher, B.; Taphoorn, M.J.; Belanger, K.; Brandes, A.A.; Marosi, C.; Bogdahn, U. Radiotherapy plus concomitant and adjuvant temozolomide for glioblastoma. *New Engl. J. Med.* **2005**, *352*, 987–996. [CrossRef]
9. Igual, M.M. Lina Stern (1878–1968) and the blood-brain barrier. A life between Geneva and Moscow. *Neurosci. Hist.* **2017**, *5*, 94–104.
10. Shen, H.H. Core Concept: Circumventing the blood–brain barrier. *Proc. Natl. Acad. Sci. USA* **2017**, *114*, 11261–11263. [CrossRef]
11. Sweeney, M.D.; Sagare, A.P.; Zlokovic, B.V. Blood-brain barrier breakdown in Alzheimer disease and other neurodegenerative disorders. *Nat. Rev. Neurol.* **2018**, *14*, 133–150. [CrossRef] [PubMed]
12. Zlokovic, B.V. The blood-brain barrier in health and chronic neurodegenerative disorders. *Neuron* **2008**, *57*, 178–201. [CrossRef] [PubMed]
13. McDannold, N.; Zhang, Y.; Power, C.; Arvanitis, C.D.; Vykhodtseva, N.; Livingstone, M. Ultrasound-mediated blood-brain barrier disruption for targeted drug delivery in the central nervous system. In *Proceedings of Micro-and Nanotechnology Sensors, Systems, and Applications VII*; Baltimore Convention Center Baltimore: Baltimore, MD, USA, 2015; p. 94670H.
14. Nitta, T.; Hata, M.; Gotoh, S.; Seo, Y.; Sasaki, H.; Hashimoto, N.; Furuse, M.; Tsukita, S. Size-selective loosening of the blood-brain barrier in claudin-5–deficient mice. *J. Cell Biol.* **2003**, *161*, 653–660. [CrossRef] [PubMed]
15. Bicker, J.; Alves, G.; Fortuna, A.; Falcão, A. Blood–brain barrier models and their relevance for a successful development of CNS drug delivery systems: A review. *Eur. J. Pharm. Biopharm.* **2014**, *87*, 409–432. [CrossRef] [PubMed]
16. DeStefano, J.G.; Jamieson, J.J.; Linville, R.M.; Searson, P.C. Benchmarking in vitro tissue-engineered blood–brain barrier models. *Fluid. Barriers CNS* **2018**, *15*, 32. [CrossRef]
17. Zeniya, S.; Kuwahara, H.; Daizo, K.; Watari, A.; Kondoh, M.; Yoshida-Tanaka, K.; Kaburagi, H.; Asada, K.; Nagata, T.; Nagahama, M. Angubindin-1 opens the blood–brain barrier in vivo for delivery of antisense oligonucleotide to the central nervous system. *J. Control. Release* **2018**, *283*, 126–134. [CrossRef] [PubMed]
18. Persidsky, Y.; Ramirez, S.H.; Haorah, J.; Kanmogne, G.D. Blood–brain barrier: Structural components and function under physiologic and pathologic conditions. *J. Neuroimmu. Pharmacol.* **2006**, *1*, 223–236. [CrossRef]
19. Daneman, R.; Prat, A. The blood–brain barrier. *Cold Spring Harbor Perspect. Biol.* **2015**, *7*, a020412. [CrossRef]
20. Lee, S.-W.; Kim, W.J.; Choi, Y.K.; Song, H.S.; Son, M.J.; Gelman, I.H.; Kim, Y.-J.; Kim, K.-W. SSeCKS regulates angiogenesis and tight junction formation in blood-brain barrier. *Nat. Med.* **2003**, *9*, 900–906. [CrossRef]
21. Abbott, N.J.; Rönnbäck, L.; Hansson, E. Astrocyte–endothelial interactions at the blood–brain barrier. *Nat. Rev. Neurosci.* **2006**, *7*, 41. [CrossRef] [PubMed]
22. Correale, J.; Villa, A. Cellular elements of the blood-brain barrier. *Neurochem. Res.* **2009**, *34*, 2067. [CrossRef]
23. Trost, A.; Lange, S.; Schroedl, F.; Bruckner, D.; Motloch, K.A.; Bogner, B.; Kaser-Eichberger, A.; Strohmaier, C.; Runge, C.; Aigner, L. Brain and retinal pericytes: Origin, function and role. *Front. Cell. Neurosci.* **2016**, *10*, 20. [CrossRef] [PubMed]

24. Armulik, A.; Genové, G.; Mäe, M.; Nisancioglu, M.H.; Wallgard, E.; Niaudet, C.; He, L.; Norlin, J.; Lindblom, P.; Strittmatter, K. Pericytes regulate the blood–brain barrier. *Nature* **2010**, *468*, 557–561. [CrossRef] [PubMed]
25. Hori, S.; Ohtsuki, S.; Hosoya, K.i.; Nakashima, E.; Terasaki, T. A pericyte-derived angiopoietin-1 multimeric complex induces occludin gene expression in brain capillary endothelial cells through Tie-2 activation in vitro. *J. Neurochem.* **2004**, *89*, 503–513. [CrossRef]
26. Zenker, D.; Begley, D.; Bratzke, H.; Rübsamen-Waigmann, H.; von Briesen, H. Human blood-derived macrophages enhance barrier function of cultured primary bovine and human brain capillary endothelial cells. *J. Physiol.* **2003**, *551*, 1023–1032. [CrossRef] [PubMed]
27. Schiera, G.; Bono, E.; Raffa, M.P.; Gallo, A.; Pitarresi, G.L.; Di Liegro, I.; Savettieri, G. Synergistic effects of neurons and astrocytes on the differentiation of brain capillary endothelial cells in culture. *J. Cell. Mol. Med.* **2003**, *7*, 165–170. [CrossRef]
28. Oberoi, R.K.; Parrish, K.E.; Sio, T.T.; Mittapalli, R.K.; Elmquist, W.F.; Sarkaria, J.N. Strategies to improve delivery of anticancer drugs across the blood–brain barrier to treat glioblastoma. *Neuro-oncology* **2015**, *18*, 27–36. [CrossRef] [PubMed]
29. Chen, Z.; Shi, T.; Zhang, L.; Zhu, P.; Deng, M.; Huang, C.; Hu, T.; Jiang, L.; Li, J. Mammalian drug efflux transporters of the ATP binding cassette (ABC) family in multidrug resistance: A review of the past decade. *Cancer Lett.* **2016**, *370*, 153–164. [CrossRef] [PubMed]
30. Cho, H.; Lee, H.-Y.; Han, M.; Choi, J.-r.; Ahn, S.; Lee, T.; Chang, Y.; Park, J. Localized down-regulation of P-glycoprotein by focused ultrasound and microbubbles induced blood-brain barrier disruption in rat brain. *Sci. Rep.* **2016**, *6*, 31201. [CrossRef] [PubMed]
31. Ribeiro, M.; Domingues, M.; Freire, J.; Santos, N.; Castanho, M. Translocating the blood-brain barrier using electrostatics. *Front. Cell. Neurosci.* **2012**, *6*, 44. [CrossRef]
32. McGowan, J.W.; Bidwell, G.L., III; Vig, P.J. Challenges and new strategies for therapeutic peptide delivery to the CNS. *Ther. Deliv.* **2015**, *6*, 841–853. [CrossRef] [PubMed]
33. Liu, H.; Dong, K.; Zhang, W.; Summerfield, S.G.; Terstappen, G.C. Prediction of brain: Blood unbound concentration ratios in CNS drug discovery employing in silico and in vitro model systems. *Drug Discov. Today* **2018**, *23*, 1357–1372. [CrossRef] [PubMed]
34. Sharma, B.; Luhach, K.; Kulkarni, G. In vitro and in vivo models of BBB to evaluate brain targeting drug delivery. In *Brain Targeted Drug Delivery System*; Elsevier: Amsterdam, Netherlands, 2019; pp. 53–101.
35. Arvanitis, C.D.; Askoxylakis, V.; Guo, Y.; Datta, M.; Kloepper, J.; Ferraro, G.B.; Bernabeu, M.O.; Fukumura, D.; McDannold, N.; Jain, R.K. Mechanisms of enhanced drug delivery in brain metastases with focused ultrasound-induced blood–tumor barrier disruption. *Proc. Natl. Acad. Sci. USA* **2018**, *115*, E8717–E8726. [CrossRef] [PubMed]
36. Boujelben, A.; Watson, M.; McDougall, S.; Yen, Y.-F.; Gerstner, E.R.; Catana, C.; Deisboeck, T.; Batchelor, T.T.; Boas, D.; Rosen, B. Multimodality imaging and mathematical modelling of drug delivery to glioblastomas. *Interface Focus* **2016**, *6*, 20160039. [CrossRef]
37. Begley, D.J. Brain superhighways. *Sci. Transl. Med.* **2012**, *4*, 147fs129. [CrossRef]
38. Kuo, Y.-C.; Lu, C.-H. Effect of human astrocytes on the characteristics of human brain-microvascular endothelial cells in the blood–brain barrier. *Colloids Surf. B Biointerfaces* **2011**, *86*, 225–231. [CrossRef] [PubMed]
39. Patabendige, A.; Skinner, R.A.; Abbott, N.J. Establishment of a simplified in vitro porcine blood–brain barrier model with high transendothelial electrical resistance. *Brain Res.* **2013**, *1521*, 1–15. [CrossRef]
40. Mensch, J.; Jaroskova, L.; Sanderson, W.; Melis, A.; Mackie, C.; Verreck, G.; Brewster, M.E.; Augustijns, P. Application of PAMPA-models to predict BBB permeability including efflux ratio, plasma protein binding and physicochemical parameters. *Int. J. Pharm.* **2010**, *395*, 182–197. [CrossRef]
41. Burek, M.; Salvador, E.; Förster, C.Y. Tissue-based in vitro and ex vivo models for blood–brain barrier permeability studies. In *Concepts and Models for Drug Permeability Studies*; Elsevier: Amsterdam, Netherlands, 2016; pp. 343–356.
42. Kaisar, M.A.; Abhyankar, V.V.; Cucullo, L. In Vitro BBB Models: Working with Static Platforms and Microfluidic Systems. In *Blood-Brain Barrier*; Springer: New York, NY, USA, 2019; pp. 55–70.
43. Mendes, B.; Marques, C.; Carvalho, I.; Costa, P.; Martins, S.; Ferreira, D.; Sarmento, B. Influence of glioma cells on a new co-culture in vitro blood–brain barrier model for characterization and validation of permeability. *Int. J. Pharm.* **2015**, *490*, 94–101. [CrossRef]

44. Brown, R.C.; Morris, A.P.; O'Neil, R.G. Tight junction protein expression and barrier properties of immortalized mouse brain microvessel endothelial cells. *Brain Res.* **2007**, *1130*, 17–30. [CrossRef]
45. Souza, G.R.; Molina, J.R.; Raphael, R.M.; Ozawa, M.G.; Stark, D.J.; Levin, C.S.; Bronk, L.F.; Ananta, J.S.; Mandelin, J.; Georgescu, M.-M. Three-dimensional tissue culture based on magnetic cell levitation. *Nat. Nanotechnol.* **2010**, *5*, 291–296. [CrossRef]
46. Brown, J.A.; Codreanu, S.G.; Shi, M.; Sherrod, S.D.; Markov, D.A.; Neely, M.D.; Britt, C.M.; Hoilett, O.S.; Reiserer, R.S.; Samson, P.C. Metabolic consequences of inflammatory disruption of the blood-brain barrier in an organ-on-chip model of the human neurovascular unit. *J. Neuroinflamm.* **2016**, *13*, 306. [CrossRef]
47. Maoz, B.M.; Herland, A.; FitzGerald, E.A.; Grevesse, T.; Vidoudez, C.; Pacheco, A.R.; Sheehy, S.P.; Park, T.-E.; Dauth, S.; Mannix, R. A linked organ-on-chip model of the human neurovascular unit reveals the metabolic coupling of endothelial and neuronal cells. *Nat. Biotechnol.* **2018**, *36*, 865. [CrossRef]
48. Campisi, M.; Shin, Y.; Osaki, T.; Hajal, C.; Chiono, V.; Kamm, R.D. 3D self-organized microvascular model of the human blood-brain barrier with endothelial cells, pericytes and astrocytes. *Biomaterials* **2018**, *180*, 117–129. [CrossRef] [PubMed]
49. Adriani, G.; Ma, D.; Pavesi, A.; Goh, E.; Kamm, R. Modeling the blood-brain barrier in a 3D triple co-culture microfluidic system. In Proceedings of the 2015 37th Annual International Conference of the IEEE Engineering in Medicine and Biology Society (EMBC), Milan, Italy, 25–29 August 2015; pp. 338–341.
50. Van Der Helm, M.W.; Van Der Meer, A.D.; Eijkel, J.C.; van den Berg, A.; Segerink, L.I. Microfluidic organ-on-chip technology for blood-brain barrier research. *Tissue Barriers* **2016**, *4*, e1142493. [CrossRef]
51. Bickel, U. How to measure drug transport across the blood-brain barrier. *NeuroRx* **2005**, *2*, 15–26. [CrossRef] [PubMed]
52. Pardridge, W.M. CSF, blood-brain barrier, and brain drug delivery. *Expert Opin. Drug Deliv.* **2016**, *13*, 963–975. [CrossRef]
53. Liu, Z.; Liu, J.; Wang, S.; Liu, S.; Zhao, Y. Neuronal uptake of serum albumin is associated with neuron damage during the development of epilepsy. *Exp. Ther. Med.* **2016**, *12*, 695–701. [CrossRef]
54. Choi, C.; Kim, H.M.; Shon, J.; Park, J.; Kim, H.-T.; Oh, S.-H.; Kim, N.K.; Kim, O.J. Additional increased effects of mannitol-temozolomide combined treatment on blood-brain barrier permeability. *Biochem. Biophys. Res. Commun.* **2018**, *497*, 769–775. [CrossRef] [PubMed]
55. Rodriguez, A.; Tatter, S.; Debinski, W. Neurosurgical techniques for disruption of the blood–brain barrier for glioblastoma treatment. *Pharmaceutics* **2015**, *7*, 175–187. [CrossRef]
56. NIH. Clinicaltrials.gov. Available online: https://clinicaltrials.gov/ct2/results?cond=brain+tumor&term=mannitol&cntry=&state=&city=&dist= (accessed on 20 April 2019).
57. Garg, P.; Pandey, S.; Seonwoo, H.; Yeom, S.; Choung, Y.-H.; Cho, C.-S.; Choung, P.-H.; Chung, J.H. Hyperosmotic polydixylitol for crossing the blood brain barrier and efficient nucleic acid delivery. *Chem. Commun.* **2015**, *51*, 3645–3648. [CrossRef]
58. Carpentier, A.; Canney, M.; Vignot, A.; Reina, V.; Beccaria, K.; Horodyckid, C.; Karachi, C.; Leclercq, D.; Lafon, C.; Chapelon, J.-Y. Clinical trial of blood-brain barrier disruption by pulsed ultrasound. *Sci. Transl. Med.* **2016**, *8*, re342–re343. [CrossRef]
59. Aryal, M.; Arvanitis, C.D.; Alexander, P.M.; McDannold, N. Ultrasound-mediated blood–brain barrier disruption for targeted drug delivery in the central nervous system. *Adv. Drug Deliv. Rev.* **2014**, *72*, 94–109. [CrossRef]
60. Marty, B.; Larrat, B.; Van Landeghem, M.; Robic, C.; Robert, P.; Port, M.; Le Bihan, D.; Pernot, M.; Tanter, M.; Lethimonnier, F. Dynamic study of blood–brain barrier closure after its disruption using ultrasound: A quantitative analysis. *J. Cereb. Blood Flow Metab.* **2012**, *32*, 1948–1958. [CrossRef]
61. Chu, P.-C.; Chai, W.-Y.; Tsai, C.-H.; Kang, S.-T.; Yeh, C.-K.; Liu, H.-L. Focused ultrasound-induced blood-brain barrier opening: Association with mechanical index and cavitation index analyzed by dynamic contrast-enhanced magnetic-resonance imaging. *Sci. Rep.* **2016**, *6*, 33264. [CrossRef]
62. Sirsi, S.; Borden, M. Microbubble compositions, properties and biomedical applications. *Bubble Sci. Eng. Technol.* **2009**, *1*, 3–17. [CrossRef]
63. Tsai, H.-C.; Tsai, C.-H.; Chen, W.-S.; Inserra, C.; Wei, K.-C.; Liu, H.-L. Safety evaluation of frequent application of microbubble-enhanced focused ultrasound blood-brain-barrier opening. *Sci. Rep.* **2018**, *8*, 17720. [CrossRef]

64. Horodyckid, C.; Canney, M.; Vignot, A.; Boisgard, R.; Drier, A.; Huberfeld, G.; François, C.; Prigent, A.; Santin, M.D.; Adam, C. Safe long-term repeated disruption of the blood-brain barrier using an implantable ultrasound device: A multiparametric study in a primate model. *J. Neurosurg.* **2017**, *126*, 1351–1361. [CrossRef]
65. Miller, D.L. Particle gathering and microstreaming near ultrasonically activated gas-filled micropores. *J. Acoust. Soc. Am.* **1988**, *84*, 1378–1387. [CrossRef]
66. Sboros, V. Response of contrast agents to ultrasound. *Adv. Drug Deliv. Rev.* **2008**, *60*, 1117–1136. [CrossRef]
67. Xia, C.-y.; Liu, Y.-h.; Wang, P.; Xue, Y.-x. Low-frequency ultrasound irradiation increases blood–tumor barrier permeability by transcellular pathway in a rat glioma model. *J. Mol. Neurosci.* **2012**, *48*, 281–290. [CrossRef]
68. Hersh, D.S.; Nguyen, B.A.; Dancy, J.G.; Adapa, A.R.; Winkles, J.A.; Woodworth, G.F.; Kim, A.J.; Frenkel, V. Pulsed ultrasound expands the extracellular and perivascular spaces of the brain. *Brain Res.* **2016**, *1646*, 543–550. [CrossRef] [PubMed]
69. Willmann, J.r.K.; Cheng, Z.; Davis, C.; Lutz, A.M.; Schipper, M.L.; Nielsen, C.H.; Gambhir, S.S. Targeted microbubbles for imaging tumor angiogenesis: Assessment of whole-body biodistribution with dynamic micro-PET in mice. *Radiology* **2008**, *249*, 212–219. [CrossRef]
70. Castano, A.P.; Demidova, T.N.; Hamblin, M.R. Mechanisms in photodynamic therapy: Part one—photosensitizers, photochemistry and cellular localization. *Photodiagn. Photodyn. Ther.* **2004**, *1*, 279–293. [CrossRef]
71. Yang, X.; Zhang, Y.; Zhao, K.; Zhao, Y.; Liu, Y.; Gong, H.; Luo, Q.; Zhu, D. Skull optical clearing solution for enhancing ultrasonic and photoacoustic imaging. *IEEE Trans. Med. Imaging* **2016**, *35*, 1903–1906. [CrossRef]
72. McGinity, M.; Floyd, J.R.; McGinity, J.; Zhang, F. Implant compositions for the unidirectional delivery of drugs to the brain. *Drug Dev. Ind. Pharm.* **2017**, *43*, 1421–1429. [CrossRef] [PubMed]
73. Bourdillon, P.; Boissenot, T.; Goldwirt, L.; Nicolas, J.; Apra, C.; Carpentier, A. Incomplete copolymer degradation of in situ chemotherapy. *J. Mater. Sci. Mater. Med.* **2018**, *29*, 25. [CrossRef]
74. Perry, J.; Chambers, A.; Spithoff, K.; Laperriere, N. Gliadel wafers in the treatment of malignant glioma: A systematic review. *Curr. Oncol.* **2007**, *14*, 189–194. [CrossRef]
75. Asano, K.; Kurose, A.; Kamataki, A.; Kato, N.; Ogawa, K.; Katayama, K.; Kakuta, K.; Fumoto, T.; Ohkuma, H. Importance and accuracy of intraoperative frozen section diagnosis of the resection margin for effective carmustine wafer implantation. *Brain Tumor Pathol.* **2018**, *35*, 131–140. [CrossRef]
76. Murai, S.; Ichikawa, T.; Kurozumi, K.; Shimazu, Y.; Oka, T.; Otani, Y.; Shimizu, T.; Date, I. Quantitative analysis of brain edema in patients with malignant glioma treated with BCNU wafers. *J. Clin. Neurosci.* **2016**, *33*, 148–153. [CrossRef]
77. Kuramitsu, S.; Motomura, K.; Natsume, A.; Wakabayashi, T. Double-edged sword in the placement of Carmustine (BCNU) wafers along the eloquent area: A case report. *NMC Case Rep. J.* **2015**, *2*, 40–45. [CrossRef]
78. Chaichana, K.L.; Kone, L.; Bettegowda, C.; Weingart, J.D.; Olivi, A.; Lim, M.; Quinones-Hinojosa, A.; Gallia, G.L.; Brem, H. Risk of surgical site infection in 401 consecutive patients with glioblastoma with and without carmustine wafer implantation. *Neurol. Res.* **2015**, *37*, 717–726. [CrossRef] [PubMed]
79. Sippl, C.; Ketter, R.; Bohr, L.; Kim, Y.J.; List, M.; Oertel, J.; Urbschat, S. MiRNA-181d Expression Significantly Affects Treatment Responses to Carmustine Wafer Implantation. *Neurosurgery* **2018**. [CrossRef]
80. Yavuz, B.; Zeki, J.; Coburn, J.M.; Ikegaki, N.; Levitin, D.; Kaplan, D.L.; Chiu, B. In vitro and in vivo evaluation of etoposide-silk wafers for neuroblastoma treatment. *J. Control. Release* **2018**, *285*, 162–171. [CrossRef] [PubMed]
81. Zembko, I.; Ahmed, I.; Farooq, A.; Dail, J.; Tawari, P.; Wang, W.; Mcconville, C. Development of disulfiram-loaded poly (lactic-co-glycolic acid) wafers for the localised treatment of glioblastoma multiforme: A comparison of manufacturing techniques. *J. Pharm. Sci.* **2015**, *104*, 1076–1086. [CrossRef] [PubMed]
82. Tseng, Y.-Y.; Yang, T.-C.; Wang, Y.-C.; Lee, W.-H.; Chang, T.-M.; Kau, Y.-C.; Liu, S.-J. Targeted concurrent and sequential delivery of chemotherapeutic and antiangiogenic agents to the brain by using drug-loaded nanofibrous membranes. *Int. J. Nanomed.* **2017**, *12*, 1265–1276. [CrossRef]
83. Mehta, A.I.; Linninger, A.; Lesniak, M.S.; Engelhard, H.H. Current status of intratumoral therapy for glioblastoma. *J. Neuro-Oncol.* **2015**, *125*, 1–7. [CrossRef]
84. Teicher, B.A.; Fricker, S.P. CXCL12 (SDF-1)/CXCR4 pathway in cancer. *Clin. Cancer Res.* **2010**, *16*, 2927–2931. [CrossRef]
85. Cornelison, R.C.; Brennan, C.E.; Kingsmore, K.M.; Munson, J.M. Convective forces increase CXCR4-dependent glioblastoma cell invasion in GL261 murine model. *Sci. Rep.* **2018**, *8*, 17057. [CrossRef]

86. Hayward, S.L.; Wilson, C.L.; Kidambi, S. Hyaluronic acid-conjugated liposome nanoparticles for targeted delivery to CD44 overexpressing glioblastoma cells. *Oncotarget* **2016**, *7*, 34158–34171. [CrossRef]
87. Tripathy, K.; Das, B.; Singh, A.K.; Misra, A.; Misra, S.; Misra, S.S. Prognostic significance of epidermal growth factor receptor in patients of glioblastoma multiforme. *J. Clin. Diagn. Res. JCDR* **2017**, *11*, EC05. [CrossRef]
88. Kim, D.G.; Kim, K.H.; Seo, Y.J.; Yang, H.; Marcusson, E.G.; Son, E.; Lee, K.; Sa, J.K.; Lee, H.W.; Nam, D.-H. Anti-miR delivery strategies to bypass the blood-brain barrier in glioblastoma therapy. *Oncotarget* **2016**, *7*, 29400–29411. [CrossRef]
89. Saucier-Sawyer, J.K.; Seo, Y.-E.; Gaudin, A.; Quijano, E.; Song, E.; Sawyer, A.J.; Deng, Y.; Huttner, A.; Saltzman, W.M. Distribution of polymer nanoparticles by convection-enhanced delivery to brain tumors. *J. Control. Release* **2016**, *232*, 103–112. [CrossRef] [PubMed]
90. Zhang, C.; Nance, E.A.; Mastorakos, P.; Chisholm, J.; Berry, S.; Eberhart, C.; Tyler, B.; Brem, H.; Suk, J.S.; Hanes, J. Convection enhanced delivery of cisplatin-loaded brain penetrating nanoparticles cures malignant glioma in rats. *J. Control. Release* **2017**, *263*, 112–119. [CrossRef] [PubMed]
91. Zhu, Y.; Jiang, Y.; Meng, F.; Deng, C.; Cheng, R.; Zhang, J.; Feijen, J.; Zhong, Z. Highly efficacious and specific anti-glioma chemotherapy by tandem nanomicelles co-functionalized with brain tumor-targeting and cell-penetrating peptides. *J. Control. Release* **2018**, *278*, 1–8. [CrossRef]
92. Valiante, S.; Falanga, A.; Cigliano, L.; Iachetta, G.; Busiello, R.A.; La Marca, V.; Galdiero, M.; Lombardi, A.; Galdiero, S. Peptide gh625 enters into neuron and astrocyte cell lines and crosses the blood–brain barrier in rats. *Int. J. Nanomed.* **2015**, *10*, 1885–1898.
93. Byeon, H.J.; Lee, S.; Min, S.Y.; Lee, E.S.; Shin, B.S.; Choi, H.-G.; Youn, Y.S. Doxorubicin-loaded nanoparticles consisted of cationic-and mannose-modified-albumins for dual-targeting in brain tumors. *J. Control. Release* **2016**, *225*, 301–313. [CrossRef]
94. Voigt, N.; Henrich-Noack, P.; Kockentiedt, S.; Hintz, W.; Tomas, J.; Sabel, B.A. Surfactants, not size or zeta-potential influence blood–brain barrier passage of polymeric nanoparticles. *Eur. J. Pharm. Biopharm.* **2014**, *87*, 19–29. [CrossRef]
95. Kreuter, J.; Shamenkov, D.; Petrov, V.; Ramge, P.; Cychutek, K.; Koch-Brandt, C.; Alyautdin, R. Apolipoprotein-mediated transport of nanoparticle-bound drugs across the blood–brain barrier. *J. Drug Target.* **2002**, *10*, 317–325. [CrossRef]
96. Hülsermann, U.; Hoffmann, M.M.; Massing, U.; Fricker, G. Uptake of apolipoprotein E fragment coupled liposomes by cultured brain microvessel endothelial cells and intact brain capillaries. *J. Drug Target.* **2009**, *17*, 610–618. [CrossRef]
97. Warren, K.E. Beyond the Blood: Brain Barrier: The Importance of CNS Pharmacokinetics for the Treatment of CNS Tumors, including DIPG. *Front. Oncol.* **2018**, *8*, 239. [CrossRef]
98. Greenberg, M.L.; Fisher, P.G.; Freeman, C.; Korones, D.N.; Bernstein, M.; Friedman, H.; Blaney, S.; Hershon, L.; Zhou, T.; Chen, Z. Etoposide, vincristine, and cyclosporin A with standard-dose radiation therapy in newly diagnosed diffuse intrinsic brainstem gliomas: A pediatric oncology group phase I study. *Pediatric Blood Cancer* **2005**, *45*, 644–648. [CrossRef]
99. Zhang, E.Y.; Knipp, G.T.; Ekins, S.; Swaan, P.W. Structural biology and function of solute transporters: Implications for identifying and designing substrates. *Drug Metab. Rev.* **2002**, *34*, 709–750. [CrossRef]
100. Jensen, R.L.; Chkheidze, R. The Role of Glucose Transporter-1 (GLUT-1) in Malignant Gliomas. In *Tumors of the Central Nervous System, Volume 1*; Springer: Dordrecht, Netherlands, 2011; pp. 99–108.
101. Li, J.; Zhou, L.; Ye, D.; Huang, S.; Shao, K.; Huang, R.; Han, L.; Liu, Y.; Liu, S.; Ye, L. Choline-Derivate-Modified Nanoparticles for Brain-Targeting Gene Delivery. *Adv. Mater.* **2011**, *23*, 4516–4520. [CrossRef]
102. Wu, D.; Qin, M.; Xu, D.; Wang, L.; Liu, C.; Ren, J.; Zhou, G.; Chen, C.; Yang, F.; Li, Y. A Bioinspired Platform for Effective Delivery of Protein Therapeutics to the Central Nervous System. *Adv. Mater.* **2019**. [CrossRef]
103. Johnsen, K.B.; Burkhart, A.; Melander, F.; Kempen, P.J.; Vejlebo, J.B.; Siupka, P.; Nielsen, M.S.; Andresen, T.L.; Moos, T. Targeting transferrin receptors at the blood-brain barrier improves the uptake of immunoliposomes and subsequent cargo transport into the brain parenchyma. *Sci. Rep.* **2017**, *7*, 10396. [CrossRef]
104. Lam, F.C.; Morton, S.W.; Wyckoff, J.; Han, T.-L.; Hwang, M.K.; Maffa, A.; Balkanska-Sinclair, E.; Yaffe, M.B.; Floyd, S.R.; Hammond, P.T. Enhanced efficacy of combined temozolomide and bromodomain inhibitor therapy for gliomas using targeted nanoparticles. *Nat. commun.* **2018**, *9*, 1991. [CrossRef]

105. Qian, W.; Qian, M.; Wang, Y.; Huang, J.; Chen, J.; Ni, L.; Huang, Q.; Liu, Q.; Gong, P.; Hou, S. Combination Glioma Therapy Mediated by a Dual-Targeted Delivery System Constructed Using OMCN–PEG–Pep22/DOX. *Small* **2018**, *14*, 1801905. [CrossRef]
106. Tosi, G.; Vilella, A.; Veratti, P.; Belletti, D.; Pederzoli, F.; Ruozi, B.; Vandelli, M.A.; Zoli, M.; Forni, F. Exploiting bacterial pathways for BBB crossing with PLGA nanoparticles modified with a mutated form of diphtheria toxin (CRM197): In vivo experiments. *Mol. Pharm.* **2015**, *12*, 3672–3684. [CrossRef]
107. Chai, Z.; Hu, X.; Wei, X.; Zhan, C.; Lu, L.; Jiang, K.; Su, B.; Ruan, H.; Ran, D.; Fang, R.H. A facile approach to functionalizing cell membrane-coated nanoparticles with neurotoxin-derived peptide for brain-targeted drug delivery. *J. Control. Release* **2017**, *264*, 102–111. [CrossRef]
108. Thom, G.; Burrell, M.; Haqqani, A.S.; Yogi, A.; Lessard, E.; Brunette, E.; Delaney, C.; Baumann, E.; Callaghan, D.; Rodrigo, N. Enhanced delivery of galanin conjugates to the brain through bioengineering of the anti-transferrin receptor antibody OX26. *Mol. Pharm.* **2018**, *15*, 1420–1431. [CrossRef]
109. Huwyler, J.; Wu, D.; Pardridge, W.M. Brain drug delivery of small molecules using immunoliposomes. *Proc. Natl. Acad. Sci. USA* **1996**, *93*, 14164–14169. [CrossRef]
110. Dixit, S.; Novak, T.; Miller, K.; Zhu, Y.; Kenney, M.E.; Broome, A.-M. Transferrin receptor-targeted theranostic gold nanoparticles for photosensitizer delivery in brain tumors. *Nanoscale* **2015**, *7*, 1782–1790. [CrossRef]
111. Ying, X.; Wen, H.; Lu, W.-L.; Du, J.; Guo, J.; Tian, W.; Men, Y.; Zhang, Y.; Li, R.-J.; Yang, T.-Y. Dual-targeting daunorubicin liposomes improve the therapeutic efficacy of brain glioma in animals. *J. Control. Release* **2010**, *141*, 183–192. [CrossRef]
112. Hersh, D.S.; Wadajkar, A.S.; Roberts, N.B.; Perez, J.G.; Connolly, N.P.; Frenkel, V.; Winkles, J.A.; Woodworth, G.F.; Kim, A.J. Evolving drug delivery strategies to overcome the blood brain barrier. *Curr. Pharm. Des.* **2016**, *22*, 1177–1193. [CrossRef]
113. Yu, Y.J.; Zhang, Y.; Kenrick, M.; Hoyte, K.; Luk, W.; Lu, Y.; Atwal, J.; Elliott, J.M.; Prabhu, S.; Watts, R.J. Boosting brain uptake of a therapeutic antibody by reducing its affinity for a transcytosis target. *Sci. Transl. Med.* **2011**, *3*, ra44–ra84. [CrossRef]
114. Niewoehner, J.; Bohrmann, B.; Collin, L.; Urich, E.; Sade, H.; Maier, P.; Rueger, P.; Stracke, J.O.; Lau, W.; Tissot, A.C. Increased brain penetration and potency of a therapeutic antibody using a monovalent molecular shuttle. *Neuron* **2014**, *81*, 49–60. [CrossRef]
115. Razpotnik, R.; Novak, N.; Čurin Šerbec, V.; Rajcevic, U. Targeting malignant brain tumors with antibodies. *Front. Immunol.* **2017**, *8*, 1181. [CrossRef]
116. Gedeon, P.C.; Schaller, T.H.; Chitneni, S.K.; Choi, B.D.; Kuan, C.-T.; Suryadevara, C.M.; Snyder, D.J.; Schmittling, R.J.; Szafranski, S.E.; Cui, X. A rationally designed fully human EGFRvIII: CD3-targeted bispecific antibody redirects human T cells to treat patient-derived intracerebral malignant glioma. *Clin. Cancer Res.* **2018**, *24*, 3611–3631. [CrossRef]
117. Alaofi, A.; On, N.; Kiptoo, P.; Williams, T.; Miller, D.; Siahaan, T. Comparison of linear and cyclic HAV peptides in modulating the blood-brain barrier permeability: Impact on delivery of molecules to the brain. *J. Pharm. Sci.* **2016**, *105*, 797–807. [CrossRef]
118. Bocsik, A.; Walter, F.R.; Gyebrovszki, A.; Fülöp, L.; Blasig, I.; Dabrowski, S.; Ötvös, F.; Tóth, A.; Rákhely, G.; Veszelka, S. Reversible opening of intercellular junctions of intestinal epithelial and brain endothelial cells with tight junction modulator peptides. *J. Pharm. Sci.* **2016**, *105*, 754–765. [CrossRef]
119. On, N.H.; Kiptoo, P.; Siahaan, T.J.; Miller, D.W. Modulation of blood–brain barrier permeability in mice using synthetic E-cadherin peptide. *Mol. Pharm.* **2014**, *11*, 974–981. [CrossRef]
120. Sauer, R.-S.; Krug, S.M.; Hackel, D.; Staat, C.; Konasin, N.; Yang, S.; Niedermirtl, B.; Bosten, J.; Günther, R.; Dabrowski, S. Safety, efficacy, and molecular mechanism of claudin-1-specific peptides to enhance blood–nerve–barrier permeability. *J. Control. Release* **2014**, *185*, 88–98. [CrossRef]
121. Wong, V.; Gumbiner, B.M. A synthetic peptide corresponding to the extracellular domain of occludin perturbs the tight junction permeability barrier. *J. Cell Biol.* **1997**, *136*, 399–409. [CrossRef] [PubMed]
122. Kim, D.-G.; Bynoe, M.S. A2A adenosine receptor regulates the human blood-brain barrier permeability. *Mol. Neurobiol.* **2015**, *52*, 664–678. [CrossRef] [PubMed]
123. Krug, S.M.; Hayaishi, T.; Iguchi, D.; Watari, A.; Takahashi, A.; Fromm, M.; Nagahama, M.; Takeda, H.; Okada, Y.; Sawasaki, T. Angubindin-1, a novel paracellular absorption enhancer acting at the tricellular tight junction. *J. Control. Release* **2017**, *260*, 1–11. [CrossRef] [PubMed]

124. Banks, D.B.; Chan, G.N.; Evans, R.A.; Miller, D.S.; Cannon, R.E. Lysophosphatidic acid and amitriptyline signal through LPA1R to reduce P-glycoprotein transport at the blood–brain barrier. *J. Cereb. Blood Flow Metab.* **2018**, *38*, 857–868. [CrossRef]
125. On, N.H.; Savant, S.; Toews, M.; Miller, D.W. Rapid and reversible enhancement of blood–brain barrier permeability using lysophosphatidic acid. *J. Cereb. Blood Flow Metab.* **2013**, *33*, 1944–1954. [CrossRef]
126. Miller, D.; Toews, M.; Savant, S.; Mayhan, B. Phospholipid Receptors as Targets for Enhancing Drug Permeability to Selected Tissues. U.S. Patent 12/446,284, 26 May 2011.
127. Sun, Z.; Worden, M.; Thliveris, J.A.; Hombach-Klonisch, S.; Klonisch, T.; van Lierop, J.; Hegmann, T.; Miller, D.W. Biodistribution of negatively charged iron oxide nanoparticles (IONPs) in mice and enhanced brain delivery using lysophosphatidic acid (LPA). *Nanomed. Nanotechnol. Biol. Med.* **2016**, *12*, 1775–1784. [CrossRef]
128. Wang, Z.; Cai, X.-j.; Qin, J.; Xie, F.-J.; Han, N.; Lu, H.-y. The role of histamine in opening blood-tumor barrier. *Oncotarget* **2016**, *7*, 31299–31310. [CrossRef]
129. Xie, Z.; Shen, Q.; Xie, C.; Lu, W.; Peng, C.; Wei, X.; Li, X.; Su, B.; Gao, C.; Liu, M. Retro-inverso bradykinin opens the door of blood–brain tumor barrier for nanocarriers in glioma treatment. *Cancer Lett.* **2015**, *369*, 144–151. [CrossRef]
130. Azad, T.D.; Pan, J.; Connolly, I.D.; Remington, A.; Wilson, C.M.; Grant, G.A. Therapeutic strategies to improve drug delivery across the blood-brain barrier. *Neurosurg. Focus* **2015**, *38*, E9. [CrossRef]
131. Liu, L.-b.; Liu, X.-b.; Ma, J.; Liu, Y.-h.; Li, Z.-q.; Ma, T.; Zhao, X.-h.; Xi, Z.; Xue, Y.-x. Bradykinin increased the permeability of BTB via NOS/NO/ZONAB-mediating down-regulation of claudin-5 and occludin. *Biochem. Biophys. Res. Commun.* **2015**, *464*, 118–125. [CrossRef]
132. Li, X.; Tsibouklis, J.; Weng, T.; Zhang, B.; Yin, G.; Feng, G.; Cui, Y.; Savina, I.N.; Mikhalovska, L.I.; Sandeman, S.R. Nano carriers for drug transport across the blood–brain barrier. *J. Drug Target.* **2017**, *25*, 17–28. [CrossRef] [PubMed]
133. Wei, X.; Chen, X.; Ying, M.; Lu, W. Brain tumor-targeted drug delivery strategies. *Acta Pharm. Sin. B* **2014**, *4*, 193–201. [CrossRef] [PubMed]
134. Huang, Y.; Liu, W.; Gao, F.; Fang, X.; Chen, Y. c (RGDyK)-decorated Pluronic micelles for enhanced doxorubicin and paclitaxel delivery to brain glioma. *Int. J. Nanomed.* **2016**, *11*, 1629–1641
135. Park, J.-S.; Kim, I.-K.; Han, S.; Park, I.; Kim, C.; Bae, J.; Oh, S.J.; Lee, S.; Kim, J.H.; Woo, D.-C. Normalization of tumor vessels by Tie2 activation and Ang2 inhibition enhances drug delivery and produces a favorable tumor microenvironment. *Cancer Cell* **2016**, *30*, 953–967. [CrossRef]
136. Guo, P.; Moses-Gardner, A.; Huang, J.; Smith, E.R.; Moses, M.A. ITGA2 as a potential nanotherapeutic target for glioblastoma. *Sci. Rep.* **2019**, *9*, 6195. [CrossRef]
137. Angara, K.; Rashid, M.H.; Shankar, A.; Ara, R.; Iskander, A.; Borin, T.F.; Jain, M.; Achyut, B.R.; Arbab, A.S. Vascular mimicry in glioblastoma following anti-angiogenic and anti-20-HETE therapies. *Histol. Histopathol.* **2017**, *32*, 917–928.
138. Aboody, K.; Najbauer, J.; Danks, M. Stem and progenitor cell-mediated tumor selective gene therapy. *Gene Ther.* **2008**, *15*, 739–752. [CrossRef]
139. Frank, R.T.; Najbauer, J.; Aboody, K.S. Concise review: Stem cells as an emerging platform for antibody therapy of cancer. *Stem Cells* **2010**, *28*, 2084–2087. [CrossRef] [PubMed]
140. Frank, R.T.; Edmiston, M.; Kendall, S.E.; Najbauer, J.; Cheung, C.-W.; Kassa, T.; Metz, M.Z.; Kim, S.U.; Glackin, C.A.; Wu, A.M. Neural stem cells as a novel platform for tumor-specific delivery of therapeutic antibodies. *PLoS ONE* **2009**, *4*, e8314. [CrossRef]
141. Ahmed, A.U.; Thaci, B.; Alexiades, N.G.; Han, Y.; Qian, S.; Liu, F.; Balyasnikova, I.V.; Ulasov, I.Y.; Aboody, K.S.; Lesniak, M.S. Neural stem cell-based cell carriers enhance therapeutic efficacy of an oncolytic adenovirus in an orthotopic mouse model of human glioblastoma. *Mol. Ther.* **2011**, *19*, 1714–1726. [CrossRef] [PubMed]
142. Aboody, K.S.; Najbauer, J.; Metz, M.Z.; D'apuzzo, M.; Gutova, M.; Annala, A.J.; Synold, T.W.; Couture, L.A.; Blanchard, S.; Moats, R.A. Neural stem cell–mediated enzyme/prodrug therapy for glioma: Preclinical studies. *Sci. Transl. Med.* **2013**, *5*, ra159–ra184. [CrossRef]
143. Sheets, K.T.; Bagó, J.R.; Hingtgen, S.D. Delivery of Cytotoxic Mesenchymal Stem Cells with Biodegradable Scaffolds for Treatment of Postoperative Brain Cancer. In *Targeted Drug Delivery*; Springer: New York, NY, USA, 2018; pp. 49–58.

144. Read, T.-A.; Farhadi, M.; Bjerkvig, R.; Olsen, B.R.; Rokstad, A.M.; Huszthy, P.C.; Vajkoczy, P. Intravital microscopy reveals novel antivascular and antitumor effects of endostatin delivered locally by alginate-encapsulated cells. *Cancer Res.* **2001**, *61*, 6830–6837. [PubMed]
145. Kauer, T.M.; Figueiredo, J.-L.; Hingtgen, S.; Shah, K. Encapsulated therapeutic stem cells implanted in the tumor resection cavity induce cell death in gliomas. *Nat. Neurosci.* **2012**, *15*, 197–204. [CrossRef]
146. Xue, J.; Zhao, Z.; Zhang, L.; Xue, L.; Shen, S.; Wen, Y.; Wei, Z.; Wang, L.; Kong, L.; Sun, H. Neutrophil-mediated anticancer drug delivery for suppression of postoperative malignant glioma recurrence. *Nat. Nanotechnol.* **2017**, *12*, 692–700. [CrossRef] [PubMed]
147. Challis, R.C.; Kumar, S.R.; Chan, K.Y.; Challis, C.; Beadle, K.; Jang, M.J.; Kim, H.M.; Rajendran, P.S.; Tompkins, J.D.; Shivkumar, K. Systemic AAV vectors for widespread and targeted gene delivery in rodents. *Nat. Protoc.* **2019**, *14*, 379–414. [CrossRef] [PubMed]
148. Kotterman, M.A.; Schaffer, D.V. Engineering adeno-associated viruses for clinical gene therapy. *Nat. Rev. Genet.* **2014**, *15*, 445–451. [CrossRef]
149. Zhang, H.; Yang, B.; Mu, X.; Ahmed, S.S.; Su, Q.; He, R.; Wang, H.; Mueller, C.; Sena-Esteves, M.; Brown, R. Several rAAV vectors efficiently cross the blood–brain barrier and transduce neurons and astrocytes in the neonatal mouse central nervous system. *Mol. Ther.* **2011**, *19*, 1440–1448. [CrossRef]
150. Choudhury, S.R.; Hudry, E.; Maguire, C.A.; Sena-Esteves, M.; Breakefield, X.O.; Grandi, P. Viral vectors for therapy of neurologic diseases. *Neuropharmacology* **2017**, *120*, 63–80. [CrossRef]
151. Bedbrook, C.N.; Deverman, B.E.; Gradinaru, V. Viral strategies for targeting the central and peripheral nervous systems. *Annu. Rev. Neurosci.* **2018**, *41*, 323–348. [CrossRef] [PubMed]
152. Deverman, B.E.; Pravdo, P.L.; Simpson, B.P.; Kumar, S.R.; Chan, K.Y.; Banerjee, A.; Wu, W.-L.; Yang, B.; Huber, N.; Pasca, S.P. Cre-dependent selection yields AAV variants for widespread gene transfer to the adult brain. *Nat. Biotechnol.* **2016**, *34*, 204–209. [CrossRef]
153. Montgomery, K.L.; Iyer, S.M.; Christensen, A.J.; Deisseroth, K.; Delp, S.L. Beyond the brain: Optogenetic control in the spinal cord and peripheral nervous system. *Sci. Transl. Med.* **2016**, *8*, rv335–rv337. [CrossRef] [PubMed]
154. Organization, W.H. Available online: www.who.int/features/qa/42/en/index.html (accessed on 19 April 2019).
155. NIH. Available online: https://clinicaltrials.gov/ct2/results?cond=brain+cancer&term=mannitol&cntry=&state=&city=&dist= (accessed on 19 April 2019).
156. Shibahara, I.; Hanihara, M.; Watanabe, T.; Dan, M.; Sato, S.; Kuroda, H.; Inamura, A.; Inukai, M.; Hara, A.; Yasui, Y. Tumor microenvironment after biodegradable BCNU wafer implantation: Special consideration of immune system. *J. Neuro-Oncol.* **2018**, *137*, 417–427. [CrossRef] [PubMed]
157. Vogelbaum, M.A.; Brewer, C.; Barnett, G.H.; Mohammadi, A.M.; Peereboom, D.M.; Ahluwalia, M.S.; Gao, S. First-in-human evaluation of the Cleveland Multiport Catheter for convection-enhanced delivery of topotecan in recurrent high-grade glioma: Results of pilot trial 1. *J. Neurosurg.* **2018**, *130*, 476–485. [CrossRef]
158. Sagnella, S.M.; Trieu, J.; Brahmbhatt, H.; MacDiarmid, J.A.; MacMillan, A.; Whan, R.M.; Fife, C.M.; McCarroll, J.A.; Gifford, A.J.; Ziegler, D.S. Targeted doxorubicin-loaded bacterially derived nano-cells for the treatment of neuroblastoma. *Mol. Cancer Therap.* **2018**, *17*, 1012–1023. [CrossRef]
159. Miranda-Filho, A.; Piñeros, M.; Soerjomataram, I.; Deltour, I.; Bray, F. Cancers of the brain and CNS: Global patterns and trends in incidence. *Neuro-Oncology* **2016**, *19*, 270–280. [CrossRef] [PubMed]

pharmaceutics

MDPI

Review

Antibodies for the Treatment of Brain Metastases, a Dream or a Reality?

Marco Cavaco, Diana Gaspar, Miguel ARB Castanho * and Vera Neves *

Instituto de Medicina Molecular, Faculdade de Medicina, Universidade de Lisboa, Av. Prof. Egas Moniz, 1649-028 Lisboa, Portugal

* Correspondence: macastanho@medicina.ulisboa.pt (M.A.R.B.C.); veraneves@medicina.ulisboa.pt (V.N.)

Received: 19 November 2019; Accepted: 28 December 2019; Published: 13 January 2020

check for updates

Abstract: The incidence of brain metastases (BM) in cancer patients is increasing. After diagnosis, overall survival (OS) is poor, elicited by the lack of an effective treatment. Monoclonal antibody (mAb)-based therapy has achieved remarkable success in treating both hematologic and non-central-nervous system (CNS) tumors due to their inherent targeting specificity. However, the use of mAbs in the treatment of CNS tumors is restricted by the blood–brain barrier (BBB) that hinders the delivery of either small-molecules drugs (sMDs) or therapeutic proteins (TPs). To overcome this limitation, active research is focused on the development of strategies to deliver TPs and increase their concentration in the brain. Yet, their molecular weight and hydrophilic nature turn this task into a challenge. The use of BBB peptide shuttles is an elegant strategy. They explore either receptor-mediated transcytosis (RMT) or adsorptive-mediated transcytosis (AMT) to cross the BBB. The latter is preferable since it avoids enzymatic degradation, receptor saturation, and competition with natural receptor substrates, which reduces adverse events. Therefore, the combination of mAbs properties (e.g., selectivity and long half life) with BBB peptide shuttles (e.g., BBB translocation and delivery into the brain) turns the therapeutic conjugate in a valid approach to safely overcome the BBB and efficiently eliminate metastatic brain cells.

Keywords: adsorptive-mediated transcytosis; antibody fragments; blood–brain barrier; brain metastases; monoclonal antibodies; peptide shuttles

1. Brain Metastases

Brain metastases (BM) account for significant morbidity and mortality. The exact incidence is unknown [1,2]. Based on various studies, investigators estimate that BM occurs in 10%–20% of adult patients with cancer [3]. Nevertheless, the incidence might be higher, and it is increasing due to prolonged life expectancy, increased resistance to cancer therapies, and improved imaging techniques. In addition, the increased patient survival by treating primary tumors may increase the number of patients that will develop more aggressive BM, or that are resistant to therapy. Among the different cancer types, lung cancer (19.9%), breast cancer (15.2%), and melanoma (6.9%) are the most common primary tumors developing BM [4]. After diagnosis, overall survival (OS) is poor. However, early diagnosis, improved systemic therapies, and multimodality treatments have significantly increased patients' survival [5].

1.1. BM Pathophysiology

The pathophysiology of BM is complex and involves a multi-step process constituted of two major stages (Figure 1) [6]. The first stage is tumor migration, which includes (i) metastatic clone progression, due to tumor cells' ability to degrade extracellular matrix (ECM); (ii) intravasation (transendothelial migration of cancer cells into vessels); (iii) dissemination (spread of tumor cells via bloodstream); (iv) extravasation (transendothelial migration of cancer cells into tissues). The second stage corresponds to tumor colonization.

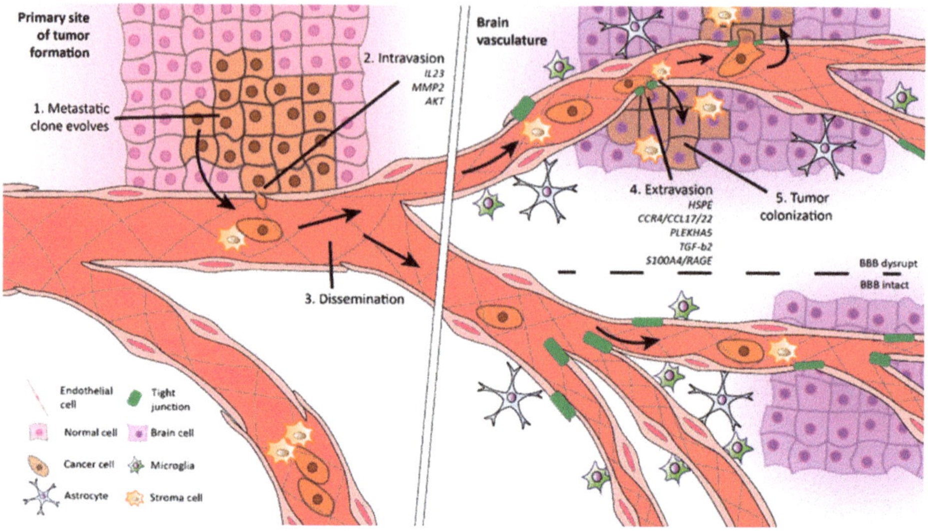

Figure 1. Steps in the formation of brain metastases (BM). Metastases formation begins in the microenvironment of the primary tumor with 1. metastatic clones developing, degrading the extracellular matrix (ECM), and suffering an epithelial–mesenchymal transition (EMT) to further detach from the connective tissue. 2. Subsequently, tumor cells invade and enter the circulation (intravation). 3. The dissemination within the vascular system drives tumor cells to distant sites, like the brain. 4. Then, they extravasate across the blood–brain barrier (BBB) and enter the brain parenchyma due to the release of proteolytic enzymes and cellular interactions. 5. Once inside the brain, cancer cells colonize the tissue and develop secondary tumors.

The cells presented in the primary tumor are heterogeneous. Among others, the tumor microenvironment is composed of cancer stem cells (CSCs), partially differentiated progenitor cells, and fully differentiated end-stage cells [6]. Recent findings attribute to CSCs the primary responsibility for enhanced malignancy since they can complete the two stages of metastases formation (Figure 1) [7]. However, during cancer progression, other cells undergo an epithelial–mesenchymal transition (EMT), changing their plasticity by morphological and phenotypical conversions [8,9]. EMT enables non-CSCs to resemble a CSC state. Thus, they acquire the ability to invade and colonize distant sites, creating secondary niches that may progress to a secondary tumor [10]. Therefore, in the end, within the tumor microenvironment, all cells are malignant. Nevertheless, the development of distal metastases only occurs in <0.1% of disseminated cancer cells. Thus, although the formation of metastases represents a major threat, it is considered highly inefficient [8,11].

1.2. BBB Physiology

BBB is a complex system composed of a structurally distinct and continuous endothelial cell layer separating two brain compartments, namely, the blood and extracellular fluid. Its components include an endothelial cell layer, adjoined by tight cell-to-cell junction proteins, and pinocytic vesicles [12]. All together, they contribute to the selective permeability of the barrier, allowing brain homeostasis. The BBB is also dynamic. It responds to regulatory signals from both the blood and the brain [13], being the main portal into the brain of gaseous molecules, such as O_2 and CO_2, ions, nutrients, hormones, and water (Figure 2). Hydrophobic compounds (<500 Da) diffuse across the endothelium membrane.

Carrier-mediated transport (CMT) is responsible for the transport of glucose and amino acid residues. While water-soluble molecules (e.g., ions) cross the BBB through ion channels. On the other hand, macromolecules (proteins and peptides) transport rely on endocytic vesicles, which involve either receptor-mediated transport (RMT) or adsorptive-mediated transport (AMT) [14,15].

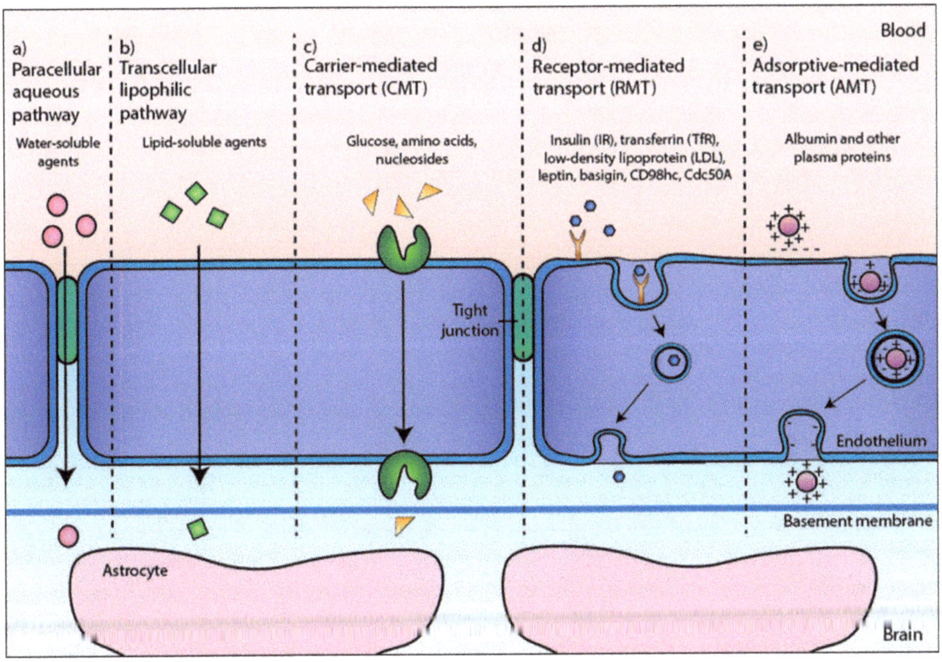

Figure 2. Pathways across the blood–brain barrier (BBB). Representation of the BBB formed by the endothelial cells and their interaction with astrocytes. Different translocation routes are presented. (**a**) Tight junctions usually restrict the penetration of water-soluble compounds. (**b**) The large surface area of the lipid membranes of the endothelium offers an efficient diffuse route for lipid-soluble agents. (**c**) Several transport proteins (carriers) are present in the endothelium for glucose (Gluc-1), amino acids, nucleosides, and other substances. (**d**) Large molecules such as antibodies, lipoproteins, proteins, and peptides can only transverse the BBB by receptor-mediated transport (RMT). (**e**) The transport of native plasma proteins or peptides is limited, but cationization can increase their uptake by adsorptive-mediated transport (AMT).

Nevertheless, the BBB interacts with the metastatic cells in an unidentified way. The BBB is hypothesized to create a unique brain microenvironment and to influence metastatic colonization [16]. The increased permeability of tumor-associated endothelial cells, due to tumor penetration into the brain, permits leakage of proteins and water into brain parenchyma. The mechanism described is responsible for the edema often associated with BM [17]. Microglia and macrophages influence tumor proliferation and invasion by secreting multiple cytokines, growth factors, enzymes, and reactive oxygen species (ROS). Other immune cells may also participate in the BBB translocation. However, the exact mechanisms are debatable [18–21]. The BBB structure may be affected momentarily during cancer cells' invasion; however, in other non-cancer-related central-nervous system (CNS) pathologies, only in advanced disease stages, the dysfunction is usually significant.

1.3. BM Treatment

Major developments have been made in understanding brain function, metastases progression, and the development of medical technologies. However, in many cases, the major drawback in BM treatment is the inefficient drug delivery into the brain [22,23]. The BBB remains the most significant obstacle to the efficient delivery of small-molecule drugs (sMDs) and therapeutic proteins (TPs) [24]. In addition, some authors also attributed to the therapeutic resistance of metastatic cells the responsivity for therapy inefficiency. According to them, during cancer progression, some survival pathways, such as the PI3K/AKT/mTOR are activated in specific cells, which contributes to the poor response of metastatic cells [25,26]. In the end, both mechanisms might be contributing to therapeutic failure.

The first-line approach to treat BM includes surgery, stereotactic radiosurgery (SRS), and whole-brain radiation therapy (WBRT) [27–29]. However, both systemic and intracranial disease control are also possible with the improved systemic therapies that have begun to offer greater potential for specific cancer types and genotypes. Thus, the management of BM has become increasingly individualized [1]. Depending on the histology and systemic disease status, physicians may consider all the available therapies. For instance, recommendations suggest better disease management by using a multidisciplinary modality in patients with BM from breast cancer, melanoma, and specific genotypes of non-small cell lung cancer (NSCLC) (e.g., epidermal growth factor receptor (EGFR) gene mutations, translocations in the anaplastic lymphoma kinase (ALK) gene). Whenever possible, patients may enroll in clinical trials for a novel or existing therapy. Still, given the paucity of effective treatment options, BM elimination represents an unmet clinical need [30].

2. Strategies to Overcome the BBB

The BBB restricts the delivery of therapeutics to the brain. Overall, 98% of sMDs and probably all TPs cannot cross the barrier by free diffusion [31–33]. In the last decade, intense investigation allowed the discovery of new strategies to increase brain penetration of existing therapeutics (invasive, pharmacological, and physiological) [34–36]. Ideally, the translocation should not compromise the BBB integrity. However, some of the current strategies do not meet this criterion, namely, the invasive and pharmacological approaches.

2.1. Invasive Approach

This strategy allows drugs to flow directly from the systemic circulation into the brain by BBB disruption using different methodologies. The most important are: (1) osmotic disruption, due to the administration of hypertonic solutions (e.g., mannitol) causing cells shrinking based on cell dehydration [37]; (2) ultrasound methods, which rely on transcranial delivery of low-frequency ultrasound waves resulting in the opening of tight junctions [38]; and (3) pharmacological agents, such as bradykinin-like compounds (e.g., histamine, bradykinin) that disrupt tight junctions by stimulating B_2 receptors presented in endothelial cells and transiently increasing cytosolic Ca^{2+} [39]. The costs, anesthetic administration, and hospitalization are significant drawbacks for all these approaches. Also, the disruption of the BBB may increase tumor dissemination, as well as irreversible neuropathological changes due to the entry of unwanted substances [34].

2.2. Pharmacological Approach

The pharmacological approach relies on the observation that some molecules freely enter the brain owing to their molecular weight (<500 Da), charge (low hydrogen bonding capabilities), and lipophilicity [40]. Thus, researchers started modifying, through medicinal chemistry, molecules that are active against CNS diseases or BM to enable them to get into the brain [31]. Although it has enormous potential, the modifications may result in loss of pharmacological activity. In addition, the new molecule may become a substrate for the efflux pumps by increasing drugs' lipophilicity, which decreases brain accumulation [34].

2.3. Physiological Approach

This natural strategy exploits the various transporters and receptors expressed at the BBB, as well as the physiological properties of the BBB (e.g., charge and lipid composition) (Figure 2) [41–43]. These translocation mechanisms are fundamental for the uptake of essential substances to maintain brain homeostasis. They can be classified into: (1) CMT, which are responsible for the cross of glucose (glucose transporter—GLUT1), amino acids (large neutral amino acid transporter—LAT1, and cationic amino acid transporter—CAT1), and nucleosides (nucleobase transporter—NBT); (2) RMT, fundamental for large molecules translocation, such as transferrin (transferrin receptor—TfR), insulin (insulin receptor—IR), low-density lipoprotein (lipoprotein receptor-mediated protein—LRP), leptin (leptin receptor—LEPR), and fragment crystallizable (Fc) fragment of immunoglobulin G (IgG) (Fc fragment of IgG receptor transporter α—FCGRT); and finally, (3) AMT that drives albumin and other plasma proteins to brain [44].

CMT is an interesting transporter due to the easy coupling of endogenous substrates to sMDs [45]. Besides, it is also possible to perform direct modification of sMDs to resemble CMT natural substrates. The changes allow drugs to be recognized and transported across the BBB [46]. Nevertheless, the molecules generated: (1) must mimic that of the endogenous CMT substrate; (2) should not affect CMT physiological function; and (3) must maintain its pharmacological activity. So far, targeting nanocarriers to CMTs have been the best example of the strategy's success. However, the application of this approach was only possible for small molecules, as revised in Witt et al. [47].

Another promising strategy to develop molecules that can efficiently cross the BBB is the RMT. These molecules are known as Trojan Horses and can be either peptides or antibodies [48]. TfR and IR are the most important BBB receptors explored by researchers. Pardridge et al. have extensively documented the use of antibodies targeting these receptors [49–51]. The in vivo studies demonstrated an accumulation of different anti-TfR monoclonal antibodies (mAbs) in the brain tissue and a distinct biodistribution. The high affinity of antibodies towards these receptors is, however, a limitation since results in weak receptor dissociation. Consequently, the high-affinity antibodies follow the lysosomes pathway during intracellular trafficking leading to its degradation [52].

Yu et al. elegantly solved the problem by reducing the affinity of anti-TfR mAbs [53]. Next, the group developed a bispecific therapeutic antibody with a low affinity for TfR and a high affinity for the enzyme β-secretase (BACE1), an Alzheimer's disease drug target. Relevant results were obtained from the evaluation of the bispecific antibody efficacy in non-human primates. The brain accumulation was significantly higher than control, and the amyloid β-peptide presence in the brain and serum reduced considerably [54]. Similarly to TfR, exciting studies targeting IR have been developed. Pardridge et al. have shown a total of 4% brain uptake 3 h after intravenous administration in Rhesus monkeys. In the treatment of Parkinson's, stroke, metachromatic leukodystrophy, and Sanfilippo type A syndrome, some therapeutic drugs have been linked to the mAb and successfully translocated across the BBB [49,55].

Another interesting strategy to deliver drugs into the brain exploiting RMT is the use of nanoparticles (NPs) coupled with mAbs or peptides that recognize these receptors [43,56,57]. NPs are colloidal carriers of natural or synthetic origin with a size varying from 1 to 1000 nm. They are a fascinating system due to their modulating capacity concerning shape, size, hydrophobicity, coating, chemistry, and surface charge [58]. In addition, they also have a high capacity of drug payload, the relatively few mAbs or peptides to achieve high levels of drug targeting, protection of the encapsulated drug, and the ability to provide a controlled release of the drug [59,60].

Although the considerable achievements accomplished, the drawback of these RMT systems is related to the competition with natural substrates, which may affect brain homeostasis; and may result in receptors' saturation due to the high affinity of antibodies [34,61,62]. To overcome these limitations, recently, more attention has been given to AMT. The concept of AMT through the BBB began with the observation that polycationic proteins' brain uptake did not involve binding to the endothelial cell surface [63]. Electrostatic interaction between positively charged substances and negatively charged BBB drives the translocation. The vesicles created allow BBB cross and, consequently, brain accumulation [64]. Lack of selectivity of these systems and possible BBB disruption were the major concerns highlighted by researchers. However, the recent proof-of-concept given by the use of cell-penetrating peptides (CPPs) in BBB translocation (BBB peptide-shuttle) has launched a new interest in this strategy. These peptides have demonstrated a natural selectivity towards negatively charged membranes and the ability to translocate large cargoes without BBB damage both in vitro and in vivo models [65–67]. Therefore, the optimization of BBB peptide shuttles based-systems in the delivery of sMDs and TPs, using AMT, will be an area of intense investigation during the next decade.

3. Therapeutic Antibodies for BM Elimination

TPs are the standard of care in a number of therapeutic areas [68]. They are protein manufactured for biopharmaceutical use and include, for instance, mAbs, peptides, growth factors, cytokines, and enzymes [69]. Their production is relatively easy and relies mostly on either simple purification or recombinant DNA technology. Throughout the following sections, the description of antibodies' activity and their therapeutic value only concern human antibodies. The activity of antibodies is species dependent. Thus, some features presented might not be accurate for non-human antibodies.

3.1. Monoclonal Antibodies

mAbs represent the fastest growing class of TPs. Currently, over 50 therapeutic antibodies are on the market [70]. They are complex molecules consisting of homodimers of variable and constant regions (Figure 3) [71]. The former has antigen specificity owing to the presence of complementary determining regions (CDRs). On the other hand, the Fc domain is responsible for the long half-life of antibodies, due to antibody recycling after interaction to the neonatal Fc receptor (FcRn) [72]; and immune activation (complement-dependent cytotoxicity—CDC; and antibody-dependent cellular cytotoxicity—ADCC) by engaging Fcγ receptors (FcγRs) on immune cells (e.g., neutrophils, natural killer cells, monocytes) [73]. High therapeutic tolerability and low risk-to-benefit ratios favor the use of therapeutic antibodies. Thus, their exquisite specificity, high binding affinity, long half-life, low toxicity, and versatility are characteristics that contributed to antibodies' success [74,75]. Additionally, the low number of drug–drug interactions between mAbs and sMDs increased their combination in many therapeutic regimens [70].

The mechanism of action of mAbs differs depending on the molecule engineered (Figure 3). They can target soluble mediators (e.g., cytokines) to inhibit their binding to receptors and, consequently, inhibit signaling; or they can target membrane receptors either inducing or antagonizing signaling (e.g., programmed-cell death ligant-1—PD-L1; or human epidermal receptor-2—HER2, respectively) [76]. In addition, the presence of the Fc domain allows immune stimulation (CDC and/or ADCC). CDC is related to complement activation. Complement is one of the first mediators of the immune response to pathogens and cells. After binding, antibodies activate the classical complement cascade. Thus, releasing cytokines (e.g., anaphylatoxins and opsonins) and forming the membrane attack complex (MAC), which lead to cell lysis and phagocytosis [77]. On the other hand, ADCC occurred due to the interaction of the Fc domain with FcγRs on effector immune cells (e.g., neutrophils—FcγRI, natural killer cells—FcγIIIA; or monocytes—FcγIIIB). After recognition of an antibody-coated target cell, effector cells engage the release of granzymes and perforins [78]. The consequence is cell death. The magnitude of the stimulation of either CDC or ADCC depends on the IgG subset (IgG 1–4). For instance, IgG2 and IgG4 do not activate both mechanisms. Therefore, they are designed primarily for signaling blockage. Oppositely, IgG1 and IgG3 strongly activate both CDC and ADCC. Owing to its short half-life due to a low FcRn affinity, IgG3 does not have the therapeutic value of other IgG subsets [79].

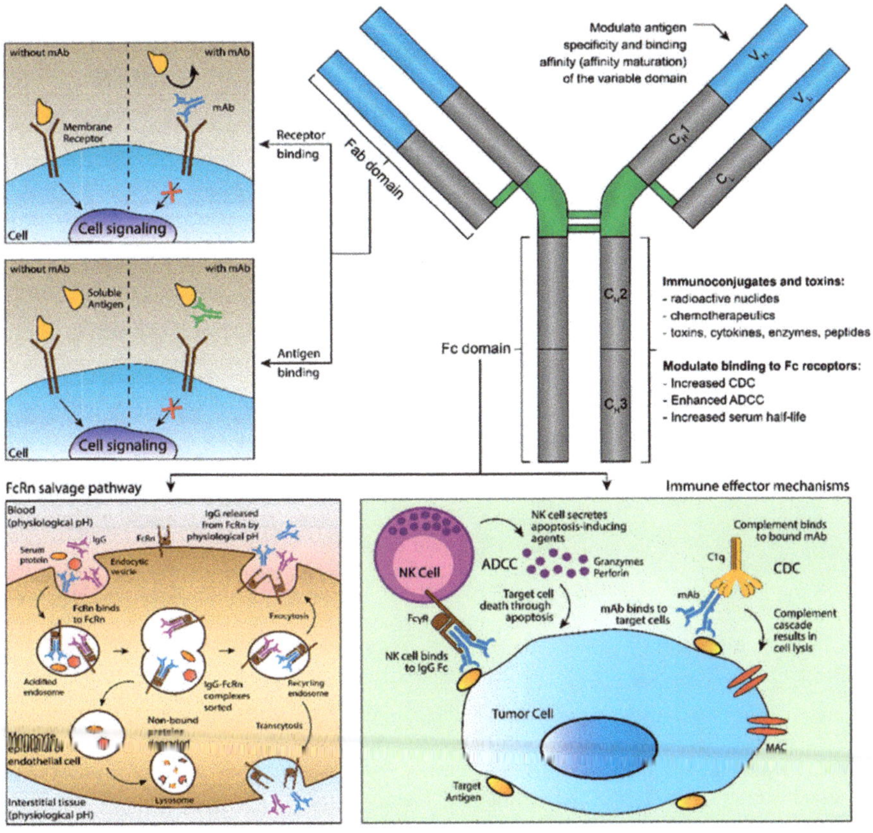

Figure 3. Main functions of therapeutic monoclonal antibodies (mAbs). mAbs have two antigen-binding fragments (Fabs) and one constant fragment crystallizable (Fc). The variable domain of the Fab confers specificity and binding affinity to either membrane receptors or soluble antigens. The Fc domain binds neonatal Fc receptor (FcRn), prolonging the half-life of mAbs; and connects immunoglobulin G (IgG) antibodies to immune effector mechanisms (antibody-dependent cell cytotoxicity—ADCC; and complement-dependent cytotoxicity—CDC) by engaging Fcγ receptors (FcγR) on immune cells, promoting cell lysis.

3.2. Therapeutic Value

The first mAb approved was muromonab-CD3 in the prevention of transplant rejection. Ever since, mAbs have been introduced in a number of therapeutic regimens in a wide range of conditions, such as organ transplantation (e.g., basiliximab and belatacept), inflammatory diseases (e.g., adalimumab and tocilizumab), and cancer (e.g., trastuzumab and cetuximab) [70]. The use of antibodies is increasing and improved mAb-based strategies will appear on the market in response to current therapeutic challenges (Figure 4). In particular, antibody research focused on the development of antibody fragments (e.g., single-chain Fv—scFv, single-domain antibody—sdAb, antigen-binding fragments—Fab); antibody–drug conjugates (ADC) (e.g., trastuzumab emtansine); fusion proteins (e.g., etanercept); and intrabodies [80–82]. Although physicians use mAbs in a variety of conditions, their applicability in the treatment of CNS diseases and BM remains challenging. Nevertheless, there are some mAb-based systems already approved (Table 1) or in investigation (Table 2).

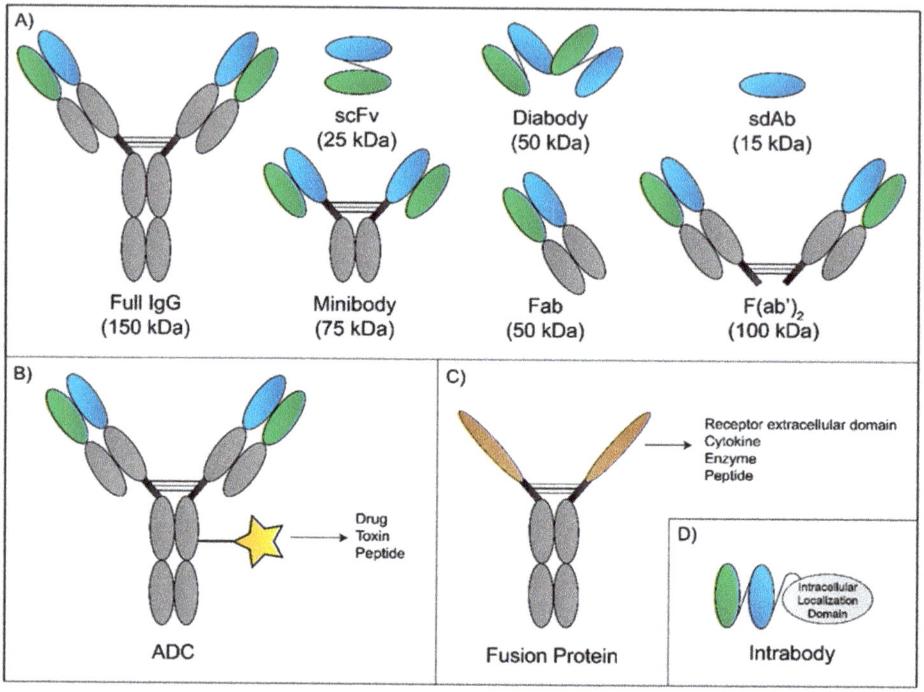

Figure 4. Novel or innovative monoclonal antibody (mAb) strategies. Schematic representation of different antibody formats currently in research. (**A**) An intact IgG molecule alongside with various antibody fragments and their respective molecular weight. (**B**) Antibody-drug conjugates (ADC) are usually intact IgG molecules linked to a drug, toxin, or peptide to increase the cargo selectivity. (**C**) Fusion proteins are biopharmaceutical molecules where the binding domains can be derived from a receptor extracellular domain, cytokine, enzyme, and peptide. Depending on the IgG molecule, the Fc region is capable of FcγR and C1q binding, potentially enabling the fusion protein to initiate antibody-dependent cell-mediated cytotoxicity (ADCC) and complement-dependent cytotoxicity (CDC). (**D**) The scheme shows an intrabody made of a variable region of the light and heavy chain constituting the antigen-binding domain and an intracellular location domain to allow nuclear binding.

3.3. *CNS Diseases*

Multiple sclerosis and episodic headache are the only neurologic pathologies where mAbs have been administrated, which started with the approval of natalizumab [83] and erenumab [84], respectively. Natalizumab is an IgG4 mAb targeting α4β1-integrin (very late activation antigen-4—VLA-4), which is present on the surface of leukocytes. After binding to VLA-4, the mAb inhibits the interaction between VLA-4 and vascular cell adhesion molecule-1 (VCAM-1). Consequently, reducing the adhesion, attachment, and migration of leukocytes across the BBB into the CNS [85]. In a pivotal phase III trial (AFFIRM), natalizumab reduced clinical relapse at one-year by 68% and the risk of continuous progression of disability by 42–54% over two years [86]. Nevertheless, the therapeutic effect observed occurs due to a peripheral action, instead of a direct antibody penetration into the brain. Erenumab is an IgG2, which targets the calcitonin gene-related peptide (CGRP) receptor [87]. The mAb competes with the binding of CGRP and inhibits its function at the CGRP receptor. The CGRP receptors are located at relevant sites to migraine pathophysiology, such as the trigeminal ganglion and the paraventricular structures. The BBB does not protect these regions. Thus, erenumab also exerts action at the periphery and not at the brain. In phase III STRIVE clinical trial, erenumab was able to significantly reduce the number of migraine days per month by 3.2 versus 1.8 in the placebo group. Efficacy was sustained up to one year [88].

A different strategy was applied in the management of Alzheimer's disease. In this case, instead of antibodies acting at the periphery, researchers are using mAbs and re-engineered antibody fragments targeting natural brain portals (e.g., TfR and IR) [44]. Pardridge et al. reported for the first time an anti-β-amyloid (Aβ) scFv fused to mAbs targeting either TfR or IR [89]. The brain uptake of the molecule was of 0.88% ID/brain. The value is within the boundaries of the brain uptake of other drugs that are active in the brain [44,90,91]. However, the mAb demonstrated a high affinity towards the receptor. Thus, receptor dissociation was a major limitation. The consequence might be antibody degradation since the molecules, instead of crossing the BBB, follow the lysosomal pathway. Yu et al. elegantly solved the problem by reducing the affinity of these mAb [53]. In addition, the group further developed a bispecific therapeutic antibody with a low affinity for TfR and a high affinity for another Alzheimer's disease drug target, the BACE1 [54]. The brain accumulation of the molecule in non-human primates was significantly higher than control, and the amyloid β-peptide presence in the brain and serum reduced considerably. The drawback of these RMT systems is the competition with natural substrates, which may affect brain homeostasis; and receptor saturation due to the high affinity of the antibodies engineered [44].

3.4. Brain Metastases

In the treatment of BM, the reality is different. In the recruitment phase of clinical trials, an active exclusion of patients presenting BM occurs [92,93]. Therefore, to date, no clinical trial supports the use of mAbs in the management of BM. The lack of information concerning efficacy and safety are the main reasons. Consequently, the standard cancer regimens with well-established antibody-based treatments cannot be applied to patients with BM [94]. Their use by physicians represents an off-label use. Nevertheless, a class of mAbs related to immunotherapy (e.g., nivolumab, pembrolizumab) is showing promising results. Several studies suggest their role in the elimination of metastatic tumors, such as in the brain. Still, more data is necessary to approve these antibodies in the treatment of metastatic brain cancers [95]. Another promising field is the radionuclide therapy. In the last years, numerous papers have been published reporting its success in brain tumors and metastases [96].

New therapeutic targets in metastatic progression in either primary or secondary tumors have driven intense research into the development of mAb-based systems [97,98], as they offer effective targeted treatment with low adverse events. However, the lack of specificity and poor BBB penetration render them ineffective. It is therefore imperative to find strategies that allow antibody translocation across the BBB. For instance, the use of antibody fragments to reduce their molecular weight. Or modify the mAbs to contain a translocation moiety, such as a CPPs. CPPs are effective in the delivery of large cargoes across cell membranes and even across the BBB (BBB peptide shuttles) [66,99]. Similar to the Trojan horse approach they engage interaction with BECs and BBB translocation, the main advantage being that CPP does not require receptors in the majority of the cases, thus reducing the toxicity of the system significantly.

Table 1. Antibody therapeutics that are approved or in review for marketing for cancer or central-nervous-system (CNS) diseases.

<table>
<tr><th>Name</th><th>Brand Name</th><th>Type</th><th>Target</th><th>First Indication Approved or Reviewed</th><th>Approval Status (EU|USA)</th></tr>
<tr><td>[fam-]trastuzumab deruxtecan</td><td>-</td><td>Humanized IgG1 ADC</td><td>HER2</td><td>HER2+ metastatic breast cancer</td><td>N.A.|In review</td></tr>
<tr><td>Ado-trastuzumab emtansine</td><td>Kadcyla</td><td>Humanized IgG1 ADC</td><td>HER2</td><td>Breast cancer</td><td>2013|2012</td></tr>
<tr><td>Alemtuzumab</td><td>Lemtrada</td><td>Humanized IgG1</td><td>CD52</td><td>Multiple sclerosis; chronic myeloid leukemia#</td><td>2013; 2001 #|2014;2001 #</td></tr>
<tr><td>Atezolizumab</td><td>Tecentriq</td><td>Humanized IgG1</td><td>PD-L1</td><td>Bladder cancer</td><td>2017|2016</td></tr>
<tr><td>Avelumab</td><td>Bavencio</td><td>Human IgG1</td><td>PD-L1</td><td>Merkel cell carcinoma</td><td>2017|2017</td></tr>
<tr><td>Bevacizumab</td><td>Avastin</td><td>Humanized IgG1</td><td>VEGF</td><td>Colorectal cancer</td><td>2005|2004</td></tr>
<tr><td>Blinatumomab</td><td>Blincyto</td><td>Murine bispecific tandem scFv</td><td>CD19, CD3</td><td>Acute lymphoblastic leukemia</td><td>2015|2014</td></tr>
<tr><td>Brentuximab vedotin</td><td>Adcentris</td><td>Chimeric IgG1 ADC</td><td>CD30</td><td>Hodgkin lymphoma, systemic anaplastic large cell lymphoma</td><td>2012|2011</td></tr>
<tr><td>Cemiplimab</td><td>Libtayo</td><td>Human mAb</td><td>PD-1</td><td>Cutaneous squamous cell carcinoma</td><td>2019|2018</td></tr>
<tr><td>Cetuximab</td><td>Erbitux</td><td>Chimeric IgG1</td><td>EGFR</td><td>Colorectal cancer</td><td>2004|2004</td></tr>
<tr><td>Daclizumab</td><td>Zinbryta; Zenapax</td><td>Humanized IgG1</td><td>IL-2R</td><td>Multiple sclerosis #; prevention of kidney transplant rejection #</td><td>2016 #; 1999 #|2016 #; 1997 #</td></tr>
<tr><td>Daratumumab</td><td>Darzalex</td><td>Human IgG1</td><td>CD38</td><td>Multiple myeloma</td><td>2016|2015</td></tr>
<tr><td>Dinutuximab</td><td>Unituxin</td><td>Chimeric IgG1</td><td>GD2</td><td>Neuroblastoma</td><td>2015|2015</td></tr>
<tr><td>Durvalumab</td><td>IMFINZI</td><td>Human IgG1</td><td>PD-L1</td><td>Bladder cancer</td><td>2018|2017</td></tr>
<tr><td>Edrecolomab</td><td>Panorex</td><td>Murine IgG2a</td><td>EpCAM</td><td>Colon cancer</td><td>1995 *#|NA</td></tr>
<tr><td>Elotuzumab</td><td>Empliciti</td><td>Humanized IgG1</td><td>SLAMF7</td><td>Multiple myeloma</td><td>2016|2015</td></tr>
<tr><td>Enfortumab vedotin</td><td>-</td><td>Human IgG1 ADC</td><td>Nectin-4</td><td>Urothelial cancer</td><td>N.A.|In review</td></tr>
<tr><td>Eptinezumab</td><td>-</td><td>Humanized IgG1</td><td>CGRP</td><td>Migraine prevention</td><td>N.A.|In review</td></tr>
<tr><td>Erenumab</td><td>Aimovig</td><td>Human IgG2</td><td>CGRP</td><td>Migraine prevention</td><td>2018|2018</td></tr>
<tr><td>Fremanezumab</td><td>Ajovy</td><td>Humanized IgG2</td><td>CGRP</td><td>Migraine prevention</td><td>2019|2018</td></tr>
<tr><td>Galcanezumab</td><td>Emgality</td><td>Humanized IgG4</td><td>CGRP</td><td>Migraine prevention</td><td>2018|2018</td></tr>
<tr><td>Gemtuzumab ozogamicin</td><td>Mylotarg</td><td>Humanized IgG4 ADC</td><td>CD33</td><td>Acute myeloid leukemia</td><td>2018|2017; 2000 #</td></tr>
<tr><td>Ibritumomab tiuxetan</td><td>Zevalin</td><td>Murine IgG1</td><td>CD20</td><td>Non-Hodgkin lymphoma</td><td>2004|2002</td></tr>
<tr><td>Idarucizumab</td><td>Praxbind</td><td>Humanized Fab</td><td>Dabigatran</td><td>Reversal of dabigatran-induced anticoagulation</td><td>2015|2015</td></tr>
<tr><td>Inotuzumab ozogamicin</td><td>BESPONSA</td><td>Humanized IgG4 ADC</td><td>CD22</td><td>Acute lymphoblastic leukemia</td><td>2017|2017</td></tr>
<tr><td>Ipilimumab</td><td>Yervoy</td><td>Human IgG1</td><td>CTLA-4</td><td>Metastatic melanoma</td><td>2011|2011</td></tr>
<tr><td>Isatuximab</td><td>-</td><td>Humanized IgG1</td><td>CD38</td><td>Multiple myeloma</td><td>In review|In review</td></tr>
<tr><td>Natalizumab</td><td>Tysabri</td><td>Humanized IgG4</td><td>α4 integrin</td><td>Multiple sclerosis</td><td>2006|2004</td></tr>
<tr><td>Nebacumab</td><td>Centoxin</td><td>Human IgM</td><td>Endotoxin</td><td>Gran-negative sepsis</td><td>1991 *#|N.A.</td></tr>
<tr><td>Necitumumab</td><td>Portrazza</td><td>Human IgG1</td><td>EGFR</td><td>Non-small cell lung cancer</td><td>2015|2015</td></tr>
<tr><td>Nivolumab</td><td>Opdivo</td><td>Human IgG4</td><td>PD1</td><td>Melanoma, non-small cell lung cancer</td><td>2015|2014</td></tr>
</table>

Table 1. *Cont.*

Name	Brand Name	Type	Target	First Indication Approved or Reviewed	Approval Status (EU|USA)
Obinutuzumab	Gazyva	Humanized IgG1 Glycoengineered	CD20	Chronic lymphocytic leukemia	2014|2013
Ocrelizumab	OCREVUS	Humanized IgG1	CD20	Multiple sclerosis	2018|2017
Ofatumumab	Arzerra	Human IgG1	CD20	Chronic lymphocytic leukemia	2010|2009
Olaratumab	Lartruvo	Human IgG1	PDGFRα	Soft tissue sarcoma	2016|2016
Panitumumab	Vectibix	Human IgG2	EGFR	Colorectal cancer	2007|2006
Pembrolizumab	Keytruda	Humanized IgG4	PD1	Melanoma	2015|2014
Pertuzumab	Perjeta	Humanized IgG1	HER2	Breast cancer	2013|2012
Polatuzumab vedotin	Polivy	Humanized IgG1 ADC	CD79b	Diffuse large B-cell lymphoma	In review|2019
Ramucirumab	Cyramza	Human IgG1	VEGFR2	Gastric cancer	2014|2014
Rituximab	MabThera	Chimeric IgG1	CD20	Non-Hodgkin lymphoma	1998|1997
Sacituzumab govitecan	-	Humanized IgG1	TROP-2	Triple-negative breast cancer	N.A.|In review
Tafasitamab	-	Humanized IgG1	CD19	Diffuse large B-cell lymphoma	N.A.|In review
Tositumomab-I131	Bexxar	Murine IgG2a	CD20	Non-Hodgkin lymphoma	N.A.|2003 #
Trastuzumab	Heceptin	Humanized IgG1	HER2	Breast cancer	2000|1998

ADC, Antibody–drug conjugate; CGRP, Calcitonin gene-related peptide; CTLA-4, Cytotoxic T-lymphocyte-associated protein 4; EGFR, Epidermal growth factor receptor; EpCAM, Epithelial cellular adhesion molecule; Fab, Fragment antigen-binding; HER2, Human epidermal growth factor receptor-2; mAb, Monoclonal antibody; PD1, Programmed cell death protein-1; PDGFR, Platelet-derived growth factor receptor; PD-L1, Programmed death-Ligand 1; scFv, Single-chain fragment variable; SLAMF7, Signaling lymphocytic activation molecule F7; TROP-2, Tumor-associated calcium signal transducer 2; VEGF, Vascular endothelial growth factor; N.A. Not approved; * Country-specific approval; # Withdrawn or marketing discontinued. Adapted from [100].

Table 2. Antibody therapeutics that are in investigation for cancer or central-nervous-system (CNS) diseases.

Name	Type	Target	Clinical Indications	Most Advanced Phase
(vic-)trastuzumab duocarmazine	Humanized IgG1 ADC	HER2	Breast cancer	Phase 3
[125I]-mAb 425	Human mAb	EGFR	Glioblastoma	Phase 2
[131I]-BC-2 mAb	Human mAb	Tenascin	Glioblastoma	Phase 1/2
[131I]-chTNT-1/B MAb	Human mAb	DNA-histone H1 complex	Glioblastoma	Phase 1/2
[131I]-SGMIB anti-HER2 VHH1	Humanized VHH	HER2	Breast cancer	Phase 1
[188Re]-labeled Nimotuzumab	Humanized mAb	EGFR	Glioblastoma	Phase 1
[211At]-labeled 81C6 mAb	Human mAb	Tenascin	Glioblastoma	Phase 1/2
131I-omburtamab	Murine mAb, radio-labeled	BT-H3	Neuroblastoma central nervous system/leptomenigeal metastases	Phase 2/3
68-Ga-NOTA-anti-HER2 VHH1	Humanized VHH	HER2	Brain metastases of breast carcinoma	Phase 2
ABBV-8E12	Human mAb	Tau protein	Alzheimer's disease	Phase 2
Aducanumab	Human IgG1	Amyloid beta	Alzheimer's disease	Phase 3
AL002	Human mAb	TREM2 receptor	Alzheimer's disease	Phase 1
AL003	Human mAb	SIGLEC-3	Alzheimer's disease	Phase 1
ALX-0651	Humanized VHH	CXCR4	-	Phase 1
Andecaliximab	Humanized IgG4	MMP9	Gastric or gastroesophageal junction adenocarcinoma	Phase 3
BAT8001	Humanized IgG1 ADC	HER2	Breast cancer	Phase 3
BCD-100	Human mAb	PD-1	Melanoma	Phase 2/3
Bemarituzumab	Humanized IgG1	FGFR2b	Gastric or gastroesophageal junction adenocarcinoma	Phase 3
BIB092	Human mAb	Tau protein	Alzheimer's disease	Phase 2
biotin-coupled BC-4 + Avidin + [90Y]-Biotin	Human mAb	Tenascin	Glioblastoma	Phase 1/2
Bispecific nb-derived CAR-T cells	Bispecific Humanized tandem VH	CD19/CD20	Refractory/relapsed B-cell lymphoma	Phase 1
Camrelizumab	Humanized IgG4	PD-1	Hodgkin's lymphoma, hepatocellular carcinoma	Phase 3
Carotuximab	Chimeric IgG1	Endoglin	Angiosarcoma	Phase 3
Crenezumab	Humanized IgG4	Amyloid beta	Alzheimer's disease	Phase 3
CS1001	Human mAb	PD-L1	Non-small cell lung cancer	Phase 3
Depatuxizumab mafodotin	IgG1 ADC	EGFR	Glioblastoma	Phase 2b/3
Donanemab	Humanized IgG1 mAb	Amyloid beta	Alzheimer's disease	Phase 2
Enfortumab vedotin	Human IgG1 ADC	Nectin-4	Urothelial cancer	Phase 3
Eptinezumab	Humanized IgG1	CGRP	Episodic migraines	Phase 3
Gantenerumab	Human IgG1	Amyloid beta	Alzheimer's disease	Phase 3

Table 2. *Cont.*

Name	Type	Target	Clinical Indications	Most Advanced Phase
I-131-BC8	Murine IgG1, radio-labeled	CD45	Ablation of bone marrow to hematopoietic cell transplantation in AML patients	Phase 3
IBI308	Human mAb	PD-1	Squamous cell non-small cell lung cancer	Phase 3
Isatuximab	Humanized IgG1	CD38	Multiple myeloma	Phase 3
JNJ-63733657	Human mAb	Tau protein	Alzheimer's disease	Phase 1
KN035	mAb single domain	PD-L1	Bile tract carcinoma	Phase 3
L19IL2 + L19TNF	scFv conjugates	Fibronectin extra-domain B	Melanoma	Phase 3
Loncastuximab tesirine	Humanized IgG1 ADC	CD19	Diffuse large B-cell lymphoma	Phase 2
Margetuximab	Chimeric IgG1	HER2	Breast cancer	Phase 3
Mirvetuximab soravtansine	IgG1 ADC	Folate receptor 1	Ovarian cancer	Phase 3
Naxitamab	Humanized mAb	GD2	High-risk neuroblastoma and refractory osteomedullary disease	Phase 3
Opicinumab	Human mAb	LINGO-1	Multiple sclerosis, acute optic neuritis	Phase 2
Oportuzumab monatox	Humanized scFv immunotoxin	EpCAM	Bladder cancer	Phase 3
Polatuzumab vedotin	Humanized IgG1 ADC	CD79b	Diffuse large B-cell lymphoma	Phase 3
Relatlimab	Human mAb	LAG-3	Melanoma	Phase 2/3
Rovalpituzumab tesirine	Humanized IgG1 ADC	DLL3	Small cell lung cancer	Phase 3
Semorinemab	Humanized IgG4	Tau protein	Alzheimer's disease	Phase 2
Solanezumab	Humanized IgG1 mAb	Monomers	Alzheimer's disease	Phase 3
Spartalizumab	Humanized IgG4	PD1	Melanoma	Phase 3
Tisdelizumab	Humanized mAb	PD1	Non-small cell lung cancer, Hodgkin's lymphoma	Phase 3
Trastuzumab deruxtecan	Humanized ADC	HER2	Breast cancer, HER2+ gastric or gastroesophageal junction adenocarcinoma	Phase 3
Tremelimumab	Human IgG2	CTLA4	Non-small cell lung, head & neck, urothelial cancer	Phase 3
TSR-042	Humanized mAb	PD1	Ovarian cancer	Phase 3
Ublituximab	Chimeric IgG1	CD20	Chronic lymphocytic leukemia	Phase 3
Utomilumab	Human IgG2	CD137	Diffuse large B-cell lymphoma	Phase 3
XMAB-5574	Humanized IgG1	CD19	Diffuse large B-cell lymphoma	Phase 2/3
Zagotenemab	Human mAb	Tau protein	Alzheimer's disease	Phase 2
Zolbetuximab	Chimeric IgG1	Claudin-18.2	Gastric or gastroesophageal junction adenocarcinoma	Phase 3

ADC, Antibody–drug conjugate; AML, Acute myeloid leukemia; CGRP, Calcitonin gene-related peptide; CTLA-4, Cytotoxic T-lymphocyte-associated protein 4; CXCR4, Chemokine receptor type 4; DLL3, Delta-like protein 3; EGFR, Epidermal growth factor receptor; EpCAM, Epithelial cellular adhesion molecule; HER2, Human epidermal growth factor receptor-2; LAG-3, Lymphocyte-activation gene 3; mAb, Monoclonal antibody; MMP-9, Matrix metalloproteinase-9; PD1, Programmed cell death protein-1; PD-L1, Programmed death-ligand 1; scFv, single-domain fragment variable; TREM2, Triggering receptor expressed on myeloid cells 2. Adapted from [96,101–104].

4. The Role of BBB Peptide-Shuttles

CPPs are short peptides (less than 30 amino residues) capable of crossing cell membranes without causing significant membrane damage [99]. They represent a broad group of peptides with different physicochemical properties. Accordingly, they can be: (1) cationic, which comprises peptides with highly positive charges at physiological pH; (2) amphipathic, that contains both polar (hydrophilic) and nonpolar (hydrophobic) regions of amino acids; and (3) hydrophobic, the less studied class, which are CPPs mainly containing nonpolar residues, resulting in a low net charge. Naturally occurring proteins and peptides are the principal sources of CPPs [105,106], however, to optimize the peptides' properties, fully engineered peptides have been designed, based on computational modeling [107].

The specific internalization mechanism of CPPs is unclear [108,109]. The peptide's concentration, the cargo conjugated, the physicochemical properties, and molecular weight are features affecting the efficiency of cellular entry, as well as the internalization pathway followed. Nevertheless, energy-dependent endocytic pathways, which include clathrin-mediated endocytosis, caveolin-mediated endocytosis, and macropinocytosis, are considered the main translocation mechanisms [110]. The intensive research using peptides and the development of technologies that allowed their conjugation to TPs (e.g., recombinant DNA technology) resulted in the capacity of cargo-transportation not only across cell membranes, especially epithelia, but also the endothelial BBB (BBB peptide shuttle).

Human Immunodeficiency Virus Trans-activator of transduction (TAT) peptide was the first peptide demonstrating translocation properties [111]. Subsequently, many other peptides, like SynB, Penetratin, Angiopep-2, dNP2, and PepH3 were studied with relevant results [112]. Despite the variation in length and amino acids' sequence, these peptides share common features. Among others, their amphipathic nature, net positive charge, theoretical hydrophobicity, helical moment, as well as the ability to interact with lipid membranes. The mechanism by which these peptide shuttles cross the BBB and mediate cargo translocation is not fully understood and may vary according to the concentration, cell type, and the cargo of interest [113]. Direct membrane permeation, RMT, and AMT are the three principal possibilities. The latter constitutes an advantage compared with others since it avoids enzymatic degradation, problems related to endosomal escape, receptor saturation, and toxicity, among others. The process is based on the electrostatic binding of positive charge peptide-shuttle to negative charge proteoglycans (Figure 2), forming a vesicle that transports the system across the endothelial cells layer [112].

The efficiency of large proteins delivery, such as antibodies or fusion proteins across cell membranes by CPPs, has been intensively studied, mainly in vitro. However, the delivery to the brain by peptide shuttles was not. Schwarze et al. performed the first in vivo study using BBB peptide shuttles. In their work, they successfully delivered a 120 kDa β-galactosidase fused to TAT into the brain [114]. These results showed that the direct delivery of proteins into the brain was possible. Afterward, others conjugated TAT to B-cell lymphoma-extra-large (TAT-Bcl-xL), glial cell-derived neurotrophic factor (TAT-GDNF), NR2B9c (TAT-NR2B9c), and c-Jun N-terminal kinase-1 (TAT-JNK1) fusion proteins, and evaluated their concentration in the brain [115]. In addition, the use of rabies virus glycoprotein (RVG) fused to brain-derived neurotrophic factor (RVG-BDNF), and fibroblast growth factor-4 (FGF4) fused to suppressor of cytokine signaling-3 (FGF4-SOCS3) also validated the use of these peptide shuttles [116,117].

Angiopep-2 is a 19-amino acid peptide, derived from the Kunitz domain, which binds to LRP1 and efficiently penetrates the BBB via RMT. In the study performed by Demeule et al., the translocation of angiopep-2 in an in vitro BBB model was found to be seven-fold higher than of aprotinin, an LRP1 natural ligand with BBB translocation properties [118]. Furthermore, the apparent distribution of the peptide shuttle in vivo was far greater than both transferrin and aprotinin, confirming the BBB translocation capabilities of angiopep-2. To further challenge the peptide, researchers conjugated it to an anti-HER2 mAb to investigate the ability of cargo translocation across the BBB. HER2+ breast cancer patients demonstrate a high incidence of BM. The low concentration of mAb in the brain provides a "sanctuary site" for tumor proliferation. Nevertheless, in this study, after carotid artery administration, 60% of the molecule was localized in the brain, demonstrating high brain accumulation. Besides, increased survival was reported compared with control [119].

dNP2 is an amphiphilic human-derived CNS-permeable peptide shuttle. To evaluate the abilities of cargo translocation across the BBB, Lim et al. conjugated the peptide with the cytoplasmic domain of CTLA4 (ctCTLA-4) [120]. CTLA-4 is an immune regulatory receptor expressed on the surface of T cells and often associated with susceptibility to multiple sclerosis. Thus, proving the autoimmune pathogenesis of the disease. In Lim et al. study, the administration of the dNP2-conjugated ctCTLA-4 protein successfully controlled autoimmune effector T-cell responses in an experimental autoimmune encephalomyelitis (EAE) model, an experimental mouse model of multiple sclerosis. The exact mechanism of BBB translocation and cell internalization were not determined. However, due to dNP2 properties, BBB crossing was considered to be AMT, and cellular uptake of the fusion protein through lipid-raft mediated endocytosis [120].

The most recent peptide shuttle reported is PepH3, a cationic peptide derived from Dengue virus type-2 capsid protein (DEN2C) [121]. In Neves et al. study, the in vitro BBB transmigration after 24 h was 67.2%. Furthermore, in an in vivo model, the peptide showed a brain biodistribution of 0.31% after 5 minutes. Although the exact mechanism is not fully described, studies with endocytosis inhibitors reveal that the PepH3 mechanism is consistent with the AMT. The peptide BBB translocation capacity was also evaluated in conjugation to anti-β-amyloid protein 42 (bAP42) sdAb (anti-bAP42 sdAb) [122]. bAP42 is an amyloid precursor protein fragment that plays a significant role in the formation of "senile plaques" characteristic of Alzheimer's disease. Through conjugation of PepH3 to an antibody fragment, the investigators expected the increase in antibody concentration in the brain, followed by binding to bAP42, decreasing plaques formation. Also, the complex may be detected in peripheral blood aiding in the diagnosis of disease. Interestingly, the PepH3-anti-bAP42 sdAb conjugate showed a brain accumulation of 1.5% after the same period (5 minutes), showing that cargoes do influence the translocation mechanisms of BBB peptide shuttles [122].

The results obtained with these fusion proteins are promising. Nevertheless, their high elimination rates prevent the clinical use of such peptide shuttles-based systems. Peptides and small proteins are rapidly metabolized by serum/tissue proteases and easily eliminated by glomerular filtration [123]. Since they do not present a high circulation time, BBB translocation is also affected. By increasing circulation time, the brain concentration the conjugates will improve, due to a higher contact frequency and interaction time. Thus, increasing the half-life of such systems will hypothetically augment brain accumulation. Several strategies are available. The most interesting methods explore the human serum albumin protein (HSA) and the FcRn mediated recycling [124].

In the peptide shuttles-based approaches mentioned, researchers used mainly antibody fragments of a full IgG as therapeutic agents [120,122]. However, antibodies are proteins also used as pharmacokinetic enhancers owing to their Fc region. Thus, a strategy combining these properties with the translocation capabilities of peptide shuttles will improve the half-life of the system and increase brain accumulation. Depending on the target selected, the strategy may be used to eliminate BM or CNS diseases. Since antibodies are large proteins and BBB translocation is complicated, instead of a full antibody, the engineering of a minibody comprising the antibodies' minimal domains (Fc fragment and variable region) to keep the antibody properties is preferable. In this, a reduction from 150 kDa for 25–80 kDa is achieved (Figure 4). These molecular weights are similar to the fusion proteins studied, increasing the interest of the strategy proposed. Another approach is the nanobodies (smallest Ab-derivative) which reveal promising results in preclinical and clinical studies in the elimination of brain tumors [102,125,126].

5. Conclusions and Future Perspectives

Our understanding of cancer biology and technological advances in cancer diagnosis and therapy improved significantly over the past decades. One of the landmarks for reversing the worldwide increase in cancer incidence and mortality was the development of more effective, tumor-specific, and less toxic anti-cancer drugs. The use of targeted therapy of human cancers using mAb-based systems has revolutionized cancer therapy. They are currently being used as the first choice to treat some of the most frequent metastatic cancers, such as HER2+ breast cancers or colorectal cancers. More recently, the efficacy demonstrated by antibodies inhibiting immune checkpoints has extended their use in other tumor types. In addition, they have also been introduced in many therapeutic protocols in combination with sMDs. For the treatment of cancer, the Food and Drug Administration (FDA) and the European Medicines Agency (EMA) have approved 32 mAbs-based systems. Interestingly, between 2012 and 2017 the number has doubled. Therefore, the value of mAbs in cancer therapy is undisputable.

Despite the excitement of mAbs-based systems in cancer treatment, their use in brain cancers and BMs are limited since they are unable to cross the BBB. Their molecular weight and hydrophilic nature difficult brain accumulation. Therefore, to create BM specific mAbs-based systems, they need to be modified to get across the BBB. Many strategies have been employed for mAbs to become a reality in the treatment of BMs. The most promising uses BBB peptide shuttles conjugated to mAbs-based systems. These peptides use endogenous routes, such as RMT or AMT to cross the BBB. Although RMT has been by far the most exploited route, it presents several disadvantages, namely, the receptor's saturation and natural ligand competition. Therefore, with the advent of many peptide shuttles exploring AMT, the crossing of the BBB without interfering with brain homeostasis has become a reality. Consequently, the number of studies involving BBB peptide shuttles mAb-based systems are increasing in the literature.

In conclusion, BBB peptide shuttles mAb-based systems are being designed and studied with some limitations. However, the attractive results within different studies validate their application. Consequently, in the near future, it is expected a significant increase in the number of molecules conjugated to BBB peptide shuttles pushing the use of antibodies for the treatment of BMs into a reality.

Author Contributions: M.C., D.G., M.A.R.B.C., and V.N. contributed to the structuring of the present review, to the literature search, and to the writing of the manuscript. All authors have read and agreed to the published version of the manuscript.

Funding: This research was funded by the Portuguese Funding Agency, Fundação para a Ciência e Tecnologia, FCT IP, grants PD/BD/128281/2017, PTDC/BBB-BQB/1693/2014 and PTDC/BBB-NAN/1578/2014.

Conflicts of Interest: The authors declare no conflict of interest.

References

1. Lin, X.; DeAngelis, L.M. Treatment of Brain Metastases. *J. Clin. Oncol.* **2015**, *33*, 3475–3484. [CrossRef] [PubMed]
2. Cagney, D.N.; Martin, A.M.; Catalano, P.J.; Redig, A.J.; Lin, N.U.; Lee, E.Q.; Wen, P.Y.; Dunn, I.F.; Bi, W.L.; Weiss, S.E.; et al. Incidence and prognosis of patients with brain metastases at diagnosis of systemic malignancy: A population-based study. *Neuro-Oncology* **2017**, *19*, 1511–1521. [CrossRef]
3. Nayak, L.; Lee, E.Q.; Wen, P.Y. Epidemiology of Brain Metastases. *Curr. Oncol. Rep.* **2012**, *14*, 48–54. [CrossRef] [PubMed]
4. Tabouret, E.; Chinot, O.; Metellus, P.; Tallet, A.; Viens, P.; GonÇAlves, A. Recent Trends in Epidemiology of Brain Metastases: An Overview. *Anticancer Res.* **2012**, *32*, 4655–4662. [PubMed]
5. Eichler, A.F.; Loeffler, J.S. Multidisciplinary Management of Brain Metastases. *Oncology* **2007**, *12*, 884–898. [CrossRef] [PubMed]
6. Massagué, J.; Obenauf, A.C. Metastatic Colonization. *Nature* **2016**, *529*, 298–306. [CrossRef]
7. Ren, D.; Zhu, X.; Kong, R.; Zhao, Z.; Sheng, J.; Wang, J.; Xu, X.; Liu, J.; Cui, K.; Zhang, X.H.F.; et al. Targeting Brain-Adaptive Cancer Stem Cells Prohibits Brain Metastatic Colonization of Triple-Negative Breast Cancer. *Cancer Res.* **2018**, *78*, 2052–2064. [CrossRef]
8. van Zijl, F.; Krupitza, G.; Mikulits, W. Initial steps of metastasis: Cell invasion and endothelial transmigration. *Mutat. Res.* **2011**, *728*, 23–34. [CrossRef]

9. Jeevan, D.S.; Cooper, J.B.; Braun, A.; Murali, R.A.J.; Jhanwar-Uniyal, M. Molecular Pathways Mediating Metastases to the Brain via Epithelial-to-Mesenchymal Transition: Genes, Proteins, and Functional Analysis. *Anticancer Res.* **2016**, *36*, 523–532.
10. Mani, S.A.; Guo, W.; Liao, M.-J.; Eaton, E.N.; Ayyanan, A.; Zhou, A.Y.; Brooks, M.; Reinhard, F.; Zhang, C.C.; Shipitsin, M.; et al. The epithelial–mesenchymal transition generates cells with properties of stem cells. *Cell* **2008**, *133*, 704–715. [CrossRef]
11. Celià-Terrassa, T.; Kang, Y. Metastatic niche functions and therapeutic opportunities. *Nat. Cell Biol.* **2018**, *20*, 868–877. [CrossRef] [PubMed]
12. Serlin, Y.; Shelef, I.; Knyazer, B.; Friedman, A. Anatomy and Physiology of the Blood–brain Barrier. *Semin. Cell Dev. Biol.* **2015**, *38*, 2–6. [CrossRef] [PubMed]
13. Strazielle, N.; Ghersi-Egea, J.F. Physiology of Blood–Brain Interfaces in Relation to Brain Disposition of Small Compounds and Macromolecules. *Mol. Pharm.* **2013**, *10*, 1473–1491. [CrossRef] [PubMed]
14. Sharif, Y.; Jumah, F.; Coplan, L.; Krosser, A.; Sharif, K.; Tubbs, R.S. Blood brain barrier: A review of its anatomy and physiology in health and disease. *Clin. Anat.* **2018**, *31*, 812–823. [CrossRef]
15. Daneman, R.; Prat, A. The blood–brain barrier. *Cold Spring Harb. Perspect. Biol.* **2015**, *7*, a020412. [CrossRef]
16. Weil, R.J.; Palmieri, D.C.; Bronder, J.L.; Stark, A.M.; Steeg, P.S. Breast Cancer Metastasis to the Central Nervous System. *Am. J. Pathol.* **2005**, *167*, 913–920. [CrossRef]
17. Lesniak, M.S.; Brem, H. Targeted therapy for brain tumours. *Nat. Rev. Drug Discov.* **2004**, *3*, 499–508. [CrossRef]
18. Green, C.E.; Liu, T.; Montel, V.; Hsiao, G.; Lester, R.D.; Subramaniam, S.; Gonias, S.L.; Klemke, R.L. Chemoattractant Signaling between Tumor Cells and Macrophages Regulates Cancer Cell Migration, Metastasis and Neovascularization. *PLoS ONE* **2009**, *4*, e6713. [CrossRef]
19. Gonzalez, H.; Hagerling, C.; Werb, Z. Roles of the immune system in cancer: From tumor initiation to metastatic progression. *Genes Dev.* **2018**, *32*, 1267–1284. [CrossRef]
20. You, H.; Baluszek, S.; Kaminska, B. Immune Microenvironment of Brain Metastases—Are Microglia and Other Brain Macrophages Little Helpers? *Front. Immunol.* **2019**, *10*, 1941. [CrossRef]
21. Leibold, A.T.; Monaco, G.N.; Dey, M. The role of the immune system in brain metastasis. *Curr. Neurobiol.* **2019**, *10*, 33–48. [PubMed]
22. Di Lorenzo, R.; Ahluwalia, M.S. Targeted therapy of brain metastases: Latest evidence and clinical implications. *Ther. Adv. Med. Oncol.* **2017**, *9*, 781–796. [CrossRef] [PubMed]
23. Franchino, F.; Rudà, R.; Soffietti, R. Mechanisms and Therapy for Cancer Metastasis to the Brain. *Front. Oncol.* **2018**, *8*, 161. [CrossRef] [PubMed]
24. Pardridge, W.M. Blood–brain barrier delivery. *Drug Discov. Today* **2007**, *12*, 54–61. [CrossRef] [PubMed]
25. Kabraji, S.; Ni, J.; Lin, N.U.; Xie, S.; Winer, E.P.; Zhao, J.J. Drug Resistance in HER2-Positive Breast Cancer Brain Metastases: Blame the Barrier or the Brain? *Clin. Cancer Res. Off. J. Am. Assoc. Cancer Res.* **2018**, *24*, 1795–1804. [CrossRef]
26. Dong, R.; Ji, J.; Liu, H.; He, X. The evolving role of trastuzumab emtansine (T-DM1) in HER2-positive breast cancer with brain metastases. *Crit. Rev. Oncol. Hematol.* **2019**, *143*, 20–26. [CrossRef]
27. Nieder, C.; Grosu, A.L.; Gaspar, L.E. Stereotactic radiosurgery (SRS) for brain metastases: A systematic review. *Radiat. Oncol.* **2014**, *9*, 155. [CrossRef]
28. Protopapa, M.; Kouloulias, V.; Nikoloudi, S.; Papadimitriou, C.; Gogalis, G.; Zygogianni, A. From Whole-Brain Radiotherapy to Immunotherapy: A Multidisciplinary Approach for Patients with Brain Metastases from NSCLC. *J. Oncol.* **2019**, *2019*, 12. [CrossRef]
29. Ulahannan, D.; Khalifa, J.; Faivre-Finn, C.; Lee, S.M. Emerging treatment paradigms for brain metastasis in non-small-cell lung cancer: An overview of the current landscape and challenges ahead. *Ann. Oncol.* **2017**, *28*, 2923–2931. [CrossRef]
30. Hardesty, D.A.; Nakaji, P. The Current and Future Treatment of Brain Metastases. *Front. Surg.* **2016**, *3*, 30. [CrossRef]
31. Papademetriou, I.T.; Porter, T. Promising approaches to circumvent the blood–brain barrier: Progress, pitfalls and clinical prospects in brain cancer. *Ther. Deliv.* **2015**, *6*, 989–1016. [CrossRef] [PubMed]
32. Teleanu, D.M.; Chircov, C.; Grumezescu, A.M.; Volceanov, A.; Teleanu, R.I. Blood–brain Delivery Methods Using Nanotechnology. *Pharmaceutics* **2018**, *10*, 269. [CrossRef] [PubMed]
33. Dong, X. Current Strategies for Brain Drug Delivery. *Theranostics* **2018**, *8*, 1481–1493. [CrossRef] [PubMed]

34. Gabathuler, R. Approaches to transport therapeutic drugs across the blood–brain barrier to treat brain diseases. *Neurobiol. Dis.* **2010**, *37*, 48–57. [CrossRef]
35. Bellettato, C.M.; Scarpa, M. Possible strategies to cross the blood–brain barrier. *Ital. J. Pediatrics* **2018**, *44*, 131. [CrossRef]
36. Pandey, P.K.; Sharma, A.K.; Gupta, U. Blood brain barrier: An overview on strategies in drug delivery, realistic in vitro modeling and in vivo live tracking. *Tissue Barriers* **2015**, *4*, e1129476. [CrossRef]
37. Fortin, D.; Gendron, C.; Boudrias, M.; Garant, M.-P. Enhanced chemotherapy delivery by intraarterial infusion and blood–brain barrier disruption in the treatment of cerebral metastasis. *Cancer* **2007**, *109*, 751–760. [CrossRef]
38. Kinoshita, M.; McDannold, N.; Jolesz, F.A.; Hynynen, K. Targeted delivery of antibodies through the blood–brain barrier by MRI-guided focused ultrasound. *Biochem. Biophys. Res. Commun.* **2006**, *340*, 1085–1090. [CrossRef]
39. Borlongan, C.V.; Emerich, D.F. Facilitation of drug entry into the CNS via transient permeation of blood brain barrier: Laboratory and preliminary clinical evidence from bradykinin receptor agonist, Cereport. *Brain Res. Bull.* **2003**, *60*, 297–306. [CrossRef]
40. Lipinski, C.A.; Lombardo, F.; Dominy, B.W.; Feeney, P.J. Experimental and computational approaches to estimate solubility and permeability in drug discovery and development settings1PII of original article: S0169-409X(96)00423-1. The article was originally published in Advanced Drug Delivery Reviews 23 (1997) 3–25.1. *Adv. Drug Deliv. Rev.* **2001**, *46*, 3–26. [CrossRef]
41. Tajes, M.; Ramos-Fernández, E.; Weng-Jiang, X.; Bosch-Morató, M.; Guivernau, B.; Eraso-Pichot, A.; Salvador, B.; Fernàndez-Busquets, X.; Roquer, J.; Muñoz, F.J. The blood–brain barrier: Structure, function and therapeutic approaches to cross it. *Molec. Membr. Biol.* **2014**, *31*, 152–167. [CrossRef] [PubMed]
42. Parodi, A.; Rudzińska, M.; Deviatkin, A.A.; Soond, S.M.; Baldin, A.V.; Zamyatnin, A.A., Jr. Established and Emerging Strategies for Drug Delivery Across the Blood–brain Barrier in Brain Cancer. *Pharmaceutics* **2019**, *11*, 245. [CrossRef] [PubMed]
43. Pulgar, V.M. Transcytosis to Cross the Blood Brain Barrier, New Advancements and Challenges. *Front. Neurosci.* **2019**, *12*, 1019. [CrossRef] [PubMed]
44. Neves, V.; Aires-da-Silva, F.; Corte-Real, S.; Castanho, M.A.R.B. Antibody Approaches To Treat Brain Diseases. *Trends Biotechnol.* **2015**, *34*, 36–48. [CrossRef]
45. Khan, N.U.; Miao, T.; Ju, X.; Guo, Q.; Han, L. 6—Carrier-Mediated Transportation through BBB. In *Brain Targeted Drug Delivery System*; Gao, H., Gao, X., Eds.; Academic Press: New York, NY, USA, 2019; pp. 129–158. [CrossRef]
46. Pavan, B.; Dalpiaz, A.; Ciliberti, N.; Vertuani, S. Progress in Drug Delivery to the Central Nervous System by the Prodrug Approach. *Molecules* **2008**, *13*, 1035–1065. [CrossRef]
47. Witt, K.A.; Gillespie, T.J.; Huber, J.D.; Egleton, R.D.; Davis, T.P. Peptide drug modifications to enhance bioavailability and blood–brain barrier permeability. *Peptides* **2001**, *22*, 2329–2343. [CrossRef]
48. Pardridge, W.M. Drug transport across the blood–brain barrier. *J. Cereb Blood Flow. Metab.* **2012**, *32*, 1959–1972. [CrossRef]
49. Coloma, M.J.; Lee, H.J.; Kurihara, A.; Landaw, E.M.; Boado, R.J.; Morrison, S.L.; Pardridge, W.M. Transport Across the Primate Blood–brain Barrier of a Genetically Engineered Chimeric Monoclonal Antibody to the Human Insulin Receptor. *Pharm. Res.* **2000**, *17*, 266–274. [CrossRef]
50. Pardridge, W. Blood–brain barrier drug targeting: The future of brain drug development. *Mol. Interv.* **2003**, *3*, 90–105. [CrossRef]
51. Zhang, Y.; Pardridge, W.M. Blood–brain barrier targeting of BDNF improves motor function in rats with middle cerebral artery occlusion. *Brain Res.* **2006**, *1111*, 227–229. [CrossRef]
52. Bien-Ly, N.; Yu, Y.J.; Bumbaca, D.; Elstrott, J.; Boswell, C.A.; Zhang, Y.; Luk, W.; Lu, Y.; Dennis, M.S.; Weimer, R.M.; et al. Transferrin receptor (TfR) trafficking determines brain uptake of TfR antibody affinity variants. *J. Exp. Med.* **2014**, *211*, 233–244. [CrossRef]
53. Yu, Y.J.; Zhang, Y.; Kenrick, M.; Hoyte, K.; Luk, W.; Lu, Y.; Atwal, J.; Elliott, J.M.; Prabhu, S.; Watts, R.J.; et al. Boosting Brain Uptake of a Therapeutic Antibody by Reducing Its Affinity for a Transcytosis Target. *Sci. Transl. Med.* **2011**, *3*, 84ra44. [CrossRef] [PubMed]

54. Yu, Y.J.; Atwal, J.K.; Zhang, Y.; Tong, R.K.; Wildsmith, K.R.; Tan, C.; Bien-Ly, N.; Hersom, M.; Maloney, J.A.; Meilandt, W.J.; et al. Therapeutic bispecific antibodies cross the blood–brain barrier in nonhuman primates. *Sci. Transl. Med.* **2014**, *6*, 261ra154. [CrossRef] [PubMed]
55. Boado, R.J.; Zhang, Y.; Zhang, Y.; Pardridge, W.M. Humanization of anti-human insulin receptor antibody for drug targeting across the human blood–brain barrier. *Biotechnol. Bioeng.* **2007**, *96*, 381–391. [CrossRef] [PubMed]
56. Vieira, D.B.; Gamarra, L.F. Multifunctional Nanoparticles for Successful Targeted Drug Delivery across the Blood–brain Barrier, Molecular Insight of Drug Design. *IntechOpen* **2018**. [CrossRef]
57. Tang, W.; Fan, W.; Lau, J.; Deng, L.; Shen, Z.; Chen, X. Emerging blood–brain-barrier-crossing nanotechnology for brain cancer theranostics. *Chem. Soc. Rev.* **2019**, *48*, 2967–3014. [CrossRef]
58. Saraiva, C.; Praça, C.; Ferreira, R.; Santos, T.; Ferreira, L.; Bernardino, L. Nanoparticle-mediated brain drug delivery: Overcoming blood–brain barrier to treat neurodegenerative diseases. *J. Control. Release* **2016**, *235*, 34–47. [CrossRef]
59. Loureiro, J.A.; Gomes, B.; Coelho, M.A.N.; Carmo Pereira, M.d.; Rocha, S. Targeting nanoparticles across the blood–brain barrier with monoclonal antibodies. *Nanomedicine* **2014**, *9*, 709–722. [CrossRef]
60. Furtado, D.; Björnmalm, M.; Ayton, S.; Bush, A.I.; Kempe, K.; Caruso, F. Overcoming the Blood–Brain Barrier: The Role of Nanomaterials in Treating Neurological Diseases. *Adv. Mater.* **2018**, *30*, 1801362. [CrossRef]
61. Abdul Razzak, R.; Florence, G.J.; Gunn-Moore, F.J. Approaches to CNS Drug Delivery with a Focus on Transporter-Mediated Transcytosis. *Int. J. Mol. Sci.* **2019**, *20*, 3108. [CrossRef]
62. Mäger, I.; Meyer, A.H.; Li, J.; Lenter, M.; Hildebrandt, T.; Leparc, G.; Wood, M.J.A. Targeting blood–brain-barrier transcytosis—Perspectives for drug delivery. *Neuropharmacology* **2017**, *120*, 4–7. [CrossRef] [PubMed]
63. Bickel, U. Antibody delivery through the blood–brain barrier. *Adv. Drug Deliv. Rev.* **1995**, *15*, 53–72. [CrossRef]
64. Kumagai, A.; Eisenberg, J.; Pardridge, W. Adsorptive-mediated transcytosis of cationized albumin and a beta-endorphin-cationized albumin chimeric peptide by isolated brain capillaries. Model system of blood–brain barrier transport. *J. Biol. Chem.* **1987**, *262*, 15414–15419.
65. Hervé, F.; Ghinea, N.; Scherrmann, J.M. CNS Delivery Via Adsorptive Transcytosis. *AAPS J.* **2008**, *10*, 455–472. [CrossRef]
66. Oller-Salvia, B.; Sanchez-Navarro, M.; Giralt, E.; Teixido, M. Blood–brain barrier shuttle peptides: An emerging paradigm for brain delivery. *Chem. Soc. Rev.* **2016**, *45*, 4690–4707. [CrossRef]
67. McCully, M.; Sanchez-Navarro, M.; Teixido, M.; Giralt, E. Peptide Mediated Brain Delivery of Nano- and Submicroparticles: A Synergistic Approach. *Curr. Pharm. Des.* **2018**, *24*, 1366–1376. [CrossRef]
68. Lagassé, H.A.D.; Alexaki, A.; Simhadri, V.L.; Katagiri, N.H.; Jankowski, W.; Sauna, Z.E.; Kimchi-Sarfaty, C. Recent advances in (therapeutic protein) drug development. *F1000Research* **2017**, *6*, 113. [CrossRef]
69. Lee, J.I.; Zhang, L.; Men, A.Y.; Kenna, L.A.; Huang, S.M. CYP-Mediated Therapeutic Protein-Drug Interactions. *Clin. Pharm.* **2010**, *49*, 295–310. [CrossRef]
70. Cavaco, M.; Goncalves, J. Interactions between therapeutic proteins and small molecules: The shared role of perpetrators and victims. *Clin. Pharm. Ther.* **2016**, *102*, 649–661. [CrossRef]
71. Moorthy, B.S.; Xie, B.; Moussa, E.M.; Iyer, L.K.; Chandrasekhar, S.; Panchal, J.P.; Topp, E.M. Structure of Monoclonal Antibodies. In *Biobetters: Protein Engineering to Approach the Curative*; Rosenberg, A., Demeule, B., Eds.; Springer: New York, NY, USA, 2015; pp. 81–89. [CrossRef]
72. Roopenian, D.C.; Akilesh, S. FcRn: The neonatal Fc receptor comes of age. *Nat. Rev. Immunol.* **2007**, *7*, 715–725. [CrossRef]
73. Weiner, L.M.; Surana, R.; Wang, S. Antibodies and cancer therapy: Versatile platforms for cancer immunotherapy. *Nat. Rev. Immunol.* **2010**, *10*, 317–327. [CrossRef] [PubMed]
74. Dobson, C.L.; Devine, P.W.A.; Phillips, J.J.; Higazi, D.R.; Lloyd, C.; Popovic, B.; Arnold, J.; Buchanan, A.; Lewis, A.; Goodman, J.; et al. Engineering the surface properties of a human monoclonal antibody prevents self-association and rapid clearance in vivo. *Sci. Rep.* **2016**, *6*, 38644. [CrossRef] [PubMed]
75. Rodgers, K.R.; Chou, R.C. Therapeutic monoclonal antibodies and derivatives: Historical perspectives and future directions. *Biotechnol. Adv.* **2016**, *34*, 1149–1158. [CrossRef] [PubMed]
76. Shuptrine, C.; Surana, R.; Weiner, L.M. Monoclonal Antibodies for the Treatment of Cancer. *Semin. Cancer Biol.* **2012**, *22*, 3–13. [CrossRef]

77. Stoermer, K.A.; Morrison, T.E. Complement and Viral Pathogenesis. *Virology* **2011**, *411*, 362–373. [CrossRef]
78. Kubota, T.; Niwa, R.; Satoh, M.; Akinaga, S.; Shitara, K.; Hanai, N. Engineered therapeutic antibodies with improved effector functions. *Cancer Sci.* **2009**, *100*, 1566–1572. [CrossRef]
79. Irani, V.; Guy, A.J.; Andrew, D.; Beeson, J.G.; Ramsland, P.A.; Richards, J.S. Molecular properties of human IgG subclasses and their implications for designing therapeutic monoclonal antibodies against infectious diseases. *Mol. Immunol.* **2015**, *67*, 171–182. [CrossRef]
80. Chames, P.; Van Regenmortel, M.; Weiss, E.; Baty, D. Therapeutic antibodies: Successes, limitations and hopes for the future. *Br. J. Pharm.* **2009**, *157*, 220–233. [CrossRef]
81. Xenaki, K.T.; Oliveira, S.; van Bergen en Henegouwen, P.M.P. Antibody or Antibody Fragments: Implications for Molecular Imaging and Targeted Therapy of Solid Tumors. *Front. Immunol.* **2017**, *8*, 1287. [CrossRef]
82. Bates, A.; Power, C.A. David vs. Goliath: The Structure, Function, and Clinical Prospects of Antibody Fragments. *Antibodies* **2019**, *8*, 28. [CrossRef]
83. Brandstadter, R.; Katz Sand, I. The use of natalizumab for multiple sclerosis. *Neuropsychiatr. Dis. Treat.* **2017**, *13*, 1691–1702. [CrossRef] [PubMed]
84. Lipton, R.B.; Tepper, S.J.; Reuter, U.; Silberstein, S.; Stewart, W.F.; Nilsen, J.; Leonardi, D.K.; Desai, P.; Cheng, S.; Mikol, D.D.; et al. Erenumab in chronic migraine. *Neurology* **2019**, *92*, e2250. [CrossRef] [PubMed]
85. Stüve, O.; Bennett, J.L. Pharmacological Properties, Toxicology and Scientific Rationale for the use of Natalizumab (Tysabri®) in Inflammatory Diseases. *CNS Drug Rev.* **2007**, *13*, 79–95. [CrossRef] [PubMed]
86. Yaldizli, Ö.; Putzki, N. Natalizumab in the Treatment of Multiple Sclerosis. *Ther. Adv. Neurol. Disord.* **2009**, *2*, 115–128. [CrossRef]
87. Ohlsson, L.; Haanes, K.A.; Kronvall, E.; Xu, C.; Snellman, J.; Edvinsson, L. Erenumab (AMG 334), a monoclonal antagonist antibody against the canonical CGRP receptor, does not impair vasodilatory or contractile responses to other vasoactive agents in human isolated cranial arteries. *Cephalalgia* **2019**, *39*, 1745–1752. [CrossRef] [PubMed]
88. Goadsby, P.J.; Reuter, U.; Hallström, Y.; Broessner, G.; Bonner, J.H.; Zhang, F.; Sapra, S.; Picard, H.; Mikol, D.D.; Lenz, R.A. A Controlled Trial of Erenumab for Episodic Migraine. *N. Engl. J. Med.* **2017**, *377*, 2123–2132. [CrossRef] [PubMed]
89. Zhou, Q.-H.; Fu, A.; Boado, R.J.; Hui, E.K.-W.; Lu, J.Z.; Pardridge, W.M. Receptor-mediated abeta amyloid antibody targeting to Alzheimer's disease mouse brain. *Mol. Pharm.* **2011**, *8*, 280–285. [CrossRef]
90. Zuchero, Y.J.; Chen, X.; Bien-Ly, N.; Bumbaca, D.; Tong, R.K.; Gao, X.; Zhang, S.; Hoyte, K.; Luk, W.; Huntley, M.A.; et al. Discovery of Novel Blood–brain Barrier Targets to Enhance Brain Uptake of Therapeutic Antibodies. *Neuron* **2016**, *89*, 70–82. [CrossRef]
91. Sehlin, D.; Syvänen, S.; Ballanger, B.; Barthel, H.; Bischof, G.N.; Boche, D.; Boecker, H.; Bohn, K.P.; Borghammer, P.; Cross, D.; et al. Engineered antibodies: New possibilities for brain PET? *Eur. J. Nucl. Med. Mol. Imaging* **2019**. [CrossRef]
92. Camidge, D.R.; Lee, E.Q.; Lin, N.U.; Margolin, K.; Ahluwalia, M.S.; Bendszus, M.; Chang, S.M.; Dancey, J.; de Vries, E.G.E.; Harris, G.J.; et al. Clinical trial design for systemic agents in patients with brain metastases from solid tumours: A guideline by the Response Assessment in Neuro-Oncology Brain Metastases working group. *Lancet Oncol.* **2018**, *19*, e20–e32. [CrossRef]
93. Wang, A.; Komiya, T. Brain metastasis as exclusion criteria in clinical trials involving extensive-stage small cell lung cancer. *J. Cancer Res. Clin. Oncol.* **2019**, *145*, 3099–3144. [CrossRef] [PubMed]
94. Baik, C.S.; Chamberlain, M.C.; Chow, L.Q. Targeted Therapy for Brain Metastases in EGFR-Mutated and ALK-Rearranged Non-Small-Cell Lung Cancer. *J. Thorac. Oncol.* **2015**, *10*, 1268–1278. [CrossRef] [PubMed]
95. Lazaro, T.; Brastianos, P.K. Immunotherapy and targeted therapy in brain metastases: Emerging options in precision medicine. *CNS Oncol.* **2017**, *6*, 139–151. [CrossRef] [PubMed]
96. Puttemans, J.; Lahoutte, T.; D'Huyvetter, M.; Devoogdt, N. Beyond the Barrier: Targeted Radionuclide Therapy in Brain Tumors and Metastases. *Pharmaceutics* **2019**, *11*, 376. [CrossRef] [PubMed]
97. Zhang, X.; Hao, J. Development of anticancer agents targeting the Wnt/β-catenin signaling. *Am. J. Cancer Res.* **2015**, *5*, 2344–2360. [PubMed]
98. Li, Z.; Kang, Y. Emerging therapeutic targets in metastatic progression: A focus on breast cancer. *Pharmacol. Ther.* **2016**, *161*, 79–96. [CrossRef]

99. Stalmans, S.; Bracke, N.; Wynendaele, E.; Gevaert, B.; Peremans, K.; Burvenich, C.; Polis, I.; De Spiegeleer, B. Cell-Penetrating Peptides Selectively Cross the Blood–brain Barrier In Vivo. *PLoS ONE* **2015**, *10*, e0139652. [CrossRef]
100. AntibodySociety. Therapeutic Monoclonal Antibodies Approved or in Review in the EU or the US. Available online: https://www.antibodysociety.org/resources/approved-antibodies/ (accessed on 5 December 2019).
101. De Angelis, F.; Chataway, J. Novel Multiple Sclerosis Drugs in the Pipeline. *Clin. Pharmacol. Ther.* **2019**, *105*, 1082–1090. [CrossRef]
102. Jovčevska, I.; Muyldermans, S. The Therapeutic Potential of Nanobodies. *BioDrugs* **2019**. [CrossRef]
103. Cummings, J.; Lee, G.; Ritter, A.; Sabbagh, M.; Zhong, K. Alzheimer's disease drug development pipeline: 2019. *Alzheimer's Dement. (N.Y.)* **2019**, *5*, 272–293. [CrossRef]
104. Kaplon, H.; Reichert, J.M. Antibodies to watch in 2019. *mAbs* **2019**, *11*, 219–238. [CrossRef] [PubMed]
105. Guo, Z.; Peng, H.; Kang, J.; Sun, D. Cell-penetrating peptides: Possible transduction mechanisms and therapeutic applications. *Biomed. Rep.* **2016**, *4*, 528–534. [CrossRef] [PubMed]
106. Avci, F.G.; Akbulut, B.S.; Ozkirimli, E. Membrane Active Peptides and Their Biophysical Characterization. *Biomolecules* **2018**, *8*, 77. [CrossRef] [PubMed]
107. Bolintineanu, D.S.; Kaznessis, Y.N. Computational studies of protegrin antimicrobial peptides: A review. *Peptides* **2011**, *32*, 188–201. [CrossRef] [PubMed]
108. Madani, F.; Lindberg, S.; Langel, U.; Futaki, S.; Gräslund, A. Mechanisms of cellular uptake of cell-penetrating peptides. *J. Biophys. (Hindawi Publ. Corp. Online)* **2011**, *2011*, 414729. [CrossRef] [PubMed]
109. Derakhshankhah, H.; Jafari, S. Cell penetrating peptides: A concise review with emphasis on biomedical applications. *Biomed. Pharmacother.* **2018**, *108*, 1090–1096. [CrossRef]
110. Zhang, T.T.; Li, W.; Meng, G.; Wang, P.; Liao, W. Strategies for transporting nanoparticles across the blood–brain barrier. *J. Biomater. Sci.* **2016**, *4*, 219–229. [CrossRef]
111. Frankel, A.D.; Pabo, C.O. Cellular uptake of the tat protein from human immunodeficiency virus. *Cell* **1988**, *55*, 1189–1193. [CrossRef]
112. Kristensen, M.; Brodin, B. Routes for drug translocation across the blood–brain barrier: Exploiting peptides as delivery vectors. *J. Pharm. Sci.* **2017**, *S0022-3549*, 30361. [CrossRef]
113. Duchardt, F.; Fotin-Mleczek, M.; Schwarz, H.; Fischer, R.; Brock, R. A Comprehensive Model for the Cellular Uptake of Cationic Cell-penetrating Peptides. *Traffic* **2007**, *8*, 848–866. [CrossRef]
114. Schwarze, S.R.; Ho, A.; Vocero-Akbani, A.; Dowdy, S.F. In Vivo Protein Transduction: Delivery of a Biologically Active Protein into the Mouse. *Science* **1999**, *285*, 1569. [CrossRef]
115. Antoniou, X.; Borsello, T. Cell Permeable Peptides: A Promising Tool to Deliver Neuroprotective Agents in the Brain. *Pharmaceutics* **2010**, *3*, 379–392. [CrossRef] [PubMed]
116. Zou, L.-L.; Ma, J.-L.; Wang, T.; Yang, T.-B.; Liu, C.-B. Cell-Penetrating Peptide-Mediated Therapeutic Molecule Delivery into the Central Nervous System. *Curr. Neuropharmacol.* **2013**, *11*, 197–208. [CrossRef]
117. Ben-Zvi, T.; Yayon, A.; Gertler, A.; Monsonego-Ornan, E. Suppressors of cytokine signaling (SOCS) 1 and SOCS3 interact with and modulate fibroblast growth factor receptor signaling. *J. Cell Sci.* **2006**, *119*, 380. [CrossRef] [PubMed]
118. Demeule, M.; Régina, A.; Ché, C.; Poirier, J.; Nguyen, T.; Gabathuler, R.; Castaigne, J.-P.; Béliveau, R. Identification and Design of Peptides as a New Drug Delivery System for the Brain. *J. Pharm. Exp.* **2008**, *324*, 1064. [CrossRef]
119. Regina, A.; Demeule, M.; Tripathy, S.; Lachowicz, J. ANG4043, a Novel Brain-Penetrant Peptide-mAb Conjugate, Is Efficacious against HER2-Positive Intracranial Tumors in Mice. *Mol. Cancer* **2015**, *14*, 129–140. [CrossRef] [PubMed]
120. Lim, S.; Kim, W.J.; Kim, Y.H.; Lee, S.; Koo, J.H.; Lee, J.A.; Yoon, H.; Kim, D.H.; Park, H.J.; Kim, H.M.; et al. dNP2 is a blood–brain barrier-permeable peptide enabling ctCTLA-4 protein delivery to ameliorate experimental autoimmune encephalomyelitis. *Nat. Commun.* **2015**, *6*, 8244. [CrossRef]
121. Neves, V.; Aires-da-Silva, F.; Morais, M.; Gano, L.; Ribeiro, E.; Pinto, A.; Aguiar, S.; Gaspar, D.; Fernandes, C.; Correia, J.D.G.; et al. Novel peptides derived from Dengue virus capsid protein translocate reversibly the blood–brain barrier through a receptor-free mechanism. *ACS Chem. Biol.* **2017**. [CrossRef]
122. Côrte-Real, S.; Neves, V.; Oliveira, S.; Canhão, P.; Outeiro, T.; Castanho, M.; Aires da Silva, F. Antibody Molecules and Peptide Delivery Systems for Use in Alzheimer'S Disease and Related Disorders. WO2016120843A1, 29 January 2016.

123. Cavaco, M.; Castanho Miguel, A.R.B.; Neves, V. Peptibodies: An elegant solution for a long-standing problem. *Pept. Sci.* **2018**, *110*, e23095. [CrossRef]
124. Pyzik, M.; Rath, T.; Lencer, W.I.; Baker, K.; Blumberg, R.S. FcRn: The Architect Behind the Immune and Nonimmune Functions of IgG and Albumin. *J. Immunol.* **2015**, *194*, 4595. [CrossRef]
125. Samec, N.; Jovcevska, I.; Stojan, J.; Zottel, A.; Liovic, M.; Myers, M.P.; Muyldermans, S.; Šribar, J.; Križaj, I.; Komel, R. Glioblastoma-specific anti-TUFM nanobody for in-vitro immunoimaging and cancer stem cell targeting. *Oncotarget* **2018**, *9*, 17282–17299. [CrossRef] [PubMed]
126. Hu, Y.; Liu, C.; Muyldermans, S. Nanobody-Based Delivery Systems for Diagnosis and Targeted Tumor Therapy. *Front. Immunol.* **2017**, *8*, 1442. [CrossRef] [PubMed]

pharmaceutics

MDPI

Review

Nanomaterials Designed for Antiviral Drug Delivery Transport across Biological Barriers

Florina-Daniela Cojocaru [1], Doru Botezat [1], Ioannis Gardikiotis [1], Cristina-Mariana Uritu [1], Gianina Dodi [1,*], Laura Trandafir [2], Ciprian Rezus [3], Elena Rezus [4], Bogdan-Ionel Tamba [1] and Cosmin-Teodor Mihai [1,5]

1 Advanced Centre for Research-Development in Experimental Medicine, Grigore T. Popa University of Medicine and Pharmacy of Iasi, 700115 Iasi, Romania; florina.cojocaru@umfiasi.ro (F.-D.C.); doru.botezat@umfiasi.ro (D.B.); ioannis.gardikiotis@umfiasi.ro (I.G.); cristina-mariana.uritu@umfiasi.ro (C.-M.U.); bogdan.tamba@umfiasi.ro (B.-I.T.); cosmin-teodor.mihai@umfiasi.ro (C.-T.M.)

2 Pediatric Department, Grigore T. Popa University of Medicine and Pharmacy of Iasi, 700115 Iasi, Romania; laura.trandafir@umfiasi.ro

3 Department of Internal Medicine, Grigore T. Popa University of Medicine and Pharmacy, 700115 Iasi, Romania; ciprian.rezus@umfiasi.ro

4 Department of Rheumatology and Physiotherapy, Grigore T. Popa University of Medicine and Pharmacy, 700115 Iasi, Romania; elena.rezus@umfiasi.ro

5 NIRDBS—Institute of Biological Research Iasi, Department of Experimental and Applied Biology, 700107 Iasi, Romania

* Correspondence: gianina.dodi@umfiasi.ro; Tel.: +40-232-267-801

Received: 20 December 2019; Accepted: 15 February 2020; Published: 18 February 2020

check for updates

Abstract: Viral infections are a major global health problem, representing a significant cause of mortality with an unfavorable continuously amplified socio-economic impact. The increased drug resistance and constant viral replication have been the trigger for important studies regarding the use of nanotechnology in antiviral therapies. Nanomaterials offer unique physico-chemical properties that have linked benefits for drug delivery as ideal tools for viral treatment. Currently, different types of nanomaterials namely nanoparticles, liposomes, nanospheres, nanogels, nanosuspensions and nanoemulsions were studied either in vitro or in vivo for drug delivery of antiviral agents with prospects to be translated in clinical practice. This review highlights the drug delivery nanosystems incorporating the major antiviral classes and their transport across specific barriers at cellular and intracellular level. Important reflections on nanomedicines currently approved or undergoing investigations for the treatment of viral infections are also discussed. Finally, the authors present an overview on the requirements for the design of antiviral nanotherapeutics.

Keywords: nanomaterials; antivirals; drug delivery; biological barriers

1. Social Impact and Economic Burden of Viral Infectious Diseases

Today, we are living through a so-called fourth great transitional period, after the other three waves of epidemiological transitions, namely early agrarian-based settlements, early Eurasian civilizations and European expansionism (more details can be found in McMichael's description [1]). The current configuration and variety of infectious diseases closely followed the combined evolutions in demography, environment, technology, social change and behaviours. Medicine itself has created new opportunities for microbes either through blood transfusions, organ transplants, the use of hypodermic syringes or the excessive use of antibiotics thus contributing to the induction of iatrogenic effects in some treatments for infections such as hepatitis C, HIV and others [1].

In recent decades, old concerns have been reactivated at both the official and the general public levels regarding infectious diseases as a threat to public health. McMichael [1] analyzed the reflection of this issue in social media and noticed the "emergence and resurgence" of infectious diseases (determined by environmental, sociological and economic changes) and a so-called "public anxiety" set on this topic.

Despite their widespread and increasing transmission, there is still a poor understanding of global economic impact of viral diseases, which makes difficult to evaluate the societal costs and the cost-effectiveness of preventive efforts. The issue of estimating a general impact of viral infections involves several aspects that hinder this approach, namely: the variety of viral infections, the incidence of associated co-morbidities, social and psychological issues having economic repercussions (hidden costs), the variety of treatments (only direct-acting antiviral (DAA) and vaccinations can be used in the evaluation) and the presence of negative externalities [2], meaning that the disease consequences are not limited to their patients infected or potentially but also to the related families who may experience social distress as well. Also, human mobility and long-distance trade have increased; ever-larger cities, often girded with slums, have become highways for microbial traffic; poverty perpetuates vulnerability to infectious disease; and sexual practices, drug injecting, intensified food production and much modern medical technology all create new "audience" for microbial opportunism, and new management issues for public health decision makers [1].

For all these reasons, a global assessment of socio-economic impact seems practically impossible or, if contrary, it could not meet all the criteria to be relevant. Rather, the impact can be split by relevant categories of viral infections, were the literature is more precise but even so, there remain some inconsistencies regarding the unification of the methodologies from the various studies that will ensure the comparability of the data. Figure 1 presents statistical facts related to the most "burden-generator" viral infection diseases [3–7].

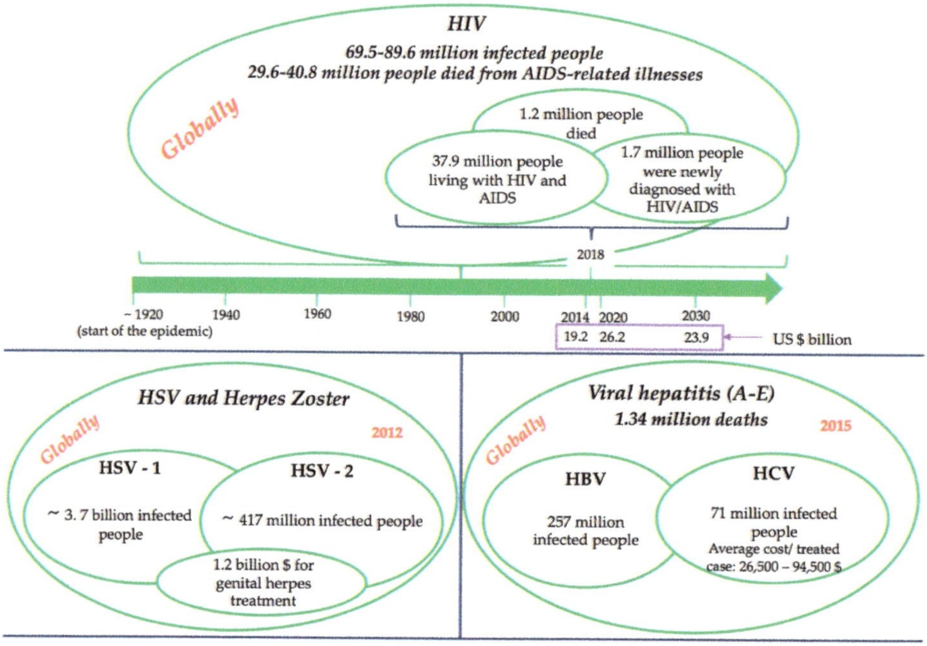

Figure 1. Global impact of viral diseases.

HIV/AIDS, considered as one of the major burdens of disease globally, became a chronic disease after the introduction of multiple antiretroviral therapy (ART), and therefore it needs to provide long-term care and support for the ill person, demanding a higher level of treatment costs for the HIV-affected households. Consequently, HIV/AIDS causes depletion of savings and productive assets, and increases the indebtedness of the HIV-affected households [8,9]. Moreover, the higher health care expenditure of the households reduces investment for nutritional food for the family members, investment for farming or business, and the children education. Death during the working age of the patient is a major factor in the economic impact of HIV/AIDS. The household level impact of HIV/AIDS includes direct costs, including medical and non-medical costs, and productivity costs such as loss of labour time, as a result of the morbidity of HIV positive household members, as well as time spent by others caring for them [10]. This evidence suggests that HIV/AIDS places significant economic pressure on households trying to pay for health care costs, and trying to make up for lost income.

The difficulty in accurately quantifying and explaining the morbidity and mortality related to viral hepatitis stems from the fact that hepatitis deaths are caused by five distinct viruses (hepatitis A–E) with different routes of transmission, or from the fact that death occurs decades after infection, and that when people die with hepatitis-related liver cancer and cirrhosis, these deaths are not always linked to the underlying infection.

Although antiviral therapy appears to be expensive (for example, average cost for a treated HCV case ranges from $26,500 to $94,500 [11]), is considered to be also cost effective when compared with other well-accepted medical interventions [12] due to sustained viral response to therapy, the cost savings and quality-of-life improvement and prolongation of life expectancy from the prevention of HCV complications. In the era of new DAAs, the statement "provide treatment to HCV-patients" generates savings compared to not provide it. Low and middle-income countries may consider HCV-treatment as a cost-saving intervention for the health system, not only in a long-term horizon, but in 5–10 years [5]. Economic impact of no treatment is higher than treatment costs itself. But still, the enthusiasm for DAA therapies, however, has been tempered by two major concerns: the price, still very high, of these medications, and, the challenges patients and clinicians face with respect to drug access in many countries.

HSV and herpes zoster (varicella-zosterian virus—VZV reactivation) are some of the most common infections in humans, with no effective treatment available at this time. The impact on the health degree varies from a very small impairment to severe, disabling forms, in most cases limiting methods are applied to local infection and reducing side effects (pain, manifestations dermis etc.). Also, specific psychological issues are related with this disease, such as negative feelings correlated to the condition following diagnosis, in particular if they have acquired the genital form of the disease. Feelings can include depression, fear of rejection, feelings of isolation, fear of being found out, and self-destructive feelings [6].

As mentioned above, all the available reports present crude estimates rather than precise measures of the economic costs of the illness, because most of the costs are calculated directly on interventions or on medical budgetary chapters, without taking into account the societal losses. Moreover, different approaches are discrepant (can be explained, at least in part, with the influence of compliance to treatment and possible under sampling of subpopulations in the data set) [7]. Also, limitations must be placed on the ability to generalize the results beyond the sample. Moreover, not only the cost matters. The costs of an intervention have to be compared with the results of interventions because the effectiveness of treatments and the efficiency can produce societal gains that must be offset by losses [13].

The usual most accurate methodology to estimate a burden (associated with a disease) is the so-called "cost-of-Illness" [10], and measures all the costs of a particular disease, including the direct, indirect, and intangible dimensions. It is widely accepted that estimating the total social cost of a disease is useful in establishing policy decisions [14]. But there are also other methods as the *cross-sectional surveys* of samples of primary and secondary care physicians, analyzing health care resource utilization or approaches based on the analysis of a large administrative data sets, such as values of spending or drugs consuming lists [15]. Other approaches are used to estimate the number of patients seeking medical treatment, the average medical expenditures (as health inputs employed per unit multiplied by number of units) and estimated national costs. These comprehensive studies can often be advantageous in allocating total national expenditures among the major diagnostic categories [16]. However, regardless of the method, such analyzes are not possible otherwise than inside countries (where the impact determined by cultural and social aspects can vary substantially). But even in the absence of global data of this nature, we still can extract from the information presented the relevant issue for the topic of this paper: the current arrangements in the management of viral infections (treatments, prevention and limitations of spread) are costly and less effective, unaffordable in some cases and burdensome for medical systems. For example, according to the current analysis of Globe Newswire Reports and Data [17], the global antiviral drugs market was valued at $49.87 billion in 2018 and is expected to reach $71.48 billion by year 2026. Sales of antivirals increased by approximately 20% each two years. Moreover, thanks to better diagnostics, innovative drugs and new therapeutics, the market is likely to witness even further future growth. However, the list of viral diseases for which antiviral therapies are available is still relatively short [18].

There are several factors that hinder the development of antiviral drugs:

- Dependence of viruses replication on host cell biosynthetic machinery [19], that leads to a limited number of virus-specific metabolic functions can be targeted by antiviral drugs without any damage to the host;
- the viruses' functions are specific to each virus, preventing the development of a broad-spectrum antivirals fighting against different viruses that cause similar symptoms. Antivirals developed for some viruses (as HSV and HIV) can treat the acute illness, but do not cure the latent infection. This leads to recurrent or chronic diseases that require treatment for longer periods of time [18].

All these limitations prompted the need for a paradigm shift. The great challenge of antiviral therapies is to move on to developing new drug formulas. This involves changing the physico-chemical and bio-pharmaceutical properties of antiviral molecules using new scientific strategies during the preparation or in dosage configuration.

2. Viruses: Types, Current Therapy and Observed Drawbacks

Viruses are sub-microscopic intracellular parasitic particles of genetic material contained in a protein coat, totally dependent by host for cell replication, showing both living and non-living characteristics [20].

Living characteristics of the viruses are represented by the high rate of multiplication (only in living host cells) and by the ability to mutate. The non-living characteristics for viruses consist in acellularity (lack of cytoplasm and organelles), the replication only by using host cell's metabolic machinery and the composition with DNA or RNA [20]. In humans, viral infections are responsible for different diseases as briefly presented in Table 1.

According to International Committee on Taxonomy of Viruses (ICTV) there are 1 phylum, 2 subphylia, six classes, 10 orders, seven suborders, 89 families, 36 subfamilies, 387 genera, 59 subgenera and 2202 species [43]. Currently, viruses are classified based on their type of nucleic acid (DNA, RNA, single-stranded, double-stranded) and their way of replication, known as Baltimore classification [44], divided as seven Baltimore classes:

- I—dsDNA viruses (e.g., adenoviruses, herpesviruses, poxviruses): enter to the host nucleus and are dependent by host cell polymerases to replicate viral genome. The virus may induce the cell to forcefully undergo cell division, which may lead to transformation of the cell and, ultimately, to cancer.

- II—ssDNA viruses (+ strand or "sense") *DNA* (e.g., parvoviruses), consists of viruses that have a single-stranded DNA genome of the same polarity as the mRNA. Excepting Parvoviruses, most of them have circular genomes and are replicating within nucleus.
- III—dsRNA viruses (e.g., reoviruses): not dependent by host replication polymerases and their replication (monocistronic) is realized into capsid (in cytoplasm).
- IV—(+)ssRNA viruses (+ strand or sense) *RNA* (e.g., picornaviruses, Togaviruses): the RNA can be directly accessed by ribosomes of the host to form proteins, and use a simple reproduction pathway (viruses with polycistronic mRNA) or a more complex transcription pathway (for which subgenomic mRNAs, ribosomal frameshifting, and proteolytic processing of polyproteins may be used).
- V—(−)ssRNA viruses (− strand or antisense) *RNA* (e.g., orthomyxoviruses, rhabdoviruses), that first must be transcribed by viral polymerases (positive-sense) before can be directly accessed by host ribosomes to form proteins.
- VI—ssRNA-RT viruses (+ strand or sense) RNA with DNA intermediate in life-cycle (e.g., retroviruses), which use the reverse transcriptase to convert the positive-sense RNA into DNA. They are using DNA to create the templates and those are spliced into host genome by integrase.
- VII—dsDNA-RT viruses DNA with RNA intermediate in life-cycle (e.g., hepadnaviruses), dsDNA viruses that replicate through a single-stranded RNA intermediate, which use a pregenome RNA as a template and conversion to DNA is done by a viral reverse transcriptase.

Table 1. Common viral infections.

Viral Infection	Viruses	References
Common cold	rhinoviruses, parainfluenza viruses, respiratory syncytial viruses coronaviruses, Influenza viruses, adenoviruses, enteroviruses, metapneumovirus, unknown	[21]
Eye infections	herpes simplex virus, adenovirus, cytomegalovirus	[22–24]
Encephalitis or meningitis	as JC virus, measles, LCM virus, arbovirus, rabies	[25]
Pneumonia	influenza virus (A and B), parainfluenza virus, respiratory syncytial virus, adenovirus, SARS coronavirus	[26]
Cardiovascular and pancreas disease	coxsackie virus;	[27–29]
Hepatitis	hepatitis viruses types A, B, C, D, E	[30–33]
Skin infections	varicella-zoster virus, human herpesvirus 6, smallpox, molluscum contagiosum, human papillomavirus, parvovirus B19, rubella, measles, coxsackie A virus	[34–38]
Gastroenteritis	adenoviruses, rotaviruses, noroviruses, astroviruses, coronaviruses	[39–41]
Sexually transmitted diseases	herpes simplex type 2, human papillomavirus HIV	[42]

There are distinct stages of viral replication (cell entry, uncoating, transcription of viral genome, translation of viral proteins, post-translational modifications and assembly of virion components) and the classes of antiviral agents that can act at each stage, the correspondence between stage of replication and classes of selective inhibitors being described in detail in reference [45] and their pharmacological properties as it follows in the next section.

Antiviral Agents Used in Nanotechnology

According to our knowledge (based on found studies) only the antivirals summarized in Figure 2 were applied in nanomedicine.

The therapy of the large HSV family of double-stranded DNA viruses, widely distributed among humans, includes highly selective and effective antivirals, from which only acyclovir (ACV) and ganciclovir (GCV) were incorporated into nanomaterials; a complete classification of HSV antivirals can be found in described in detail in references [46–50]. Briefly, ACV pharmacology can be explained as follows:

- the ACV selectivity is dependent by interaction with viral HSV thymidine kinase and DNA polymerase, therefore the cellular uptake and first phosphorilation are facilitated by HSV thymidine kinase that presents a high affinity for ACV;
- then, intracellular enzymes convert monophospate to triphosphate ACV and this form of ACV inhibits viral DNA polymerase and, to a much lesser extent, cellular DNA polymerase; ACV triphosphate is also integrated into viral DNA and acts as a chain terminator, as it binds to viral DNA polymerase and determines its irreversible inactivation by a mechanism called suicide inactivation [51–53];
- occurrence of resistance to ACV could be acquired by three mechanisms: impaired production of viral thymidine kinase (the most common), altered thymidine kinase substrate specificity (e.g., phosphorylation of thymidine but not acyclovir), or altered viral DNA polymerase (rare). Alterations in viral enzymes are caused by point mutations and base insertions or deletions in the corresponding genes [54].

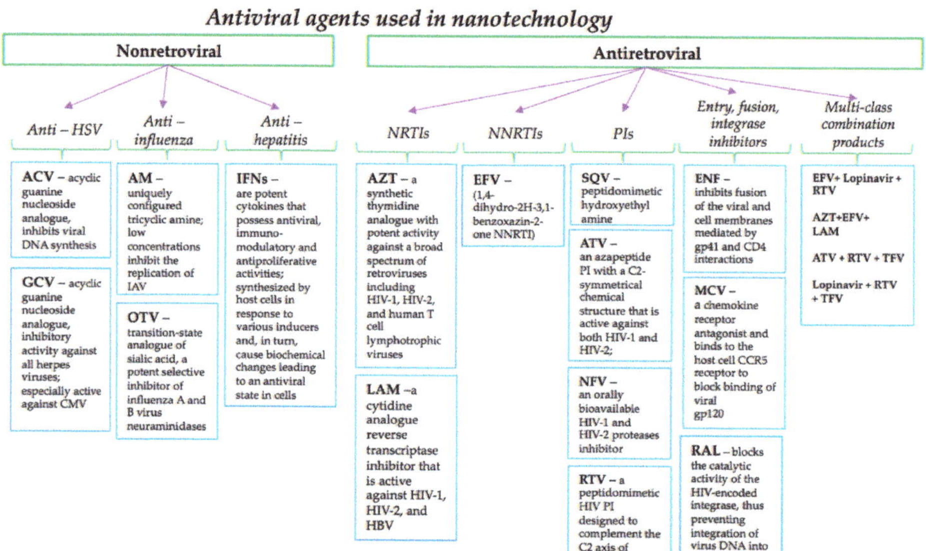

Figure 2. Summary of antiviral agents used in nanotechnology.

The mechanism of GCV inhibits viral DNA synthesis [55] as briefly explained below:

- it is monophosphorylated intracellular by viral thymidine kinase during HSV infection and by a viral phosphotransferase encoded by the UL97 gene during CMV infection, while diphosphate and GCV triphosphate forms are produced by cellular enzymes;
- CMV can become resistant to GCV by mutations in the viral phosphotransferase encoded by the UL97 gene and by mutations in viral DNA polymerase [56].

The conventional treatment (prophylaxis or therapy) of an influenza virus infection, as a major public health concern worldwide, is designed to target viral proteins and could be used, either alone or in combination [57]. These include also amantadine and neuraminidase inhibitors (zanamivir and oseltamivir), that have been encapsulated into nanoparticles as specified in Table 2. Potent vacuolar ATPase (V-ATPase) inhibitors, namely diphyllin and bafilomycin, previously shown to have broad-spectrum antiviral activity represent another possibility against influenza virus infection [58–60]. Briefly, the antiviral mechanism of amantadine is based on nterference with the viral protein, M2 (an ion channel), the protein needed for the viral particle to become uncoated once it is taken inside the cell by endocytosis [61]. Also, oseltamivir carboxylate mechanism implies a selective inhibition of influenza virus neuraminidase enzymes, which are glycoproteins found on the virion surface, very important for viral entry into uninfected cells, for the release of recently formed virus particles from infected cells, and for the further spread of the infectious virus in the body [62,63].

Hepatitis viruses have been the subject of intense study in the last years, with a special attention on therapy. As mentioned in the first section, hepatitis treatment depends upon the type of hepatitis, therefore different antivirals are considered and summarized in detail in references [64–69]. Currently, interferons (IFNs) α, β, and γ have antiviral activity, the first two being produced by nearly all cells as response to viral infections, while the third is restricted to T-lymphocytes and NK cells. IFN-induced proteins include 2′-5′-oligoadenylate [2-5(A)] synthetase and a protein kinase, either of which can inhibit protein synthesis in the presence of double-stranded RNA. The 2-5(A) synthetase produces adenylate oligomers that activate a latent cellular endoribonuclease (RNase L) to cleave both cellular and viral single-stranded RNAs.

Antiretroviral therapy—ART—refers to the treatment with HIV medicines. According to the last updated list approved by Food and Drug Administration (FDA) these drugs can be classified as can be seen in Figure 2 [70,71]. In the following paragraph the authors describe the pharmacological mechanism of the medicines for HIV treatment that were included/encapsulated/incorporated into nanomaterials. The representative nucleoside reverse transcriptase inhibitor (NRTI) zidovudine (AZT) is phosphorylated intracellular by kinases specific to AZT 5′-triphosphate, a metabolite responsible for termination in elongation of proviral DNA because it is incorporated by reverse transcriptase into nascent DNA but lacks a 3′-hydroxyl group. Resistance to AZT is associated with mutations at reverse transcriptase codons 41, 44, 67, 70, 210, 215, and 219 [72]. Also, lamivudine (LAM), another NRTI agent, enters cells by passive diffusion, and then is converted to the monophosphate by deoxycytidine kinase, and undergoes further phosphorylation by deoxycytidine monophosphate kinase and nucleoside diphosphate kinase to yield lamivudine 5′-triphosphate, which is the active anabolite [73]. Tenofovir disoproxil is a derivative of adenosine 5′-monophosphate lacking a complete ribose ring, and it is the only nucleotide analogue currently marketed for the treatment of HIV infection, being active against HIV-1, HIV-2 and HBV. After a rapid hydrolysis, tenofovir is formed, being then phosphorylated by cellular kinases to its active metabolite, tenofovir diphosphate which is a competitive inhibitor of viral reverse transcriptases and is incorporated into HIV DNA, causing chain termination [74,75].

Non-nucleoside reverse transcriptase inhibitors (NNRTIs) include a variety of chemical substrates that bind to a hydrophobic pocket in the p66 subunit of the HIV-1 reverse transcriptase and induce a conformational change in the 3D structure of the enzyme that greatly reduces its activity, and thus they act as non-competitive inhibitors. Unlike nucleoside and nucleotide reverse transcriptase inhibitors, these compounds do not require intracellular phosphorylation to attain activity [76]. Also, the binding site for NNRTIs is virus-strain-specific and the approved agents are active against HIV-1 but not HIV-2 or other retroviruses and should not be used to treat HIV-2 infection.

HIV protease inhibitors (PIs) are peptide-like chemicals that competitively inhibit the action of the virus aspartyl protease (a homodimer consisting of two 99-amino acid monomers). The preferred cleavage site for this enzyme is the N-terminal side of proline residues, especially between phenylalanine and proline. These drugs prevent proteolytic cleavage of HIV gag and pol precursor polypeptides that include essential structural (p17, p24, p9, and p7) and enzymatic (reverse transcriptase, protease, and integrase) components of the virus, preventing the metamorphosis of HIV virus particles into their mature infectious form [77,78]. Only SQV, indinavir, ATV, RTV, NFV and lopinavir are currently employed in nanotechnology research.

There are two drugs available in the entry inhibitors class: enfuvirtide—ENF and maraviroc—MCV, with different mechanisms of action, both incorporated into nanoparticles as specified in Table 2. ENF inhibits fusion of the viral and cell membranes mediated by gp41 and CD4 interactions, while MCV is a chemokine receptor antagonist and binds to the host cell CCR5 receptor to block binding of viral gp120. As such, MCV is the only approved antiretroviral drug that targets a host protein [79–81].

Raltegravir (RAL), the first approved HIV integrase inhibitor, has potent activity against both HIV-1 and HIV-2, and also retains activity against viruses that have become resistant to antiretroviral agents of other classes because of its unique mechanism of action [81–84]. RAL was encapsulated into a polymeric PLGA nano-formulation and gold nanoparticles (see Table 2).

3. Biological Barriers Security System

The first line of defence [85] that any substance encounters is the biological barriers penetration into the organism. The "security" system includes physiological barriers, such as blood-brain barrier (BBB), epithelium, stratum corneum, air-blood lung barrier [86], reproductive system barrier, etc. all of which control the extracellular and intracellular access and trafficking of foreign substances such as bacteria, viruses, fungi, and chemicals [87] but also provide selective access to "suitable candidates" such as nutrition and/or therapy molecules.

It is well established that more than one mechanism may be involved in intracellular drug delivery. The mechanisms involved in nano-based intracellular drug delivery include passive diffusion of free drug, non-specific phagocytosis of the nanocarrier, nanocarrier uptake by pinocytosis, and receptor-mediated endocytosis [88]. In this section, we will discuss in particular how to overcome biological barriers such as mucus, skin, cell membrane and BBB in antiviral therapy.

3.1. Mucus

The gastrointestinal tract, respiratory system, the urogenital cavities, eyes and mouth, are all covered with mucosal membranes. The highly adhesive mucus acts as protective layer as well as for lubrication purposes. Although large molecules cannot pass, small ones along with viruses can easily penetrate. These are also the reasons that drug delivery is so challenging. Mucus contains 95% of water along with mucin fibres, lipids, salts, cholesterol and proteins. It is continuously produced but the thickness, pH and amount differ by its position.

Two strategies have to be followed for passing through mucus, mainly depending how fast the turnover is: fast mucosal penetration or highly adhesive particles (slow turnover) to increase the drug's residence on the targeted mucosa.

Mucoadhesion can be the resultant of interactions like hydrophobic, hydrogen bonding, ionic bonding or van der Waals ones. Other possible attractive interactions can be covalent bond formation between catechols, maleimides, thiols, and acrylates with domains of mucin glycoproteins rich in cysteine. Chitosan, alginate, pectin or cellulose polymers are mostly used for achieving adhesion on mucosa. Furthermore, it is known that thiolation of the mentioned polymers develops high mucoadhesive properties.

Mucopenetration is the second strategy found to permeate the mucus layers by two potential mechanisms: active strategy characterized by the interaction with the mucus and chemically shifting the features of the mucus or their own structures and passive strategy that uses hydrophilicity enhancers to penetrate the mucus [89].

The classical HSV therapy includes daily dosing of orally administered ACV [90] that is effective in most cases, and of course problematic in other cases, for example, in long-term use of ACV patients report resistance against the drug followed by renal injury [91,92]. Another issue produced by standard HSV treatment in this case for the topical form of trifluridine (TFT) and ganciclovir (ACV analog) gels, is denoted by low retention time on vaginal and corneal mucus followed by multiple doses up to 10 times [93].

The nanotechnology field offers a great deal of drug delivery modalities in order to overpass the biological barriers, deliver efficiently the incorporated active principle in a controlled and targeted manner, reduce circulating drug levels and attenuate the renal damage. A recent review points out the nanogels based on the above-mentioned materials capabilities as an adequate example to pass through these types of biological barriers [94]. For example, the synthesized nano-drug delivery micelles based on chitosan-*g*-oligo (NiPAam) copolymers stabilised by ionotropic crosslinking by Raskin et al. [95], gave good results for delivering antiretroviral drugs (EFV) through mucosa. More studies on how ACV penetrates different barriers can be found in Table 2 and detailed in Section 5.

It is important to mention a "special barrier" namely the ocular mucus that changes completely in 5 to 8 min, making the drug absorption unfavourable. In addition, the eye is protected by blood-anterior chamber and blood-retinal barriers. In case of drug administration, a combination of penetration and adhesion of the substances is necessary. Polymers like phosphotyrosine could be the solution. It was demonstrated that if intestinal alkaline phosphatase is present, polymers can manifest a zeta potential change, thus causing their immobilization after penetration. Another kind of construct for combining adhesion and penetration is through thiolated systems with mucolytic enzymes, pH dependent. Therefore, at acid pH, the absence of disulphide bonds formation with cysteine-rich domains in mucins does not manifest mucosal adhesion unless they are near the epithelium [94].

3.2. Skin

Skin, as the largest organ of our body, protects us from microorganisms and chemicals, regulates our body temperature and maintains hydroelectrolytic balance. The two layers of the skin are the epidermis and the dermis. The first one is avascular and is composed of stratified, keratinized squamous epithelium, in four layers from bottom up: basale, spinosum, granulosum and corneum. Thick skin (palms and soles) has a fifth layer (under the most superficial corneum) called lucidum. The dermis consists of two layers (reticular and the more superficial papillary) of connective tissue of elastin and collagenous fibres and has in its component blood and lymphatic vessels network, nerves, touch receptors (Meissner corpuscles), adipocytes, phagocytes, hair follicles and sweat glands [96].

Topical, through skin drug delivery, has a local effect, requiring less drug for the targeted outcome. Transdermal therapy results in fewer side effects with no need of regular treatment but systemic distribution of the drug. Both methods of treatment have a common blockage, stratum corneum. To pass through this barrier, different approaches were developed based mostly on disrupting this structure chemically with substances like surfactants, alcohols, esters, amines, terpenes, alkanes phospholipids, or mechanically by using ultrasounds, micro needling, magnetophoresis, iontophoresis, electroporation or lasers. Excessive use, though, can damage the skin.

Analysing the literature data, we can suggest that there are different processes and mechanisms that govern the penetration of small/large molecules through skin barrier. According to Schneider et al. [97] review there are two general pathways for skin absorption: through skin appendages or through the stratum corneum and the underlying layers. The lipophilic statum corneum medium determines the first mechanism of skin penetration, namely absorption of lipophilic compounds [98]. The three transport routes of substances across stratum corneum can be classified in transcellular, intercellular and trans-appendageal pathways as defined by Liang et al. [98]. More examples are available but the conclusion is the same: "the full understanding of the penetration or absorption processes is still under evaluation" due to the challenges associated with delivering complex burdens through the skin barrie [99].

As specified in Section 2, the current antiviral therapy for HSV infection includes topical formulations of ACV that is unable permeate stratum corneum and target the virus site at the basal epidermis due to its polarity and solubility, leading to poor clinical efficacy due to delayed antiviral activity and sub-inhibitory concentrations [100]. Nanotechnology strategies [97] seem to facilitate the "admission fees" due to the rationally design and innovative functionalities of the synthesized nano-platforms as presented in Figure 3.

One more problem in dermal drug passing is related to inflammation skin pathology. The barrier is changed, and the drug resides much less on the targeted site because of fast penetration. The thermoresponsive drug delivery nanogels used in this purpose have encouraging results for overcoming the above-mentioned problems. pH sensitive nanogels can also be utilised for controlled medicinal release.

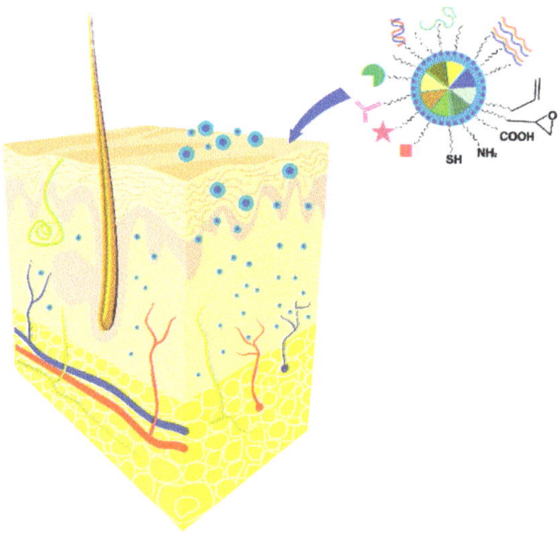

Figure 3. Schematic illustration of nano-scale carrier systems and their interactions with the dermal barrier.

Thermoresponsive nanogels can be controlled through irradiation with infrared lamp. Another way of skin penetration is the hair follicle. The stratum corneum is less intact in the lower infundibulum, so the nanoparticle's (NPs) passage is dependent on size and not on composition [94].

3.3. Cell Membrane

The cell membrane separates the content of a cell from the exterior surroundings. Besides the standard protection around the cell, the cell membrane controls what substances go in and out. The composition is based on a bilayer of phospholipids, internally hydrophobic (tails) and externally hydrophilic (heads) with different proteins and cholesterol between them. The membrane is permeable selectively, permitting only some materials to pass through its lipid layer by active (through protein pumps or vesicles) or passive (diffusion) processes of transportation. Water passes the membrane through a process called osmosis, which occurs when an imbalance of solutes appears outside and inside the cell [101].

In most of the cases, hydrophobic and small molecules can pass through diffusion. Nanoscale drug delivery systems depend upon an active mechanism (endocytosis). Over this process, the cell unit takes in ions, solid particles and molecules. There are studies that prove that positive charged nanogels can bind the membrane of the cell (negative charge) through electrostatic intercommunication. More than that, receptor-mediated endocytosis can provide a mechanism through which selectively attracted cell groups are targeted. In addition, hydrophobic nano-platforms can grow the adhesion to the membrane and the amount of drug entered in the cell [94].

3.4. Blood-Brain Barrier

The BBB is a highly selective semipermeable structure composed of five parts: the basement membrane, the astrocytes, the immune cells, the pericytes and an endothelial cell layer of capillaries. The area between basement membrane and the neurons is called Virchow-Robin space. In this region there is interstitial fluid in which reside microglia. All the above components are called neurovascular unit. The kinetics of this unit is crucial to the role of BBB and its states of illness. The capillaries of BBB compose a layer of squamous epithelial cells that fold to form a circular vessel. These cells are linked with strong connections for blocking entrance or exit of materials through central nervous system. Protein transportation facilitates the selective flow of molecules through vessel lumens, essential biomolecules being in higher concentrations, like glucose and at the same time eliminating toxins.

The BBB is a physical and metabolic "obstacle", physiologically important and active, which survey blood-brain traffic and control it, restricting the paracellular diffusion between the endothelial cells (microvessels) and the efflux pumps activity that quickly expel back into the capillary lumen a wide variety of xenobiotics. BBB integrity and function is critically influenced by what is now referred to as "the extended neurovascular unit" that incorporates not only microvascular endothelial cells and adjacent pericytes, astrocytes and neurons, but also neighbouring smooth muscle cells and microglia in the brain, and blood cells in the capillary lumen such as polymorphonuclear cells, lymphocytes and monocytes [102].

Transport at the BBB level is assured by numerous transport mechanisms that provides to the brain the necessary nutrients and also protects from the toxic xenobiotics. The main transport mechanisms are represented by free diffusion of small lipophilic substances or by active transport (carrier mediated, receptor mediated and active efflux transport).

Active efflux transport is assured by two major types of transporters that extrude metabolic waste, xenobiotics and a large number of drugs from the brain back into the blood. The first superfamily of BBB efflux transporters is the solute carrier proteins (SLC) superfamily, being represented at the level of BBB by SLC22 and SLCO (SLC21) efflux transporters. The second is the ATP-binding cassette (ABC) efflux transporter family represented by permeability glycoprotein (P-gp), breast cancer resistance protein (BCRP) and the multidrug resistance associated proteins (MRPs) [103–105].

Based on their localisation, the ABC efflux transporters prevent lipophilic and amphiphilic environmental toxic compounds or drugs, including anti-inflammatory, immunosuppressive, anti-infectious, antineoplastic drugs, some antiepileptic, antidepressant and psychotropic agents, and drug conjugates by an energy-dependent, unidirectional direct transport mechanism, from entering specific substrates [106]. The BBB's efflux machinery does an excellent job of recognizing xenobiotics, but a poor job on distinguishing between toxicants and therapeutic drugs, creating an important obstacle to treatment of brain cancer, epilepsy and neuro AIDS [107].

The penetration of the BBB for drug delivery, although challenging, captivates the interest of numerous researchers in antiviral therapy since the mechanisms by which for example HSV-1 penetrates the CNS remain unclear. The most likely routes include retrograde transport via the olfactory or trigeminal nerve fibres, occasionally leading to herpes simplex encephalitis (HSE) caused by HSV-1 [108].

Another studied virus that is involved in encephalitis and BBB disruption is HIV, known to cause severe neurological disorders and leading to HIV-related encephalitis [109] since BBB is impermeable to 98% of antiretroviral drugs [110]. The possible mechanism responsible for BBB disruption in HIV-1 encephalitis is considered a "Trojan horse" mechanism, where HIV infects specific T-lymphocytes and circulating monocytes, then entering to CNS through BBB gaps and followed by inflammatory reactions [111], but, in the last years nanotechnology has been intensely explored and several experimental attempts have been carried out in order to enhance the BBB permeability toward antiretroviral drugs, briefly described below based on the nano-based formulation composition, since it is well known that the size and surface functionalization influence transport properties within tissues:

- polymeric polybutylcyanoacrylate (PBCA) nanoparticles with two incorporeated antiretroviral drugs (AZT and lamivudine) showed a 8–20 and 10–18 fold increase in BBB permeation, by three possible mechanisms as presented by the authors: prolonged interaction interval between drug-loaded nanoparticles and brain-microvascular endothelial cells elevated the concentration gradient between blood and the brain, Polysorbate 80 covering on the periphery of nanoparticles was able to be absorbed and degraded nanoparticles improved drug absorption [112];

- spherical transferrin coated-PEGylated albumin nanoparticles encapsulating AZT prepared by ultra-emulsification method using chemical cross-linking by glutaraldehyde gained an access across the BBB through the transferrin receptor mediated endocytosis on the membrane [113];
- transferrin-conjugated quantum rod nanoparticles conjugated with saquinavir crossed an in vitro BBB model by exploiting a receptor-mediated transport [114];
- magnetic liposomal nanoformulations of azidothymidine 5′-triphosphate (the active form of azidothymidine) migrate across BBB in vitro, either directly or by a monocyte-mediated transport, under the influence of an external magnetic field [115];
- novel nanodrug consisting of an iron oxide nanoparticle coated with PMA amphiphilic polymer and functionalized with the antiretroviral peptide enfuvirtide crossed the BBB by a passive diffusion, probably mediated by the absorption of the amphiphilic coating on the cell membrane [116].

As briefly presented the preliminary results are more than encouraging. Certainly, future investigations on the mechanisms about BBB disruption are needed along with novel, innovative, safe and efficacious therapeutic approaches.

4. Nanotechnology: How Does It Face the Antiviral Therapy?

Up to now we have concluded that current antiviral therapy has not yet achieved the ideal shape and efficiency and also that the complex biological barriers are major obstacles, but can we critically say that nanotechnology could be the identified solution?

Search engine queries on "nanotechnology" generate more than 114,000 items on specialized platforms (Science Direct, for example) that represents potential and challenges in different fields from biosensors and industry-related applications up to nanomedicine and biomaterials. When the search keywords are "nanotechnology as antiviral therapy" the same engine only returns 1404 results starting from 1997.

It is a given fact that nanotechnology is defined as the application of materials on the nanometer scale. According to the literature data results, namomaterials designed with different shapes and morphologies display numerous advantages for use in antiviral therapy, namely: nanometric size that permits drug delivery through impermeable barriers [88], large surface area to volume ratios for large drug payloads incorporation [117] and improved efficacy, surface modification and/or backbone functionalization versatility that facilitates cellular membranes passage [118] or enhancing stability and bioavailability [119], virucidal activity against a series of viruses (HIV, HSV, HBV, etc.) due to biomimetic properties [120], increased specificity, improved antiviral delivery and controlled drug release to the target [121] through engineered moieties, decrease the emergence of drug resistance, personalized therapy possibility, protection of the drugs and low adverse drug side effects mainly due to the composition.

The mechanisms of nanomaterial-mediated drug delivery are determined by the chemistry, the architecture and the specific properties of each nanosystem (as presented in the schematic representation in Figure 4).

The design of new drug delivery systems for the antiviral therapy is focused on manipulating these features that are relevant in viral diseases where high drug doses are compulsory, implies high costs and the patient is depended on the administration protocol.

Lembo and Cavalli [18] present the current status up to 2010 in the nanoparticulate delivery systems in antiviral therapy area, highlighting their perspective on the challenges that must be tackled before the nanotechnology can be translated into clinical use as safe and effective antiviral formulations. Therefore, the nanoparticulate antiviral systems synthesised up to 2010 consisted mainly of micelles, polymeric NPs, solid lipid NPs (SLNs), nanostructured lipid carriers (NLCs), liposomes, nanocapsules, vesicles, dendrimers, nanogels, cyclodextrin-based systems and emulsions.

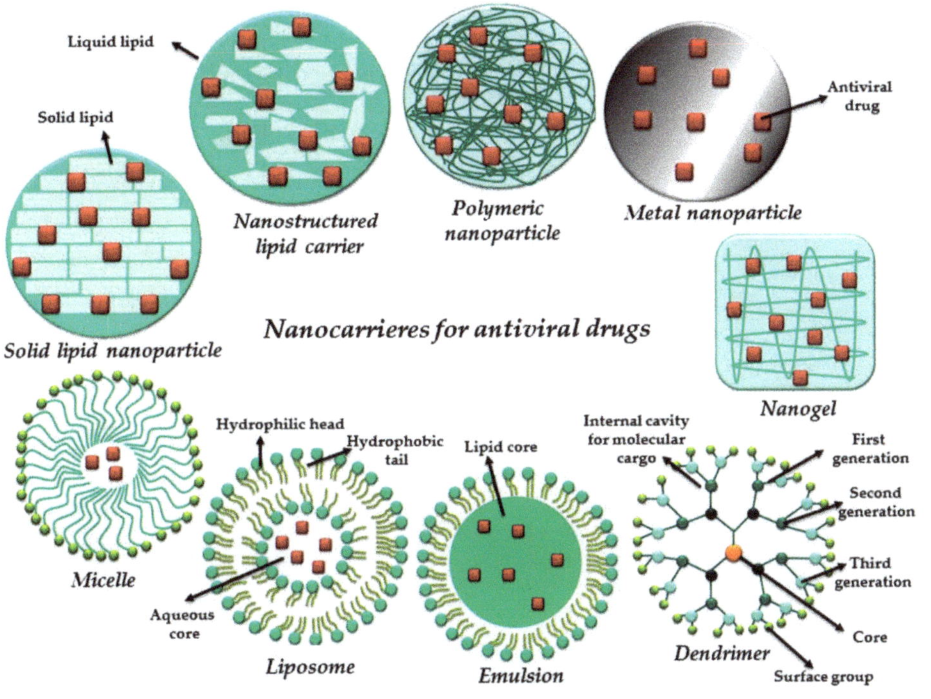

Figure 4. Nanocarriers developed for antiviral therapy.

In 2016, Liu and Chen [122] summarized in a review paper an interesting perspective of nanotechnology use in HIV/AIDS vaccine development. Their overview underline the potential of various nanomaterials and nano-architectures to be used as HIV vaccine carriers or adjuvants due to proven capabilities to improve delivery, permeability, stability, solubility and pharmacokinetics of traditional HIV vaccine approaches. The authors exhibit also the desired features of nano-carriers and adjuvants with high benefits-cost ratio.

In 2017, Milovanovic et al. [123] outlined, beside, the virus replication cycle and mechanism of actions of antiviral agents, an overview of particulate carriers for drug delivery. The review summarized several classes of the mostly considered carriers namely liposomes, micelles, microspheres, dendrimers and NPs synthesized as alternative supports for antiviral therapy. Table 2 highlights part of their summary and described based on virus classification in Section 5.

In 2019, Cao and Woodrow [124] reviewed the nanotechnology solutions used to eradicate HIV reservoirs and also the gene delivery and immunotherapy nanocarriers used in cancer with potential in HIV treatment. In Section 4. Nanocarriers for eradicating HIV reservoirs" the authors focused mainly on nanocarriers incorporating combination therapeutics developed in order to boost drug effectiveness and minimize toxicity. Several examples are presented in Table 2 and described in Section 5.

Table 2. Nano-delivery systems developed for antiviral drugs.

Nanoplatform Type	Nanoplatform Characteristics (Size, Morphology, Toxicity etc.)	Drug	Virus Type	REF
	Liposomes			
- Reverse phase evaporation	• GCV mixed with PC/CH/NaDC dissolved in chloroform/diethyl; • Spherical liposomes; • Liposome sizes of 210 ± 17 nm, ζ-potential—52.4 mV; polydisperse;	GCV	HSV	[125]
- rHDL	• rHDL-nosiheptide complex with a diameter < 30 nm;	Nosiheptide	HBV	[126]
	• rDHL-ACV palmitate complex size of 33.5 nm, around 10 times smaller than ACV-liposoms;	ACV		[127]
- cationic	• Viral gene expression reduce by 65–75% in liver after 2 days of administration at mice;	siRNA	HCV	[128]
- immunoliposomes	• Viral secretion reduced by 81% and free viral particles neutralized in vitro;	HIV gp 120 Folding inhibitor	HIV	[129–131]
	• In vivo resistance to infection has been enhanced;	anti-CCR5 siRNA		
	• Immunoliposomes diameter with average between 100 and 120 nm; really useful to deliver high concentrations of indinavir;	Indinavir		
- pegylated	• In house synthesized pis; reduced toxicity and increased adherence in vitro;	PIs		[132]
	Nanoparticles			
- HPAC	• HPAC (different concentrations)—non-cytotoxic for human epithelial cells (corneal, vaginal), HeLa cells, foreskin fibroblasts: cell viability >75%; • 99% drug loading efficiency;	ACV	HSV	[133]
- PLGA	• Three GCV pro-drugs have been separately loaded on PLGA NPs; • Uniform, spherical and smooth surface nps; • Particle size between 116 and 143 nm; ζ-potential between −13.8 and −15 mv; • Non-cytotoxic PLGA-nps (24 h and 48 h contact of three different NPs concentrations with HCEC cell);	GCV	HSV-1	[134]
- Se	• Uniformly spherical Se@AM; • Se@AM size of 70 nm, compared with SeNPs size which is 200 nm; • Se@AM—more stable than SeNPs; • Se@AM—superior antivirial effect on kidney cells treated with H1N1 and less citotoxicity (79.26% viability) than SeNPs (58.8%) or free AM (53.23%);	AM	H1N1	[135]
	• Uniformly spherical Se@OTV • Se@OTV size of 100 nm; • Se@OTV—superior antivirial effect on kidney cells treated with H1N1 and less citotoxicity (93% viability) than SeNPs (60%) or free OTV (53%);	OTV		[136]
- Ag	• Monodisperse and uniformly spherical particles; • Ag@AM size of 2 nm; highly stable NPs for more than 28 days; • NPs loaded with AM on their surface less cytotoxic (90%) than free AM (56%) or AgNPs (65%);	AM		[137]
- PEG-PLGA	• Uniformly spherical shape; • Size of NPs loaded with: diphyllin—178 nm and bafilomycin—197 nm; • Superior biocompatibility and antiviral activity for the drugs loded on NPs than the free drugs;	Diphyllin and Bafilomycin		[138]

Table 2. *Cont.*

Nanoplatform Type	Nanoplatform Characteristics (Size, Morphology, Toxicity etc.)	Drug	Virus Type	REF
- Human serum albumin + copolymers of maleic anhydride/alkyl vinyl ethers of oligo (ethylene glycol)	• Mean size of NPs in the range of 100–300 nm; • NPs surface with targeting moieties able to interact with liver cells receptors;	INFs-α		[139]
- Tf-Albumin-PEG	• Spherical particles; • NPs size between 114 and 124 nm; NPs surface had a negative charge;			[113]
- Lactoferrin	• Spherical AZT-lactoferrin particles with the diameter in the range of 50–60 nm; • The drug was intact after the preparation process; • AZT encapsulated in lactoferrin NPs is more efficient and less genotoxic (Wistar rats) compared to free AZT;	AZT		[140]
- PLGA NPs	• Polydisperse particles loaded with LAM with the size between 221 and 250 nm; ζ-potential between −4.64 and −3.65 mv; • The molecular interaction between LAM and the polymer confirmed by FTIR and DSC; • Slow degradation of NPs in simulated intestinal fluid PBS;	LAM		[141]
- Hybrid NPs (PLGA, MMA-SPM, PLA and PMMA)	• PLGA NPs size between 58–224 nm; MMA-SPM NPs size between 91–823 nm; • Almost spherical NPs; • Non-toxic NPs (male mice);	LAM+AZT		[142]
- PEO-PCL	• Spherical PEO-PCL NPs with smooth surface; • PEO-PCL size around 200 nm, PEO-PCL size around 270 nm; SQV was encapsulated into NPs;	SQV		[143]
PLGA PEG	• Spherical shape; • Size of the NPs: 125 nm for those loaded with SAHA and NFV; 118 nm for those loaded with NFV; 119 nm for those loaded with SAHA; • Low cytotoxicity effects of NPs loaded with the drugs (tested on ACH-2 cells);	SAHA NFV	HIV	[144]
- PLGA	• Spherical and smooth surface NFV NPs with mean size of 185 nm; an almost narrow distribution; ζ-potential of 28 mV;	NFV		[145]
- Lactoferrin	• NPs mean size of 45–60 nm, hydrodynamic radius of 103 nm, ζ-potential around −23 mV; polydisperse NPs; • Chemically stable NPs proved by FTIR and DSC;	EFV		[146]
- Folic acid-conjugated-P407	• NPs based on folic acid conjugated with P407 with inclusion of ATV and RTV significantly decreased the amount of HIV produced by cells in mice;	ATV+RTV		[147]
- PMA coated MNP	• Uniformly spherical NPs conjugated with ENF with size of 35.2 nm and ζ-potential around −29 mV; • Non-toxic in vitro and in vivo NPs conjugated with ENF;	ENF		[116]
- PVA-AOT	• Particles with diameters between 658 and 823 nm; ζ-potential between −12.8 and −25.3 mV;	MCV		[148]
- pMBA-Au NPs	• NPs with diameter of 1.8 nm; • Non-toxic in vitro NPs;	RAL		[130]
- PLGA	• NPs average size of 138.3 nm and ζ-potential around −13.7 mV; • Non-cytotoxic NPs loaded with drugs compared to free NPs;	EFV+Lopinavir+ RTV		[149]

Table 2. *Cont.*

Nanoplatform Type	Nanoplatform Characteristics (Size, Morphology, Toxicity etc.)	Drug	Virus Type	REF
- PLGA+Pluronic F127	• Well-defined polydisperse NPs with average size of 220 nm and −19.2 mV ζ-potential; • No adverse events and toxicity in a 14 days pharmacokinetic study on mice;	TAF+EVG	HIV	[150]
- Lactoferrin	• Average drug loaded particles size of 67 nm; • Insignificant in vitro toxicity to red blood cells; • Improved bioaviability of the three drugs;	AZT+EFV+LAM		[151]
	Dendrimers			
- PG	• Non-toxic peptide–PG conjugates in vitro; • Antiviral effect in vitro;	Peptides	IAV	[152]
- Alginate-PEG	• Dendritic structure confirmed by TEM; • Hydrodynamic diameter in the range of 601–782 nm and ζ-potential in the range of −45.8 and −65 mv; • Great biocompatibility proved by the values of cell viability: between 88% and 98% (neuro cells, Hela cells, glioma cells);	AZT	HIV	[118]
	Nanorods			
- PVP-PEG coated with Ag	• The presence of AM on the surface of Ag nanorods confirmed by FTIR; • The nanorods loaded with AM of 540 nm diameter. • Acceptable viability of cells (hela, huvecs, dendritic cells, macrophages) after 72 h contact with three different concentrations of AM nanorods;	AM	HIV	[153]
- Tf-conjugated QRs	• QR-Tf-SQV hydrodynamic diameter is 130–140 nm; • QR-Tf with different SQV concentrations non-cytotoxic after 6–48 h of contact with BMVECs;	SQV		[154]
	Nanospheres			
- Cs	• Spherical NS with an almost smooth surface; average diameter around 200 nm, polydisperse NS, ζ-potential around 40 mV; • ACV encapsulation efficiency 86% • Satisfactory Vero cell viability after contact with NS;	ACV	HSV	[155]
	Micelles			
- Cs-*g*-oligo(NiPAam)	• Copolymers self-assembled into multimicellar aggregates with hydrodynamic diameter between 330 and 436 nm and ζ-potential between +7 and +22.8 mV; • good mucoadhesion and cytocompatibility properties;	EFV	HIV	[95]
	Nanosuspensions			
- zirconium oxide beads stabilized with PVP, poloxamers and SLS	• Mean particele size around 320 nm, ζ-potential −32.8 mV; • EFV bioavailability improved after incorporation in nanosuspensions (in vivo, rabits);	EFV	HIV	[156]
	Nanoemulsions			
- Mucoadhesive NEs	• Based on triacetin-oil, tween 20-surfactant, transcutol P-cosurfactant; • 23–200 nm spherical particles; • Nontoxic and noniritant nanoplatforms (New Zealand albino rabbit);	GCV	HSV	[157]

Table 2. *Cont.*

Nanoplatform Type	Nanoplatform Characteristics (Size, Morphology, Toxicity etc.)	Drug	Virus Type	REF
	SLNs			
- Borneol	• Microemulsion-based method; • Particle size between 113 and 142 nm; ζ-potential between −15.1 and −18.3 mV; polydisperse particles;	GCV	CMV	[158]
- Stearic acid + Pluronic F68)	• Spherical SLNs with a mean diameter of 167 nm and ζ-potential around −18 mV; • SLNs loaded with AZT can successfully deliver the drug in vitro to human brain endothelial cells;	AZT	HIV	[159]
	Lipid nanoparticles			
- bryostatin-2	• The lipid nanoparticles can stimulate latent HIV and can inhibit virus spread in vitro;	NFV	HIV	[160]
- PEG and phospholipids	• Size of LNPs loaded with drug/drugs between 33 and 68 nm and incorporation efficiency between 88% and 96%;	ATV+RTV		[161]
		ATV+RTV+TFV		
- DSPC+MPEG+DSPE	• Particle diameter between 52 and 68 nm; • In vivo anti-HIV LNPS do not exhibit local reactions and animal platelet counts are within normal limits.	Liponavir+ RTV+TFV		[162]

Recently, Arca-Lafuente et al. [163] overviewed nanotechnology-based systems as reliable alternative diagnostic tools for HCV infectious disease. Even if our review does not cover screening, it is important to mention that new diagnostic methods are required in order to overcome current drawbacks of HCV under-diagnosed infection as highlighted in the above-mentioned review. The nanotechnology-based tools described in the review seem to fulfil the necessary features for HCV elimination.

With the aim of developing new personalized diagnostic tools, Farzin et al. [164] summarized current strategies and under-development tools for early diagnosis of HIV. Their review combines the use of nanomaterials such as carbon nanostructures, nanoclusters, quantum dots, metallic and metal oxide NPs as advanced structures for HIV detection with possible biosensing strategies targeting to offer innovative outlooks for developing intelligent, sensitive and specific nano-objects for in situ and real-time detection of HIV.

5. Current Overview of Nanotechnology Use in Antiviral Therapy: Virus Related

In this section the authors point out the suitability of nanomaterials (recent data) for antiviral therapy, highlighting the enhanced features pursued to overcome the identified issues as related above.

5.1. *Nanomaterials Designed for Non-Retroviral Antiviral Agents*

5.1.1. Nano-Based Antiviral Agents against Herpes Viruses

Donalisio et al. [155] have reported the preparation by a modified nano-emulsion method of chitosan nanospheres (NS) loaded with 8.5% ACV as a topical formulation against both HSV-1 and HSV-2 herpes virus strains. The main component, chitosan, a natural polycationic polysaccharide, was selected as a material for ACV release, due to its distinctive properties: hydrophilic character, in situ gelling, mucoadhesion, permeation enhancing, in addition to a low cytotoxicity, biocompatibility and bioresorbability features [165]. The obtained gel formulation based on ACV-loaded NS proved an enhanced ability to penetrate porcine skin to about 55% (at 24 h) greater than the commercial cream product (10%). IC_{50} values against HSV-1 and HSV-2 were also determined on Vero cell cultures infected with above-mentioned strains, displaying significant reduced values of 0.012 μM and 0.100 μM, respectively, when using the NS formulation as compare to 0.156 μM and 1.608 μM for free acyclovir. This nano-technological approach attests the higher efficacy of the described formulation and with promising expectations for further preclinical and clinical experiments.

Yadavalli et al. [133] have explored the potential of highly porous activated carbon (HPAC) particles as a model for restricting HSV-1 and HSV-2 from entering target cells. They have considered this material due to the charcoal surface-active that could provide antiviral effects through virion sequestration approach. Furthermore, ACV molecules adsorbed or encapsulated inside the HPAC pores revealed sustained drug release acting in a synergistic manner to obtain an enhanced therapeutic effect. The HPAC compound prove d a 40 to 60% reduction in HSV-1 and HSV-2 entry for concentrations as low as 1 mg/mL. The IC_{50} value of HPAC corresponding to HSV-1 and HSV-2 infection in prophylactic administration was found of 0.8 and 1 mg/mL, respectively, significantly lower than clinically accepted TC50 value (half maximal toxicity concentration) of HPAC. Following the promising outcomes from in vitro tests, further determinations of antiviral efficacy on in vivo studies using a murine model of ocular (HSV-1) and genital (HSV-2) infection were performed. As a result, the ACV loaded HPAC acts by capturing the virus and releasing the encapsulated drug, hindering inflammation and immune cell infiltration in targeted tissue. The strong antiviral activity of this product was assigned to the charged surface of its pores which may interact with the cell's surface, stimulating an active exchange of ions (Na^+, K^+, Ca^+, Cl^-, and OH^-), when sustained or slow release of ACV has been acquired. Moreover, these particles exhibited both prophylactic and therapeutic effects against HSV-1/HSV-2 cells, unlike the free drug that did not demonstrate a prophylactically antiviral response.

5.1.2. Nanomaterials with Antiviral Intrinsic Activity

Gold and Silver NPs Using Seaweed *Sargassum wightii* with Anti-Herpetic Activity

Biogenic gold (Au) and silver (Ag) NPs were prepared using the seaweed *Sargassum wightii* (*Sw*) and investigated for their antiviral activity against HSV-1 and HSV-2 strains [166]. The NPs synthesis resided in an eco-friendly method, previously described in the literature [167], replacing the use of different reducing agents. The obtained NPs, *Sw*-Au and *Sw*-Ag, were evaluated concerning both cytotoxic and antiviral effect, using MTT and CPE (cytopathic effect) assays on Vero cells. The results showed that cell viability ranged from 93% to 85% when the concentration ranged between 2.5 and 25 μL per sample in *Sw*-Au, and from 97% to 84.58% for concentrations of 2.5 and 1 μL per sample in *Sw*-Ag. The antiviral assay have shown a 70% decrease of CPE on both HSV-1 and HSV-2 when Vero cells were treated with 10 and 25 μL *Sw*-Au, whereas *Sw*-Ag exhibit similar reduction of CPE at a concentration of only 2.5 μL per sample. Higher concentrations of *Sw*-Ag are not accepted due to an increased cytotoxic effect. The authors claimed that the obtained results are in agreement with other published research and they inferred that functionalized metallic NPs act as antiviral agents by blocking the virus attachments and cell access, depending on particle size.

Broad-Spectrum Antiviral NPs

Cagno et al. [168] conducted a research study concerning broad-spectrum antiviral products, which usually mimic heparan sulfate proteoglycans (HSPG), as well-preserved target of "viral attachment ligands" (VALs). The antiviral effect relies on the binding mechanism of the nanoparticles to the virus surface, thus preventing virus-cell attachment. In most cases, the reversibility of these bonds is reported [169], so that by increasing the dilution viral inhibition is lost, causing those compounds not to be considered antiviral drugs. The aforementioned authors have designed antiviral nanoparticles of virucidal effect based on long and flexible linkers simulating HSPG, leading to irreversible viral deformation. Of the synthesized compounds, the most notable virucidal effect was found in the AuNPs coated with a 2:1 mixture of decanesulfonic acid (MUS) and 1-octanethiol (OT). MUS allows a multivalent binding [170] as a consequence of its structure comprising a long hydrophobic chain, sulfonic acid terminated. The enhanced activity of MUS:OT-NPs was assigned to the new construct using MUS linker that caused local distortions and then a global virus deformation, leading to irreversible loss of infectivity. The MUS:OT-NPs exhibited efficient virucidal effect against HSV-1 and HSV-2, human papilloma virus (HPV-16), respiratory syncytial virus (RSV), dengue and lenti virus.

The in vivo testing on Balb/c mice infected with RSV reveals the efficacy of MUS:OT-NPs treatment that prevented the pulmonary dissemination of the infection. These results are in agreement with previous published data [171], which assessed the relationship between the surface structure of nano-objects and their ability to cross cell membranes. Both in vivo and in vitro tests on cell cultures have proven the lack of toxicity of MUS:OT-NPs.

Lipid Nanoemulsions Encapsulating Coumestrol as Topical Treatment of Herpes Simplex

Coumestrol is an isoflavonoid-like compound having the ability to inhibit the replication of HSV-1 (both acyclovir sensitive and resistant strains) and also some strains of HSV-2 [172]. Argenta et al. [173] have designed a formulation in an effort to obtain a topical product for coumestrol delivery at the level of mucosa. In this approach, the bioactive compound was entrapped by fluid or rigid phospholipid nanoemulsions (dioleylphosphocholine, DOPC and distearoylphosphocholine, DSPC, respectively) dispersed in a hydroxyethylcellulose gel. The effectiveness of the proposed antiviral agents was tested regarding permeation and retention ability on intact and damaged porcine esophageal mucosae and for antiherpes activity on cell culture assays using Vero and GMK AH1 cell lines. The greatest performance of both coumestrol-loaded nanoemulsions NE-COU/DOPC and the same product thickened with hydroxyethylcellulose, HNE-COU/DOPC, as compared to those based on DSPC, relies largely on the physico-chemical properties of the nanoemulsion. The positively charged nanoemulsion showing highest values of ζ-potential may interact with negatively charged surface of mucosa membrane, with beneficial consequences relating to transmucosal delivery of coumestrol [174]. The length of phospholipids alkyl chain, the number of unsaturations, the lipophilic/hydrophilic balance of the active principle also contributed to the global effect, so that the fluid-state of hydrocarbon chain induced by DOPC explained the interaction between the oil-water interface and mucosa, increasing coumestrol permeation and retention. The low IC_{50} values proved an enhanced antiviral activity against HSV-1 and HSV-2 after coumestrol formulation using nanoemulsions based on DOPC, which could be considered for advanced studies in order to be introduced in therapy.

5.2. Nanomaterials Designed for Antiretroviral Drug Delivery

The huge socio-economic impact of HIV, as mentioned in Section 1, determined a continuously increased trend of studies related with finding an almost perfect treatment. Since the seven classes of antiretroviral drugs defined by FDA contain a large number of active principles, plenty of studies regarding their incorporation in different types of nanosystems can be found in the literature. Several examples are presented below.

5.2.1. NRTIs and NNRTIs

RTIs classes, NRTIs and NNRTIs, include some of the drugs often used in HIV treatment plans. Recently, Grande et al. [175] published a complex review on RTIs nanosystems for controlled drug delivery and our review complements part of the data presented in this article, emphasizing the penetration of biological barriers in vitro or in vivo by nanosystems containing RTIs.

AZT, a high bioavailable drug, has serious side effects, the most common being bone marrow suppression, toxicity for some organs, neutropenia and anaemia. Specific target drug delivery using different nanosystems is a promising solution [140]. ATZ has been incorporated in hybrid NPs based on alginate and stearic acid-poly ethylene glycol. C6 glioma, neuro brain and Hela cells have been used to study the cellular uptake and the cytotoxicity of the NPs in vitro. The results proved that these nanosystems are nontoxic and have significant brain cellular uptake, suggesting that they can be used for more complex internalization in brain cells studies [176]. In another study, sol-oil chemistry has been used to prepare small NPs lactoferrin loaded with AZT (50–60 nm in size), stable in biological simulated fluids (gastric and intestinal). Antiviral activity of NPs has been analysed using SupT1 Cells infected with HIV-193IN101 virus and the results suggested that the encapsulation of AZT in lactoferrin does not influence the drug activity. The NPs loaded with AZT and the drug alone have been orally administrated to Wistar rats of both genders, the performed assays (bone marrow micronucleus, histopathological and biochemical analysis) showing that AZT loaded in NPs is more efficient and less toxic, compared with the soluble form [140].

Lamivudine—LAM, a water-soluble drug with two main drawbacks: its half-life is only 2 h and has a deficient bioavailability, especially in paediatric patients (68%) [177]. LAM has been included in NPs based on poly(ε-caprolactone)—PCL [178,179], poly lactic-*co*-glycolic acid—PLGA [141], chitosan and sodium alginate [180], Eudragit E100 [177]. The physico-chemical characterization of the obtained NPs has shown an adequate size of the NPs and a good stability. In vitro drug release tests indicate that NPs can support the drug delivery for 24 h, indicating a less frequent administration [179,180]. Sneba et al. [177] have reported a more complex study, where LAM-polymeric non-cytotoxic NPs have been included in films for drug delivery through the buccal mucosa barrier. Four mucoadhesive polymers: polyvinyl alcohol—PVA, polyvinyl pyrrolidone—PVP, sodium carboxymethylcellulose—SCM, hydroxypropyl methylcellulose—HPMC have been used to prepare the films. Moreover, Ozturk et al. [141], obtained PLGA NPs loaded with LAM and proved that are physicochemical stable and slowly released the drug, a great property attributed to ester end-group of PLGA. Because these NPs were intended for oral administration, the authors evaluated the gastrointestinal stability of the NPs in vitro, using different fluids of biological interest with pH in the range 1.2–7.4 phosphate buffer solution, intestinal fluid phosphate buffer solution, physiological serum and distilled water; the tests have been developed at 37 °C for 24 h. The results indicated that PLGA NPs are promising intestinal targeted drug delivery system for LMA, being stable in tested media.

In clinical practice, LAM is frequently administrated together with AZT, therefore this combination of drugs is being studied for target delivery using different types of NPs. Sankar et al. [142] used PLGA, methylmethacrylate-sulfopropylmethacrylate—MMA-SPM, poly lactic acid—PLA, and poly methyl methacrylate—PMMA to prepare different types of NPs by emulsion polymerization, as drug delivery nanosystems for ATZ (52%) and LMA (58%). In vivo acute toxicity has been studied in mice; the results proving the fact that the drug doses loaded in the NPs are not toxic. ATZ-LAM PLGA NPs seemed to be the most promising nanosystems.

Efavirenz—EFV, one of the most used NNRTIs in clinical practice, is a poorly water-soluble drug, and the incorporation in different drug delivery nanosystems being a solution for this drawback. Patel et al. [156] obtained nanosuspensions—NS based on povidone polymer- polyvinylpyrrolidone (PVP) K30, poloxamers steric stabilizer (188 and 407) and an anionic electrostatic stabilizer (sodium lauryl sulphate, a steric stabilizer—SLS). Compared with the drug alone, an important improvement of saturation solubility has been noticed for the NS with EFV. The incorneporation of EFV in NS increased the absorption of the drug in rat intestine in situ, and very important the oral bioavailability in the studies on Albino rabbits. Lactoferin used to prepare NPs loaded with ATZ [140], as described above, has been used also to encapsulate EFV [146], based on the same preparation technique: sol-oil chemistry. Compared with free EFV, the encapsulation of the drug in NPs showed a reduced toxicity to peripheral blood mononuclear cells, Jurkat T cells and B16-F10 cells, an increased anti-HIV-1 activity and improved oral bioavailability and pharmacokinetic profile in studies on rats.

5.2.2. PIs

ATV. Low brain permeability and antiretroviral drug resistance are two of the most important disadvantages of ATV. This PIs drug has been encapsulated in SLNs and studied as nanosystems for brain delivery using hCMEC/D3 as a blood-brain barrier in vitro model, HCMEC/D3 being human brain microvessel endothelial cell line. Average ATV had an important increase regarding the cellular uptake once delivered through average nanosystems (around 167 nm) [159].

SQV, another anti-HIV PIs, has been incorporated in poly(ethylene oxide)—modified poly (ε-caprolactone) (PEO-PCL) NPs [143] by a solvent displacement process. Human monocyte/ macrophage (Mo/Mac) cell line—THP-1 has been used for in vitro cellular uptake assay. The drug has been successively released intracellular and a meaningful uptake of the SQV-PEO-PCL has been noticed.

NFV, used in HIV-1 and HIV-2 treatment as PI, is a promising drug that can be used also for other grave medical disorders like cancer [181]. Some studies have showed the ability of different types of NPs loaded with NFV to activate latent HIV and to restrict viral spread in vitro. Kovochich et al. [160] showed that lipid NPs–LNPs incorporated with bryostatin-2, a protein kinase C activator (LNP-Bry), can be loaded with NFV (LNP-Bry-NFV), and proved the above mentioned abilities on J-Lat Full Length Cells (10.6). Tang et al. [144] have prepared NPs based on poly(lactic-*co*-glycolic acid)-polyethylene glycol diblock copolymers and anti-CD45RO antibody conjugated with suberoylanilide hydroxamic acid (SAHA) and NFV and tested theirs in vitro properties on ACH-2 cells. More complex studies have been performed by Venkatesh et al. [145] PLGA NPs loaded with NFV have significantly enhanced the oral bioavailability of the drug studied in vivo in New Zealand rabbits, a reduced frequency of dosing being needed in this case.

The literature data reports also several studies where combinations of PIs drugs have been incorporated in nanosystems and pre-clinically evaluated. Duan et al. [161] have included separately ATV and DRV in LNPs, but only ATV-LNPs proved to form stable drug-lipid concentrations. Based on these results, the authors have developed LNPs containing ATV and RTV and also LNPs containing ATV + RTV + tenofovir (TFV—an HIV NRTIs), the last ones being prepared in a large volume for a preliminary primate pharmacokinetic study. After LNPs subcutaneously administration, the three drugs have been detected in plasma for seven days.

5.2.3. Fusion Inhibitors, Entry Inhibitors and Integrase Inhibitors

ENF is a fusion inhibitor that is incapable to cross the cerebrospinal fluid. Fiandra et al. [116] proved that by including it into a nanosystem composed from magnetic nanoparticles (MNP) synthesized by solvothermal decomposition in organic solvent followed by fluorescent labelled PMA coating could solve ENF drawback. In vitro model (co-colture of RBMVECs and astrocytes) and in vivo model (Balb/c mice) studies proved that nanoconjugated ENF could penetrate BBB.

MCV, an entry inhibitor acting as a CCR5 co-receptor antagonist, has been also included in some nanosystems in order to increase its oral bioavailability. Solid drug nanoparticles—SDNs, containing 70 wt % MVC and 30% some polymer/surfactant excipients have been prepared using the emulsion-template freeze drying technique. Monolayers of Caco-2 have been used as a human gut in vitro model in order to study the absorption behaviour of MVC SDNs and in vivo oral pharmacokinetics of the MVC solid drug nanoparticles (SDNs) has been analysed on a rat model. Both studies indicated an advanced permeability of the MCV NPs (based on PVA and sodium 1,4-bis(2-ethylhexoxy)-1,4-dioxobutane-2-sulfonate (AOT) excipients) correlated with the normal drug [148].

Bowman et al. [182] proved that small organic monovalent molecules conjugated to AuNPs acts as fusion inhibitors in vitro, while Vijayakumar et al. [183] proved that AuNPs alone acts as entry inhibitors. Moreover, integrase inhibitor, RAL, has been functionalized with a thiol group in order to link Au-NPs. In vitro cellular uptake has been tested on macrophages, human brain microendothelial cells and primary peripheral blood mononuclear cells, the results suggesting that RAL—pMBA—Au-NPs penetrate the cells and also can exhibit antiviral activity. In vivo studies performed by injection of RAL—pMBA—Au-NPs in female adult BALB mice tail proved that the studied NPs could cross the BBB [184].

6. Progress in Nanomedicine: Antiviral Nanotherapeutics Approved or under Evaluation

Nanomedicine represents a fast-revolutionizing field that faces rapidly and constantly progress assessed by the numerous nanodrugs that have entered clinical practice and also by even more being investigated in clinical trials. Table 3 presents the approved antiviral nanomedicines, from which half are vaccines.

Table 3. Examples of approved nanoplatforms for drug delivery by the FDA, EMA and other organizations (updated table, after Singh et al. [185]).

Name	Company/Approval Year/Country/Organization	Nanoplatform. Benefits	Virus	Route of Administration	REF
Epaxal®	Crucell (former Berna Biotech Ltd.); 1994 Switzerland	Virosomes (around 150 nm *spherical liposomal vesicles*)—intrinsic adjuvant properties; reduced toxicity and superior tolerability;	HAV	Intramuscular vaccine	[186,187]
Inflexal® V	Crucell (former Berna Biotech Ltd.); 1997 Switzerland	Virosomes (around 150 nm *spherical liposomal vesicles*)—biodegradable and biocompatible adjuvant systems; unwanted side effects; superior immune response;	Influenza	Intramuscular vaccine	[187,188]
PegIntron®	Schering Corporation, 2001, U.S., FDA	PEG-interferon alfa-2b (polymeric NPs) —31.000 Daltons molecules; superior protein stability;	HCV	Subcutaneous	[189]
Pegasys®	Genentech, 2002, U.S. FDA	PEG-interferon alfa-2a (polymeric NPs)—31.000 Daltons molecules; superior protein stability;	HBV, HCV	Subcutaneous	[190,191]
Influvac® Plus	BGP Pharma ULC, 2005, Canada	Virosome vaccine	Influenza	Intramuscular vaccine	[192]
VivaGel® BV	Starpharma, Australia; Mundipharma, Europe, 2019	Dendrimer (astodrimer sodium—SPL7013) incorporated in a water-based vaginal gel, acting as a targeting antiviral biofilm.	HIV, HSV	Topically Applied (Vaginal gel)	[193]

According to Singh et al. [185] review from 2017 and also to available information on the respective websites there are still several nanomedicines under evaluation, namely:

- Fluquit (STP 702) from Sirnaomics Inc. currently under preclinical evaluation, a polymer-based nanotherapeutic that incorporates siRNA and targeting the H5N1 (avian flu), H1N1 (swine flu) influenza, and newly emerging H7N9; and cervisil (STP909), a nanobased drug candidate, which incorporates siRNA for the treatment of HPV16 and HPV18;
- DermaVir from Genetic Immunity, a synthetic pathogen-like nanomedicine that incorporates single plasmid DNA expressing 15 HIV antigens that assemble to HIV-like particles; DermaVir vaccine completed Phase I/II randomized, placebo-controlled, dose-finding, double-blinded, multicenter study to assess the safety, tolerability and immune response in HIV-1-infected adults who are currently receiving anti-HIV treatment (number NCT00270205) [194];
- Doravirine (MK-1439), from Merck, a novel, next generation NNRTI described as solid drug nanoparticle formulation tested for HIV; currently doravirine completed the pharmacokinetic trial of the bioavailability of four MK-1439 nano formulations in healthy adults (number NCT02549040) [195];
- Lipid nanoparticles of ARB-001467 TKM-HBV containing three RNAi therapeutics for HBV genome targeting from Arbutus Biopharma; in 2018 the company completed the phase 2a, single blind, randomized, placebo controlled, study evaluating the safety, anti-viral activity, and pharmacokinetics (PK) following multiple doses of intravenous ARB-001467 (number NCT02631096) [196].

7. Authors' Perspective to Design Next Generation of Nano-Based Antivirals for Clinical Translation

As discussed above, nanotechnology started to be a critical player in the antiviral therapy. As mentioned by Ross et al. [99], nanotechnology frees the current therapy payloads in terms of delivery across biological complex barriers, and could resolve the low bioavailability drawback as already stated in Section 3.

Nanomaterials impart many physical, chemical and biological advantages [18,197] such as: (1) small particle size in order to facilitate drug delivery through biological barriers, (2) large surface area to volume ratios to ensure large drug payloads, (3) tunable surface charge to facilitate cellular entry across the negatively charged cellular membrane, (4) biomimetic properties which result in intrinsic antiviral assets, (5) ability to anchor targeting moieties to increase specificity to desired cell types, tissue or other compartments, (6) improved solubility and pharmacokinetic and/or pharmacodynamics properties translated in longer time to allow greater accumulation, controlled and sustained release, (7) enhanced efficiency gained either by drug molecules entrapment to protect them from physiologically hostile media, or by using surface conjugation to target drugs to specific tissues, (8) reduced toxicity and (9) multifunctionality by combining several beneficial features in a stable construct, designed to simultaneously stimulate the replication of latent virus and deliver an antiviral to the activated cell [198].

Several limitations were also acknowledged such as: (1) degradation, for example nanoparticles are degraded in the gut following oral administration, or fail to penetrate the mucus barrier and are thus minimally absorbed [199], (2) undesired interactions with biological molecules that leads to opsonization, uptake by macrophages and reduced plasma half-life [200], (3) non-specifically absorption that may induce apoptosis and disrupt cell membrane and adverse immunological responses [201], (4) large dimension for renal clearance therefore cannot be degraded within the body, and are accumulated, leading to toxicity [202], and (5) scaling up issues and high costs.

In our perspective, the "ideal" nanocarrier for efficient antiviral delivery must take into considerations several key factors namely:

- Clinical outcome, since patients need safe, effective, targeted, available and affordable therapy, as they are our inspiration;
- From the clinical perspective, the future antiviral candidates should improve the efficacy of the fused/encapsulated drug, reduce the intake frequency and time, restrict adverse side effects and reduce therapy costs;
- Design consideration for the nanoplatforms that will allow targeted delivery of the drugs in sustained released manner and improves efficacy, safety and patient convenience; therefore, from a chemist point of view, hybrid nanosystems can gather all the necessary features in terms of composition, shape and size by overcoming limitations of individual systems and offers greater advantages. Starting with the composition, the chosen materials should be biodegradable, biocompatible, and non-toxic, for example polymers are very attractive since they offer the possibility for chemical modifications over the surface or backbone. In addition to these advantages, the second component from the hybrid architecture (in the shape of potential liposomes) should offer besides advanced barrier penetration, higher encapsulation efficiency for the intended drug, which in combination with the polymeric piece will be able to modulate the release kinetics, the stability and prolong drug release. When thinking about the shape, we have in mind targeting capabilities as impact. As we already know, the shape is linked with size and surface charge and density, therefore a complex puzzle that must be solved. The surface charge and density should be carefully chosen during the nanoplatforms design through the surface modification possibility. The ideal candidate here from our perspective is PEG due to its versatility to exhibit various charges, shapes and sizes but also to enhance tolerability, reduce clearance, and lengthen circulation time. The size influences the biodistribution and the uptake rate therefore the "nominee" has to be in the submicron size range, recommended to be under 200 nm.

Taking into consideration the performance indicators of nanomedicine, we claim that the development a personalized nanomedicine is possible via a synergistically approach. Since the development of "best" viral carriers involves a multidisciplinary team, virologists should be directly implicated in the development, offering specialized support on the following matters: identification of differentially expressed moieties virus cells for targeted delivery, elucidation of the type of desired targeted and the response from the host cells to nanodelivery platforms.

Therefore, multidisciplinary research-oriented efforts have to be related also to system biology by exploring machine learning for process optimization and pharmacology in order to introduce best appropriate combination of therapeutic agents.

In 2018, an interdisciplinary team of virologists and biochemists, which developed low-cost and "cell-friendly" nanogels that can efficiently prevent viral infections, addressed these challenges [203]. Here, the flexible, nontoxic and broad-spectrum nanogels based on dendritic polyglycerol sulfate mimic cellular surface receptors where several viral families bind. The designed nanogels can multivalently interact with viral glycoproteins, shield virus surfaces, and efficiently block infection since they act as robust inhibitors for these viruses.

When thinking about the translation into the clinical practice, the nano-based future antiviral therapy must follow a specific flow-chart, starting with the optimization and scale-up practices according to the good manufacturing practice, the elaboration of suitable regulatory guidelines and finishing with the development of cost-effective and high quality formulations available worldwide. Taken into consideration all these enhanced features, the road to clinical practice still has many addressed issues in order to provide effective and safe antiviral nano-formulations to patients.

8. Conclusions

Treating or improving treatment success rate for viral diseases are fundamental responsibilities. The established potential and boosted progress of nanotechnology in antiviral therapy development generates great expectations for new therapeutic innovative strategies for attacking or eradicating viral disorders. At present, studies explored numerous and diverse nano-platforms including nanoparticles, liposomes, micelles, with different compositions, size, with single or combined entrapped drugs that may serve as potential antiviral drug delivery transporters. These nano-based systems have exhibited versatile features to improve the identified current therapy drawback. However, the clinical use of a nano-based antiviral formulation to date based on our knowledge has turned out just a few approved or under clinical trials nanoformulations, mainly vaccines, despite more than 22 years of constant efforts. It is expected in the upcoming years that part of this "success preclinical story" to be scaled-up, translated and applied for better outcome, convenience and access for patients.

Author Contributions: All authors contributed equally to this manuscript. G.D. and F.-D.C. were responsible for the review concept and design. G.D., C.-M.U., F.-D.C., C.-T.M., I.G., D.B., B.-I.T., C.R., L.T. and E.R. were responsible for draft preparation. G.D. and F.-D.C. were responsible for the review and editing of the manuscript. G.D., C.R., L.T., B.-I.T. and E.R. were responsible for visualization and funding acquisition. G.D. and F.-D.C. were responsible for the revisions of the manuscript. All authors have read and agreed to the published version of the manuscript.

Funding: This work was supported by a grant of the Ministry of Research and Innovation, CNCS-UEFISCDI, project number PN-III-P1-1.1-PD-2016-1642, within PNCDI III.

Conflicts of Interest: The authors declare no conflict of interest.

Abbreviations

AIDS: acquired immunodeficiency syndrome; **ACV**: acyclovir; **Ag**: silver; **AM**: amantadine; **ART**: antiretroviral therapy; **ATV**: atazanavir; **AZT**: zidovudine; **BMVECs**: brain microvascular endothelial cells; **Au**: gold; **CCK-8**: Cell Counting Kit-8; **CH:** cholesterol; **CMV**: cytomegalovirus; **Cs**: chitosan; **DAA:** direct-acting antiviral; **DSC**: differential scanning calorimetry; **DLS**: dynamic light-scattering; **DNA**: Deoxyribonucleic acid; **DsDNA**: double stranded DNA; **DRV**: darunavir; **DSPC**: 1,2-Distearoyl-sn-glycero-3-phosphocholine; **EFV**: efavirenz; **EVG**: elvitegravir; **FDA**: Food and Drug Administration; **FTIR**: Fourier-transform infrared spectroscopy; **GCV**: gangiclovir; **GMP**: good manufacturing practice; **AuNPs**: gold nanoparticles; **HBV**: hepatitis B virus; **HCV**: hepatitis C virus; **HSV**: herpes Simplex; **HIV**: human immunodeficiency virus; **HPAC**: highly porous activated carbon; **HPMC**: hydroxypropyl methylcellulose; **IAV**: influenza A virus; **IFNs**: interferons; **LAM**: lamivudine; **LNPs**: lipid nanoparticles; **MNPs**: magnetic nanoparticles; **MPEG**: *N*-(carbonylmethoxypolyethyleneglycol-2000); **DSPE**: 1,2-distearoyl-sn-glycero-3-phosphoethanolamine, sodium salt; **MTT**: [3-(4,5-Dimethylthiazol-2-yl)-2,5-Diphenyltetrazolium Bromide]; **MTS**: [3-(4,5-dimethylthiazol-2-yl)-5-(3-carboxymethoxyphenyl)-2-(4-sulfophenyl)-2H-tetrazolium]; **NaDC**: sodium deoxycholate; **NEs**:

nanoemuslions; **NiPAam**: *N*-isopropylacrylamide; **NLCs**: nanostructured lipid carriers; **NPs**: nanoparticles; **NRTIs**: nucleoside and nucleotide reverse transcriptase inhibitors; **NNRTIs**: non-nucleoside reverse transcriptase inhibitors; **NS**: nanospheres; **OTV**: oseltamivir; **PBS**: phosphate buffer solution; **PC**: phosphatidylcholine; **PCL**: poly(ε-caprolactone); **PEG**: poly(ethylene glycol); **PEO-PCL**: poly(ethylene oxide)-modified poly(ε-caprolactone); **PG**: polyglycerol; **PLGA**: poly lactic-*co*-glycolic acid; **PIs**: protease inhibitors; **pMBA**: p-mercaptobenzoic acid; **PVA**: polyvinyl alcohol; **PVP**: poly(vinylpyrrolidone); **QRs**: quantum rods; **RAL**: raltegravir; **REF**: reference; **rHDL**: recombinant high density lipoproteins; **RNA**: Ribonucleic acid; **RTIs**: Reverse transcriptase inhibitors; **RTV**: ritonavir; **SCM**: sodium carboxymethylcellulose; **SLNs**: solid lipid nanoparticles; **SLS**: sodium lauryl sulphate; **SAHA**: suberoylanilide hydroxamic acid; **SEM**: scanning electron microscopy; **SeNPs**: selenium nanoparticles; **Se@AM**: selenium nanoparticles with amantadine; **siRNA**: small interfering RNA; **ssDNA**: single stranded DNA; **ssRNA**: single stranded RNA; **TAF**: tenofovir alafenamide; **TEM**: transmission electron microscopy; **Tf**: transferrin; **TFV**: tenofovir; **UNAIDS**: The Joint United Nations Programme on HIV/AIDS; **VZV**: varicella-zosterian virus; **ζ-potential**: zeta potential.

References

1. McMichael, A.J. Environmental and social influences on emerging infectious diseases: Past, present and future. *Philos. Trans. R. Soc. Lond. B* **2004**, *359*, 1049–1058. [CrossRef] [PubMed]
2. Szucs, P.A.; Richman, P.B.; Mandell, M. Triage Nurse Application of the Ottawa Knee Rule. *Acad. Emerg. Med.* **2001**, *8*, 112–116. [CrossRef] [PubMed]
3. The Joint United Nations Programme on HIV/AIDS. Available online: https://www.unaids.org/en (accessed on 27 November 2019).
4. UNAIDS. Global Statistics Report 2019. Available online: https://www.unaids.org/en/resources/fact-sheet (accessed on 27 November 2019).
5. WHO. Global Hepatits Report. 2017. Available online: https://www.who.int/hepatitis/publications/global-hepatitis-report2017/en (accessed on 25 November 2019).
6. WHO. Herpes Simplex Virus Key Facts. 2017. Available online: https://www.who.int/news-room/fact-sheets/detail/herpes-simplex-virus (accessed on 20 November 2019).
7. Szucs, T.D.; Berger, K.; Fisman, D.N.; Harbarth, S. The estimated economic burden of genital herpes in the United States. An analysis using two costing approaches. *BMC Infect. Dis.* **2001**, *1*, 5. [CrossRef] [PubMed]
8. Gonzalo, T.; García Goñi, M.; Muñoz Fernández, M.A. Socio economic impact of antiretroviral treatment in HIV patients. An economic review of cost savings after introduction of HAART. *AIDS Rev.* **2009**, *11*, 79–90.
9. Tseng, A.; Seet, J.; Phillips, E.J. The evolution of three decades of antiretroviral therapy: Challenges, triumphs and the promise of the future: Three decades of antiretroviral therapy. *Br. J. Clin. Pharmacol.* **2015**, *79*, 182–194. [CrossRef]
10. Fleßa, S.; Marschall, P. Socio-Economic Impact of Antiviral Intervention. In *Antiviral Strategies*; Kräusslich, H.-G., Bartenschlager, R., Eds.; Springer: Berlin/Heidelberg, Germany, 2009; Volume 189, pp. 347–374, ISBN 978-3-540-79085-3. [CrossRef]
11. Woolston, S.L.; Kim, N. Cost and Access to Direct-Acting Antiviral Agents. *Hepatoma Res.* **2018**, *4*, 3.
12. Iyengar, S.; Tay-Teo, K.; Vogler, S.; Beyer, P.; Wiktor, S.; de Joncheere, K.; Hill, S. Prices, Costs, and Affordability of New Medicines for Hepatitis C in 30 Countries: An Economic Analysis. *PLoS Med.* **2016**, *13*, e1002032. [CrossRef]
13. Noyes, K.; Holloway, R.G. Evidence from cost-effectiveness research. *Neurotherapeutics* **2004**, *1*, 348–355. [CrossRef]
14. Byford, S. Economic Note: Cost of illness studies. *BMJ* **2000**, *320*, 1335. [CrossRef]
15. Drummond, M.F.; Sculpher, M.J.; Claxton, K.; Stoddart, G.L.; Torrance, G.W. *Methods for the Economic Evaluation of Health Care Programmes*; Oxford University Press: Oxford, UK, 2015.
16. Jo, C. Cost-of-illness studies: Concepts, scopes, and methods. *Clin. Mol. Hepatol.* **2014**, *20*, 327. [CrossRef]
17. Reports and Data. Available online: www.globenewswire.com (accessed on 10 November 2019).
18. Lembo, D.; Cavalli, R. Nanoparticulate Delivery Systems for Antiviral Drugs. *Antivir. Chem. Chemother.* **2010**, *21*, 53–70. [CrossRef] [PubMed]
19. Eisenreich, W.; Rudel, T.; Heesemann, J.; Goebel, W. How Viral and Intracellular Bacterial Pathogens Reprogram the Metabolism of Host Cells to Allow Their Intracellular Replication. *Front. Cell. Infect. Microbiol.* **2019**, *9*, 42. [CrossRef] [PubMed]

20. Van Regenmortel, M.H.V. Virus Species. In *Genetics and Evolution of Infectious Dissease*; Elsevier: Amsterdam, The Netherlands, 2011; pp. 3–19.
21. Heikkinen, T.; Järvinen, A. The common cold. *Lancet* **2003**, *361*, 51–59. [CrossRef]
22. Faith, S.C.; Durrani, A.F.; Jhanji, V. Cytomegalovirus keratitis. *Curr. Opin. Ophthalmol.* **2018**, *29*, 373–377. [CrossRef] [PubMed]
23. Kalezic, T.; Mazen, M.; Kuklinski, E.; Asbell, P. Herpetic eye disease study: Lessons learned. *Curr. Opin. Ophthalmol.* **2018**, *29*, 340–346. [CrossRef]
24. Jhanji, V.; Chan, T.C.Y.; Li, E.Y.M.; Agarwal, K.; Vajpayee, R.B. Adenoviral keratoconjunctivitis. *Surv. Ophthalmol.* **2015**, *60*, 435–443. [CrossRef]
25. Venkatesan, A.; Murphy, O.C. Viral Encephalitis. *Neurol. Clin.* **2018**, *36*, 705–724. [CrossRef]
26. Ruuskanen, O.; Lahti, E.; Jennings, L.C.; Murdoch, D.R. Viral pneumonia. *Lancet* **2011**, *377*, 1264–1275. [CrossRef]
27. Hékimian, G.; Combes, A. Myocardites. *La Revue de Médecine Interne* **2017**, *38*, 531–538. [CrossRef]
28. Rose, N.R. Viral myocarditis. *Curr. Opin. Rheumatol.* **2016**, *28*, 383–389. [CrossRef]
29. Op de Beeck, A.; Eizirik, D.L. Viral infections in type 1 diabetes mellitus—Why the β cells? *Nat. Rev. Endocrinol.* **2016**, *12*, 263–273. [CrossRef] [PubMed]
30. Thuener, J. Hepatitis A and B Infections. *Prim. Care* **2017**, *44*, 621–629. [CrossRef] [PubMed]
31. Ahmad, J.; Hepatitis, C. BMJ 358, j2861. Available online: https://bestpractice.bmj.com/topics/en-us/128 (accessed on 31 October 2019).
32. Rizzetto, M. Hepatitis D Virus: Introduction and Epidemiology. *Cold Spring Harb. Perspect. Med.* **2015**, *5*, a021576. [CrossRef] [PubMed]
33. Khuroo, M.S.; Khuroo, M.S.; Khuroo, N.S. Hepatitis E: Discovery, global impact, control and cure. *WJG* **2016**, *22*, 7030. [CrossRef] [PubMed]
34. Kennedy, P.; Gershon, A. Clinical Features of Varicella-Zoster Virus Infection. *Viruses* **2018**, *10*, 609. [CrossRef]
35. Zerboni, L.; Sen, N.; Oliver, S.L.; Arvin, A.M. Molecular mechanisms of varicella zoster virus pathogenesis. *Nat. Rev. Microbiol.* **2014**, *12*, 197–210. [CrossRef]
36. Agut, H.; Bonnafous, P.; Gautheret-Dejean, A. Human Herpesviruses 6A, 6B, and 7. *Microbiol. Spectr.* **2016**, *4*. [CrossRef]
37. Ramdass, P.; Mullick, S.; Farber, H.F. Viral Skin Diseases. *Prim. Care Clin. Off. Pract.* **2015**, *42*, 517–567. [CrossRef]
38. Schaffer, J.V.; Berger, E.M. Molluscum Contagiosum. *JAMA Dermatol.* **2016**, *152*, 1072. [CrossRef]
39. Krenzer, M.E. Viral Gastroenteritis in the Adult Population. *Crit. Care Nurs. Clin. N. Am.* **2012**, *24*, 541–553. [CrossRef]
40. Eckardt, A.J.; Baumgart, D.C. Viral Gastroenteritis in Adults. *PRI* **2011**, *6*, 54–63. [CrossRef] [PubMed]
41. Thongprachum, A.; Khamrin, P.; Maneekarn, N.; Hayakawa, S.; Ushijima, H. Epidemiology of gastroenteritis viruses in Japan: Prevalence, seasonality, and outbreak: Epidemiology of Gastroenteritis Viruses in Japan. *J. Med. Virol.* **2016**, *88*, 551–570. [CrossRef] [PubMed]
42. Cone, M.M.; Whitlow, C.B. Sexually Transmitted and Anorectal Infectious Diseases. *Gastroenterol. Clin. N. Am.* **2013**, *42*, 877–892. [CrossRef] [PubMed]
43. Virus Taxonomy: 2018b Release. Available online: https://talk.ictvonline.org/taxonomy (accessed on 30 November 2019).
44. Baltimore, D. Expression of animal virus genomes. *Bacteriol. Rev.* **1971**, *35*, 235–241. [CrossRef]
45. Brunton, L.L.; Hilal-Dandan, R.; Knollmann, B. *The Pharmacological Basis of Therapeutics*, 13th ed.; McGraw-Hill Education-Europe: New York, NY, USA, 2017; ISBN 978-1-259-58473-2.
46. Sacks, S.L.; Wilson, B. Famciclovir/Penciclovir. In *Antiviral Chemotherapy 5*; Mills, J., Volberding, P.A., Corey, L., Eds.; Springer: Boston, MA, USA, 1999; Volume 458, pp. 135–147, ISBN 978-1-4613-7150-2.
47. Hitchcock, M.J.M.; Jaffe, H.S.; Martin, J.C.; Stagg, R.J. Cidofovir, a New Agent with Potent Anti-Herpesvirus Activity. *Antivir. Chem. Chemother.* **1996**, *7*, 115–127. [CrossRef]
48. Perry, C.M.; Barman Balfour, J.A. Fomivirsen. *Drugs* **1999**, *57*, 375–380. [CrossRef]
49. Mathiesen, S.; Dam, E.; Roge, B.; Joergensen, L.B.; Laursen, A.L.; Gerstoft, J.; Clavel, F. Long-term foscarnet therapy remodels thymidine analogue mutations and alters resistance to zidovudine and lamivudine in HIV-1. *Antivir. Ther.* **2007**, *12*, 335–343.

50. Leung, D.T.; Sacks, S.L. Docosanol: A topical antiviral for herpes labialis. *Exp. Opin. Pharmacother.* **2004**, *5*, 2567–2571. [CrossRef]
51. Miller, W.H.; Miller, R.L. Phosphorylation of acyclovir diphosphate by cellular enzymes. *Biochem. Pharmacol.* **1982**, *31*, 3879–3884. [CrossRef]
52. King, D.H. History, pharmacokinetics, and pharmacology of acyclovir. *J. Am. Acad. Dermatol.* **1988**, *18*, 176–179. [CrossRef]
53. O'Brien, K.L.; Baggett, H.C.; Brooks, W.A.; Feikin, D.R.; Hammitt, L.L.; Higdon, M.M.; Howie, S.R.C.; Deloria Knoll, M.; Kotloff, K.L.; Levine, O.S.; et al. Causes of severe pneumonia requiring hospital admission in children without HIV infection from Africa and Asia: The PERCH multi-country case-control study. *Lancet* **2019**, *394*, 757–779. [CrossRef]
54. Strasfeld, L.; Chou, S. Antiviral Drug Resistance: Mechanisms and Clinical Implications. *Infect. Dis. Clin. N. Am.* **2010**, *24*, 413–437. [CrossRef]
55. McGavin, J.K.; Goa, K.L. Ganciclovir: An Update of its Use in the Prevention of Cytomegalovirus Infection and Disease in Transplant Recipients. *Drugs* **2001**, *61*, 1153–1183. [CrossRef] [PubMed]
56. Schreiber, A.; Härter, G.; Schubert, A.; Bunjes, D.; Mertens, T.; Michel, D. Antiviral treatment of cytomegalovirus infection and resistant strains. *Exp. Opin. Pharmacother.* **2009**, *10*, 191–209. [CrossRef] [PubMed]
57. De Clercq, E. Antiviral agents active against influenza a viruses. *Nat. Rev. Drug Discov.* **2006**, *5*, 1015–1025. [CrossRef] [PubMed]
58. Chen, H.-W.; Cheng, J.X.; Liu, M.-T.; King, K.; Peng, J.-Y.; Zhang, X.-Q.; Wang, C.-H.; Shresta, S.; Schooley, R.T.; Liu, Y.-T. Inhibitory and combinatorial effect of diphyllin, a v-ATPase blocker, on influenza viruses. *Antivir. Res.* **2013**, *99*, 371–382. [CrossRef]
59. Huss, M.; Wieczorek, H. Inhibitors of V-ATPases: Old and new players. *J. Exp. Biol.* **2009**, *212*, 341–346. [CrossRef]
60. Yeganeh, B.; Ghavami, S.; Kroeker, A.L.; Mahood, T.H.; Stelmack, G.L.; Klonisch, T.; Coombs, K.M.; Halayko, A.J. Suppression of influenza A virus replication in human lung epithelial cells by noncytotoxic concentrations bafilomycin A1. *Am. J. Physiol. Lung Cell. Mol. Physiol.* **2015**, *308*, L270–L286. [CrossRef]
61. Ison, M.G. Antiviral Treatments. *Clin. Chest Med.* **2017**, *38*, 139–153. [CrossRef]
62. Jefferson, T.; Jones, M.A.; Doshi, P.; Del Mar, C.B.; Hama, R.; Thompson, M.J.; Spencer, E.A.; Onakpoya, I.J.; Mahtani, K.R.; Nunan, D.; et al. Neuraminidase inhibitors for preventing and treating influenza in adults and children. *Cochrane Database Syst. Rev.* **2014**. [CrossRef]
63. De Jong, M.D.; Thanh, T.T.; Khanh, T.H.; Hien, V.M.; Smith, G.J.D.; Chau, N.V.; Cam, B.V.; Qui, P.T.; Ha, D.Q.; Guan, Y.; et al. Oseltamivir Resistance during Treatment of Influenza A (H5N1) Infection. *N. Engl. J. Med.* **2005**, *353*, 2667–2672. [CrossRef] [PubMed]
64. Te, H.S.; Randall, G.; Jensen, D.M. Mechanism of action of ribavirin in the treatment of chronic hepatitis C. *Gastroenterol. Hepatol.* **2007**, *3*, 218–225.
65. Krečmerová, M. Amino Acid Ester Prodrugs of Nucleoside and Nucleotide Antivirals. *MRMC* **2017**, *17*. [CrossRef] [PubMed]
66. De Clercq, E. Clinical Potential of the Acyclic Nucleoside Phosphonates Cidofovir, Adefovir, and Tenofovir in Treatment of DNA Virus and Retrovirus Infections. *Clin. Microbiol. Rev.* **2003**, *16*, 569–596. [CrossRef]
67. Lee, H.W.; Park, J.Y.; Ahn, S.H. An evaluation of entecavir for the treatment of chronic hepatitis B infection in adults. *Exp. Rev. Gastroenterol. Hepatol.* **2016**, *10*, 177–186. [CrossRef]
68. Matthews, S. Telbivudine for the management of chronic hepatitis B virus infection. *Clin. Ther.* **2007**, *29*, 2635–2653. [CrossRef]
69. Amarapurkar, D.N. Telbivudine: A new treatment for chronic hepatitis B. *WJG* **2007**, *13*, 6150. [CrossRef]
70. Antiretroviral Drugs Used in the Treatment of HIV Infection. Available online: https://www.fda.gov/patients/hiv-treatment/antiretroviral-drugs-used-treatment-hiv-infection (accessed on 26 October 2019).
71. FDA-Approved HIV Medicines. Available online: https://aidsinfo.nih.gov/understanding-hiv-aids/fact-sheets/21/58/fda-approved-hiv-medicines (accessed on 26 October 2019).
72. Gallant, J.E.; Gerondelis, P.Z.; Wainberg, M.A.; Shulman, N.S.; Haubrich, R.H.; St Clair, M.; Lanier, E.R.; Hellmann, N.S.; Richman, D.D. Nucleoside and nucleotide analogue reverse transcriptase inhibitors: A clinical review of antiretroviral resistance. *Antivir. Ther.* **2003**, *8*, 489–506.

73. Perry, C.M.; Faulds, D. Lamivudine: A Review of its Antiviral Activity, Pharmacokinetic Properties and Therapeutic Efficacy in the Management of HIV Infection. *Drugs* **1997**, *53*, 657–680. [CrossRef]
74. Margolis, A.M.; Heverling, H.; Pham, P.A.; Stolbach, A. A Review of the Toxicity of HIV Medications. *J. Med. Toxicol.* **2014**, *10*, 26–39. [CrossRef]
75. McConville, C.; Boyd, P.; Major, I. Efficacy of Tenofovir 1% Vaginal Gel in Reducing the Risk of HIV-1 and HSV-2 Infection. *Clin. Med. Insights Women's Health* **2014**, *7*. [CrossRef]
76. Sluis-Cremer, N.; Temiz, N.; Bahar, I. Conformational Changes in HIV-1 Reverse Transcriptase Induced by Nonnucleoside Reverse Transcriptase Inhibitor Binding. *CHR* **2004**, *2*, 323–332. [CrossRef] [PubMed]
77. Wang, Y.; Lv, Z.; Chu, Y. HIV protease inhibitors: A review of molecular selectivity and toxicity. *HIV* **2015**, 95. [CrossRef] [PubMed]
78. Latinovic, O.; Kuruppu, J.; Davis, C.; Le, N.; Heredia, A. Pharmacotherapy of HIV-1 Infection: Focus on CCR5 Antagonist Maraviroc. *Clin. Med. Ther.* **2009**, *1*, CMT.S2365. [CrossRef] [PubMed]
79. Qian, K.; Morris-Natschke, S.L.; Lee, K.-H. HIV entry inhibitors and their potential in HIV therapy. *Med. Res. Rev.* **2009**, *29*, 369–393. [CrossRef]
80. Lai, W.; Huang, L.; Chen, C. HIV entry inhibitors: Progress in development and application. *Yao Xue Xue Bao* **2010**, *45*, 131–140.
81. Hajimahdi, Z.; Zarghi, A. Progress in HIV-1 Integrase Inhibitors: A Review of their Chemical Structure Diversity. *Iran. J. Pharm. Res.* **2016**, *15*, 595–628.
82. Steigbigel, R.T.; Cooper, D.A.; Kumar, P.N.; Eron, J.E.; Schechter, M.; Markowitz, M.; Loutfy, M.R.; Lennox, J.L.; Gatell, J.M.; Rockstroh, J.K.; et al. Raltegravir with Optimized Background Therapy for Resistant HIV-1 Infection. *N. Engl. J. Med.* **2008**, *359*, 339–354. [CrossRef] [PubMed]
83. Cocohoba, J.; Dong, B.J. Raltegravir: The first HIV integrase inhibitor. *Clin. Ther.* **2008**, *30*, 1747–1765. [CrossRef]
84. Hicks, C.; Gulick, R.M. Raltegravir: The First HIV Type 1 Integrase Inhibitor. *Clin. Infect. Dis.* **2009**, *48*, 931–939. [CrossRef]
85. Avila-Olias, M.; Pegoraro, C.; Battaglia, G.; Canton, I. Inspired by nature: Fundamentals in nanotechnology design to overcome biological barriers. *Ther. Deliv.* **2013**, *4*, 27–43. [CrossRef] [PubMed]
86. Kreyling, W.G.; Hirn, S.; Möller, W.; Schleh, C.; Wenk, A.; Celik, G.; Lipka, J.; Schäffler, M.; Haberl, N.; Johnston, B.D.; et al. Air–Blood Barrier Translocation of Tracheally Instilled Gold Nanoparticles Inversely Depends on Particle Size. *ACS Nano* **2014**, *8*, 222–233. [CrossRef] [PubMed]
87. Meng, H.; Leong, W.; Leong, K.W.; Chen, C.; Zhao, Y. Walking the line: The fate of nanomaterials at biological barriers. *Biomaterials* **2018**, *174*, 41–53. [CrossRef] [PubMed]
88. Mahajan, S.; Aalinkeel, R.; Law, W.-C.; Reynolds, J.; Nair, B.B.; Sykes, D.E.; Yong, K.-T.; Roy, I.; Prasad, P.; Schwartz, S. Anti-HIV-1 nanotherapeutics: Promises and challenges for the future. *IJN* **2012**, 5301. [CrossRef] [PubMed]
89. Leyva-Gómez, G.; Piñón-Segundo, E.; Mendoza-Muñoz, N.; Zambrano-Zaragoza, M.; Mendoza-Elvira, S.; Quintanar-Guerrero, D. Approaches in Polymeric Nanoparticles for Vaginal Drug Delivery: A Review of the State of the Art. *IJMS* **2018**, *19*, 1549. [CrossRef] [PubMed]
90. Herpetic Eye Disease Study Group. Oral acyclovir for herpes simplex virus eye disease: Effect on prevention of epithelial keratitis and stromal keratitis. *Arch. Ophthalmol.* **2000**, *118*, 1030–1036.
91. Durai, R. Drug delivery approaches of an antiviral drug: A comprehensive review. *Asian J. Pharm.* **2015**, *9*. [CrossRef]
92. Wood, A.J.J.; Whitley, R.J.; Gnann, J.W. Acyclovir: A Decade Later. *N. Engl. J. Med.* **1992**, *327*, 782–789. [CrossRef]
93. Yaldiz, M.; Solak, B.; Kara, R.O.; Cosansu, N.; Erdem, M.T. Comparison of Famciclovir, Valaciclovir, and Brivudine Treatments in Adult Immunocompetent Patients with Herpes Zoster. *Am. J. Ther.* **2018**, *25*, e626–e634. [CrossRef]
94. Cuggino, J.C.; Blanco, E.R.O.; Gugliotta, L.M.; Alvarez Igarzabal, C.I.; Calderón, M. Crossing biological barriers with nanogels to improve drug delivery performance. *J. Control. Release* **2019**, *307*, 221–246. [CrossRef]
95. Raskin, M.M.; Schlachet, I.; Sosnik, A. Mucoadhesive nanogels by ionotropic crosslinking of chitosan-*g*-oligo(NiPAam) polymeric micelles as novel drug nanocarriers. *Nanomedicine* **2016**, *11*, 217–233. [CrossRef] [PubMed]

96. Heyden, R.J. Layers of the Skin. In *Anatomy and Physiology*. Available online: https://opentextbc.ca/anatomyandphysiology (accessed on 11 November 2019).
97. Schneider, M.; Stracke, F.; Hansen, S.; Schaefer, U.F. Nanoparticles and their interactions with the dermal barrier. *Dermat. Endocrinol.* **2009**, *1*, 197–206. [CrossRef] [PubMed]
98. Liang, X.; Xu, Z.; Grice, J.; Zvyagin, A.; Roberts, M.; Liu, X. Penetration of Nanoparticles into Human Skin. *CPD* **2013**, *19*, 6353–6366. [CrossRef] [PubMed]
99. Ross, K.A.; Brenza, T.M.; Binnebose, A.M.; Phanse, Y.; Kanthasamy, A.G.; Gendelman, H.E.; Salem, A.K.; Bartholomay, L.C.; Bellaire, B.H.; Narasimhan, B. Nano-enabled delivery of diverse payloads across complex biological barriers. *J. Control. Release* **2015**, *219*, 548–559. [CrossRef]
100. Prabhu, P.; Patravale, V.; Joshi, M. Nanocarriers for Effective Topical Delivery of Anti-Infectives. *CNANO* **2012**, *8*, 491–503. [CrossRef]
101. Heyden, R.J. The Cell Membrane. In *Anatomy and Physiology*. Available online: https://opentextbc.ca/anatomyandphysiology/chapter/the-cell-membrane (accessed on 11 November 2019).
102. McCaffrey, G.; Davis, T.P. Physiology and Pathophysiology of the Blood-Brain Barrier: P-Glycoprotein and Occludin Trafficking as Therapeutic Targets to Optimize Central Nervous System Drug Delivery. *J. Investig. Med.* **2012**, *60*, 1131–1140. [CrossRef]
103. Ek, C.J.; Wong, A.; Liddelow, S.A.; Johansson, P.A.; Dziegielewska, K.M.; Saunders, N.R. Efflux mechanisms at the developing brain barriers: ABC-transporters in the fetal and postnatal rat. *Toxicol. Lett.* **2010**, *197*, 51–59. [CrossRef]
104. Gottesman, M.M.; Fojo, T.; Bates, S.E. Multidrug resistance in cancer: Role of ATP–dependent transporters. *Nat. Rev. Cancer* **2002**, *2*, 48–58. [CrossRef]
105. Leandro, K.; Bicker, J.; Alves, G.; Falcão, A.; Fortuna, A. ABC transporters in drug-resistant epilepsy: Mechanisms of upregulation and therapeutic approaches. *Pharmacol. Res.* **2019**, *144*, 357–376. Available online: https://linkinghub.elsevier.com/retrieve/pii/S1043661819300726 (accessed on 18 February 2020). [CrossRef]
106. Mahringer, A.; Fricker, G. ABC transporters at the blood–brain barrier. *Exp. Opin. Drug Metab. Toxicol.* **2016**, *12*, 499–508. [CrossRef]
107. Miller, D. Regulation of ABC transporters at the blood-brain barrier. *Clin. Pharmacol. Ther.* **2015**, *97*, 395–403. [CrossRef]
108. Liu, H.; Qiu, K.; He, Q.; Lei, Q.; Lu, W. Mechanisms of Blood-Brain Barrier Disruption in Herpes Simplex Encephalitis. *J. Neuroimmune Pharmacol.* **2019**, *14*, 157–172. [CrossRef]
109. Al-Obaidi, M.M.J.; Bahadoran, A.; Wang, S.M.; Manikam, R.; Raju, C.S.; Sekaran, S.D. Disruption of the blood brain barrier is vital property of neurotropic viral infection of the central nervous system. *Acta Virol.* **2018**, *62*, 16–27. [CrossRef]
110. Sagar, V.; Pilakka-Kanthikeel, S.; Pottathil, R.; Saxena, S.K.; Nair, M. Towards nanomedicines for neuroAIDS: Nanomedicines for neuroAIDS. *Rev. Med. Virol.* **2014**, *24*, 103–124. [CrossRef]
111. Liu, Y.; Tang, X.P.; McArthur, J.C.; Scott, J.; Gartner, S. Analysis of human immunodeficiency virus type 1 gp160 sequences from a patient with HIV dementia: Evidence for monocyte trafficking into brain. *J. Neurovirol.* **2000**, *6* (Suppl. 1), S70–S81.
112. Kuo, Y.-C.; Chen, H.-H. Effect of nanoparticulate polybutylcyanoacrylate and methylmethacrylate–sulfopropylmethacrylate on the permeability of zidovudine and lamivudine across the in vitro blood–brain barrier. *Int. J. Pharm.* **2006**, *327*, 160–169. [CrossRef] [PubMed]
113. Mishra, V.; Mahor, S.; Rawat, A.; Gupta, P.N.; Dubey, P.; Khatri, K.; Vyas, S.P. Targeted brain delivery of AZT via transferrin anchored pegylated albumin nanoparticles. *J. Drug Target.* **2006**, *14*, 45–53. [CrossRef] [PubMed]
114. Jones, A.R.; Shusta, E.V. Blood–Brain Barrier Transport of Therapeutics via Receptor-Mediation. *Pharm. Res.* **2007**, *24*, 1759–1771. [CrossRef] [PubMed]
115. Saiyed, Z.; Gandhi, N.; Nair, M. Magnetic nanoformulation of azidothymidine 5′-triphosphate for targeted delivery across the blood & ndash; brain barrier. *IJN* **2010**, 157. [CrossRef]
116. Fiandra, L.; Colombo, M.; Mazzucchelli, S.; Truffi, M.; Santini, B.; Allevi, R.; Nebuloni, M.; Capetti, A.; Rizzardini, G.; Prosperi, D.; et al. Nanoformulation of antiretroviral drugs enhances their penetration across the blood brain barrier in mice. *Nanomed. Nanotechnol. Biol. Med.* **2015**, *11*, 1387–1397. [CrossRef]

117. McNeil, S.E. (Ed.) Unique Benefits of Nanotechnology to Drug Delivery and Diagnostics. In *Characterization of Nanoparticles Intended for Drug Delivery*; Humana Press: Totowa, NJ, USA, 2011; Volume 697, pp. 3–8, ISBN 978-1-60327-197-4.
118. Petros, R.A.; DeSimone, J.M. Strategies in the design of nanoparticles for therapeutic applications. *Nat. Rev. Drug Discov.* **2010**, *9*, 615–627. [CrossRef] [PubMed]
119. Goldberg, M.; Langer, R.; Jia, X. Nanostructured materials for applications in drug delivery and tissue engineering. *J. Biomater. Sci. Polym. Ed.* **2007**, *18*, 241–268. [CrossRef]
120. Gagliardi, M. Biomimetic and bioinspired nanoparticles for targeted drug delivery. *Ther. Deliv.* **2017**, *8*, 289–299. [CrossRef] [PubMed]
121. Muthu, M.S.; Singh, S. Targeted nanomedicines: Effective treatment modalities for cancer, AIDS and brain disorders. *Nanomedicine* **2009**, *4*, 105–118. [CrossRef] [PubMed]
122. Liu, Y.; Chen, C. Role of nanotechnology in HIV/AIDS vaccine development. *Adv. Drug Deliv. Rev.* **2016**, *103*, 76–89. [CrossRef]
123. Milovanovic, M.; Arsenijevic, A.; Milovanovic, J.; Kanjevac, T.; Arsenijevic, N. Nanoparticles in Antiviral Therapy. In *Antimicrobial Nanoarchitectonics*; Elsevier: Amsterdam, The Netherlands, 2017; pp. 383–410, ISBN 978-0-323-52733-0.
124. Cao, S.; Woodrow, K.A. Nanotechnology approaches to eradicating HIV reservoirs. *Eur. J. Pharm. Biopharm.* **2019**, *138*, 48–63. [CrossRef]
125. Shen, Y.; Tu, J. Preparation and ocular pharmacokinetics of ganciclovir liposomes. *AAPS J.* **2007**, *9*, E371. [CrossRef]
126. Feng, M.; Cai, Q.; Shi, X.; Huang, H.; Zhou, P.; Guo, X. Recombinant high-density lipoprotein complex as a targeting system of nosiheptide to liver cells. *J. Drug Target.* **2008**, *16*, 502–508. [CrossRef]
127. Feng, M. Liver targeting and anti-HBV activity of reconstituted HDL–acyclovir palmitate complex. *Eur. J. Pharm. Biopharm.* **2008**, *68*, 688–693. [CrossRef]
128. Kim, S.I.; Shin, D.; Lee, H.; Ahn, B.-Y.; Yoon, Y.; Kim, M. Targeted delivery of siRNA against hepatitis C virus by apolipoprotein A-I-bound cationic liposomes. *J. Hepatol.* **2009**, *50*, 479–488. [CrossRef]
129. Kim, S.-S.; Peer, D.; Kumar, P.; Subramanya, S.; Wu, H.; Asthana, D.; Habiro, K.; Yang, Y.-G.; Manjunath, N.; Shimaoka, M.; et al. RNAi-mediated CCR5 Silencing by LFA-1-targeted Nanoparticles Prevents HIV Infection in BLT Mice. *Mol. Ther.* **2010**, *18*, 370–376. [CrossRef] [PubMed]
130. Pollock, S.; Dwek, R.A.; Burton, D.R.; Zitzmann, N. N-Butyldeoxynojirimycin is a broadly effective anti-HIV therapy significantly enhanced by targeted liposome delivery. *AIDS* **2008**, *22*, 1961–1969. [CrossRef] [PubMed]
131. Gagné, J.-F.; Désormeaux, A.; Perron, S.; Tremblay, M.J.; Bergeron, M.G. Targeted delivery of indinavir to HIV-1 primary reservoirs with immunoliposomes. *BBA Biomembr.* **2002**, *1558*, 198–210. [CrossRef]
132. Clayton, R.; Ohagen, A.; Nicol, F.; Del Vecchio, A.M.; Jonckers, T.H.M.; Goethals, O.; Van Loock, M.; Michiels, L.; Grigsby, J.; Xu, Z.; et al. Sustained and specific in vitro inhibition of HIV-1 replication by a protease inhibitor encapsulated in gp120-targeted liposomes. *Antivir. Res.* **2009**, *84*, 142–149. [CrossRef]
133. Yadavalli, T.; Ames, J.; Agelidis, A.; Suryawanshi, R.; Jaishankar, D.; Hopkins, J.; Thakkar, N.; Koujah, L.; Shukla, D. Drug-encapsulated carbon (DECON): A novel platform for enhanced drug delivery. *Sci. Adv.* **2019**, *5*, eaax0780. [CrossRef]
134. Yang, X.; Shah, S.J.; Wang, Z.; Agrahari, V.; Pal, D.; Mitra, A.K. Nanoparticle-based topical ophthalmic formulation for sustained release of stereoisomeric dipeptide prodrugs of ganciclovir. *Drug Deliv.* **2015**, 1–11. [CrossRef]
135. Li, Y.; Lin, Z.; Guo, M.; Zhao, M.; Xia, Y.; Wang, C.; Xu, T.; Zhu, B. Inhibition of H1N1 influenza virus-induced apoptosis by functionalized selenium nanoparticles with amantadine through ROS-mediated AKT signaling pathways. *IJN* **2018**, *13*, 2005–2016. [CrossRef]
136. Li, Y.; Lin, Z.; Guo, M.; Xia, Y.; Zhao, M.; Wang, C.; Xu, T.; Chen, T.; Zhu, B. Inhibitory activity of selenium nanoparticles functionalized with oseltamivir on H1N1 influenza virus. *IJN* **2017**, *12*, 5733–5743. [CrossRef]
137. Li, Y.; Lin, Z.; Zhao, M.; Guo, M.; Xu, T.; Wang, C.; Xia, H.; Zhu, B. Reversal of H1N1 influenza virus-induced apoptosis by silver nanoparticles functionalized with amantadine. *RSC Adv.* **2016**, *6*, 89679–89686. [CrossRef]
138. Hu, C.-M.J.; Chen, Y.-T.; Fang, Z.-S.; Chang, W.-S.; Chen, H.-W. Antiviral efficacy of nanoparticulate vacuolar ATPase inhibitors against influenza virus infection. *IJN* **2018**, *13*, 8579–8593. [CrossRef]

139. Chiellini, E.E.; Chiellini, F.; Solaro, R. Bioerodible Polymeric Nanoparticles for Targeted Delivery of Proteic Drugs. *J. Nanosci. Nanotechnol.* **2006**, *6*, 3040–3047. [CrossRef] [PubMed]
140. Kumar, P.; Lakshmi, Y.S.; Bhaskar, C.; Golla, K.; Kondapi, A.K. Improved Safety, Bioavailability and Pharmacokinetics of Zidovudine through Lactoferrin Nanoparticles during Oral Administration in Rats. *PLoS ONE* **2015**, *10*, e0140399. [CrossRef] [PubMed]
141. Kirimlioğlu, G.Y.; Öztürk, A.A. Preparation and in vitro characterization of lamivudine loaded nanoparticles prepared by acid and/or ester terminated PLGA for effective oral anti-retroviral therapy. *JRP* **2019**, *23*, 897–913. [CrossRef]
142. Sankar, V.; Madhura Keerthi, L.; Nilaykumar, P. Formulation and In-vitro Evaluation of Zidovudine-Lamivudine Nanoparticles. *Indian J. Pharm. Educ. Res.* **2012**, *46*, 192–196.
143. Shah, L.K.; Amiji, M.M. Intracellular Delivery of Saquinavir in Biodegradable Polymeric Nanoparticles for HIV/AIDS. *Pharm. Res.* **2006**, *23*, 2638–2645. [CrossRef]
144. Tang, X.; Liang, Y.; Liu, X.; Zhou, S.; Liu, L.; Zhang, F.; Xie, C.; Cai, S.; Wei, J.; Zhu, Y.; et al. PLGA-PEG Nanoparticles Coated with Anti-CD45RO and Loaded with HDAC Plus Protease Inhibitors Activate Latent HIV and Inhibit Viral Spread. *Nanoscale Res. Lett.* **2015**, *10*, 413. [CrossRef]
145. Venkatesh, D.N.; Baskaran, M.; Karri, V.V.S.R.; Mannemala, S.S.; Radhakrishna, K.; Goti, S. Fabrication and in vivo evaluation of Nelfinavir loaded PLGA nanoparticles for enhancing oral bioavailability and therapeutic effect. *Saudi Pharm. J.* **2015**, *23*, 667–674. [CrossRef]
146. Kumar, P.; Lakshmi, Y.; Kondapi, A. An oral formulation of efavirenz-loaded lactoferrin nanoparticles with improved biodistribution and pharmacokinetic profile. *HIV Med.* **2017**, *18*, 452–462. [CrossRef]
147. Zhang, G. The mixed lineage kinase-3 inhibitor URMC-099 improves therapeutic outcomes for long-acting antiretroviral therapy. *Nanomedicine* **2016**, *12*, 109–122. [CrossRef]
148. Savage, A.C.; Tatham, L.M.; Siccardi, M.; Scott, T.; Vourvahis, M.; Clark, A.; Rannard, S.P.; Owen, A. Improving maraviroc oral bioavailability by formation of solid drug nanoparticles. *Eur. J. Pharm. Biopharm.* **2019**, *138*, 30–36. [CrossRef]
149. Shibata, A.; McMullen, E.; Pham, A.; Belshan, M.; Sanford, B.; Zhou, Y.; Goede, M.; Date, A.A.; Destache, C.J. Polymeric Nanoparticles Containing Combination Antiretroviral Drugs for HIV Type 1 Treatment. *AIDS Res. Hum. Retrovir.* **2013**, *29*, 746–754. [CrossRef] [PubMed]
150. Prathipati, P.K.; Mandal, S.; Pon, G.; Vivekanandan, R.; Destache, C.J. Pharmacokinetic and Tissue Distribution Profile of Long Acting Tenofovir Alafenamide and Elvitegravir Loaded Nanoparticles in Humanized Mice Model. *Pharm. Res.* **2017**, *34*, 2749–2755. [CrossRef] [PubMed]
151. Kumar, P.; Lakshmi, Y.S.; Kondapi, A.K. Triple Drug Combination of Zidovudine, Efavirenz and Lamivudine Loaded Lactoferrin Nanoparticles: An Effective Nano First-Line Regimen for HIV Therapy. *Pharm. Res.* **2017**, *34*, 257–268. [CrossRef] [PubMed]
152. Lauster, D.; Glanz, M.; Bardua, M.; Ludwig, K.; Hellmund, M.; Hoffmann, U.; Hamann, A.; Böttcher, C.; Haag, R.; Hackenberger, C.P.R.; et al. Multivalent Peptide-Nanoparticle Conjugates for Influenza-Virus Inhibition. *Angew. Chem. Int. Ed.* **2017**, *56*, 5931–5936. [CrossRef] [PubMed]
153. Li, W.; Balachandran, Y.L.; Hao, Y.; Hao, X.; Li, R.; Nan, Z.; Zhang, H.; Shao, Y.; Liu, Y. Amantadine Surface-Modified Silver Nanorods Improves Immunotherapy of HIV Vaccine Against HIV-Infected Cells. *ACS Appl. Mater. Interfaces* **2018**, *10*, 28494–28501. [CrossRef] [PubMed]
154. Mahajan, S.D.; Roy, I.; Xu, G.; Yong, K.-T.; Ding, H.; Aalinkeel, R.; Reynolds, J.L.; Sykes, D.E.; Nair, B.B.; Lin, E.Y.; et al. Enhancing the Delivery of Anti Retroviral Drug "Saquinavir"; Across the Blood Brain Barrier Using Nanoparticles. *CHR* **2010**, *8*, 396–404. [CrossRef] [PubMed]
155. Donalisio, M.; Leone, F.; Civra, A.; Spagnolo, R.; Ozer, O.; Lembo, D.; Cavalli, R. Acyclovir-Loaded Chitosan Nanospheres from Nano-Emulsion Templating for the Topical Treatment of Herpesviruses Infections. *Pharmaceutics* **2018**, *10*, 46. [CrossRef]
156. Patel, G.V.; Patel, V.B.; Pathak, A.; Rajput, S.J. Nanosuspension of efavirenz for improved oral bioavailability: Formulation optimization, in vitro, in situ and in vivo evaluation. *Drug Dev. Ind. Pharm.* **2014**, *40*, 80–91. [CrossRef]
157. Akhter, S.; Talegaonkar, S.; Khan, Z.I.; Jain, G.K.; Khar, R.K.; Ahmad, F.J. Assessment of Ocular Pharmacokinetics and Safety of Ganciclovir Loaded Nanoformulations. *J. Biomed. Nanotechnol.* **2011**, *7*, 144–145. [CrossRef]

158. Ren, J.; Zou, M.; Gao, P.; Wang, Y.; Cheng, G. Tissue distribution of borneol-modified ganciclovir-loaded solid lipid nanoparticles in mice after intravenous administration. *Eur. J. Pharm. Biopharm.* **2013**, *83*, 141–148. [CrossRef]
159. Chattopadhyay, N.; Zastre, J.; Wong, H.-L.; Wu, X.Y.; Bendayan, R. Solid Lipid Nanoparticles Enhance the Delivery of the HIV Protease Inhibitor, Atazanavir, by a Human Brain Endothelial Cell Line. *Pharm. Res.* **2008**, *25*, 2262–2271. [CrossRef] [PubMed]
160. Kovochich, M.; Marsden, M.D.; Zack, J.A. Activation of Latent HIV Using Drug-Loaded Nanoparticles. *PLoS ONE* **2011**, *6*, e18270. [CrossRef] [PubMed]
161. Duan, J.; Freeling, J.P.; Koehn, J.; Shu, C.; Ho, R.J.Y. Evaluation of Atazanavir and Darunavir Interactions with Lipids for Developing pH-Responsive Anti-HIV Drug Combination Nanoparticles. *J. Pharm. Sci.* **2014**, *103*, 2520–2529. [CrossRef] [PubMed]
162. Freeling, J.P.; Koehn, J.; Shu, C.; Sun, J.; Ho, R.J.Y. Anti-HIV Drug-Combination Nanoparticles Enhance Plasma Drug Exposure Duration as Well as Triple-Drug Combination Levels in Cells Within Lymph Nodes and Blood in Primates. *AIDS Res. Hum. Retrovir.* **2015**, *31*, 107–114. [CrossRef]
163. Arca-Lafuente, S.; Martínez-Román, P.; Mate-Cano, I.; Madrid, R.; Briz, V. Nanotechnology: A reality for diagnosis of HCV infectious disease. *J. Infect.* **2019**. [CrossRef]
164. Farzin, L.; Shamsipur, M.; Samandari, L.; Sheibani, S. HIV biosensors for early diagnosis of infection: The intertwine of nanotechnology with sensing strategies. *Talanta* **2020**, *206*, 120201. [CrossRef]
165. Muxika, A.; Etxabide, A.; Uranga, J.; Guerrero, P.; de la Caba, K. Chitosan as a bioactive polymer: Processing, properties and applications. *Int. J. Biol. Macromol.* **2017**, *105*, 1358–1368. [CrossRef]
166. Dhanasezhian, A.; Srivani, S.; Govindaraju, K.; Preetam, P.; Sasikala, S.; Ramesh Kumar, M.R. Anti-Herpes Simplex Virus (HSV-1 and HSV-2) activity of biogenic gold and silver nanoparticles using seaweed Sargassum wightii. *Indian J. Geo Mar. Sci.* **2019**, *48*, 1252–1257.
167. Singaravelu, G.; Arockiamary, J.S.; Kumar, V.G.; Govindaraju, K. A novel extracellular synthesis of monodisperse gold nanoparticles using marine alga, Sargassum wightii Greville. *Colloids Surf. B* **2007**, *57*, 97–101. [CrossRef]
168. Cagno, V.; Andreozzi, P.; D'Alicarnasso, M.; Jacob Silva, P.; Mueller, M.; Galloux, M.; Le Goffic, R.; Jones, S.T.; Vallino, M.; Hodek, J.; et al. Broad-spectrum non-toxic antiviral nanoparticles with a virucidal inhibition mechanism. *Nat. Mater.* **2018**, *17*, 195–203. [CrossRef]
169. Klimyte, E.M.; Smith, S.E.; Oreste, P.; Lembo, D.; Dutch, R.E. Inhibition of Human Metapneumovirus Binding to Heparan Sulfate Blocks Infection in Human Lung Cells and Airway Tissues. *J. Virol.* **2016**, *90*, 9237–9250. [CrossRef]
170. Bhatia, S.; Camacho, L.C.; Haag, R. Pathogen Inhibition by Multivalent Ligand Architectures. *J. Am. Chem. Soc.* **2016**, *138*, 8654–8666. [CrossRef]
171. Verma, A.; Uzun, O.; Hu, Y.; Hu, Y.; Han, H.-S.; Watson, N.; Chen, S.; Irvine, D.J.; Stellacci, F. Surface-structure-regulated cell-membrane penetration by monolayer-protected nanoparticles. *Nat. Mater.* **2008**, *7*, 588–595. [CrossRef]
172. Argenta, D.F.; Silva, I.T.; Bassani, V.L.; Koester, L.S.; Teixeira, H.F.; Simões, C.M.O. Antiherpes evaluation of soybean isoflavonoids. *Arch. Virol.* **2015**, *160*, 2335–2342. [CrossRef]
173. Argenta, D.F.; Bidone, J.; Koester, L.S.; Bassani, V.L.; Simões, C.M.O.; Teixeira, H.F. Topical Delivery of Coumestrol from Lipid Nanoemulsions Thickened with Hydroxyethylcellulose for Antiherpes Treatment. *AAPS PharmSciTech* **2018**, *19*, 192–200. [CrossRef]
174. Hoeller, S.; Sperger, A.; Valenta, C. Lecithin based nanoemulsions: A comparative study of the influence of non-ionic surfactants and the cationic phytosphingosine on physicochemical behaviour and skin permeation. *Int. J. Pharm.* **2009**, *370*, 181–186. [CrossRef]
175. Grande, F.; Ioele, G.; Occhiuzzi, M.A.; De Luca, M.; Mazzotta, E.; Ragno, G.; Garofalo, A.; Muzzalupo, R. Reverse Transcriptase Inhibitors Nanosystems Designed for Drug Stability and Controlled Delivery. *Pharmaceutics* **2019**, *11*, 197. [CrossRef]
176. Joshy, K.S.; George, A.; Jose, J.; Kalarikkal, N.; Pothen, L.A.; Thomas, S. Novel dendritic structure of alginate hybrid nanoparticles for effective anti-viral drug delivery. *Int. J. Biol. Macromol.* **2017**, *103*, 1265–1275. [CrossRef]

177. Sneha, R.; Vedha Hari, B.N.; Ramya Devi, D. Design of antiretroviral drug-polymeric nanoparticles laden buccal films for chronic HIV therapy in paediatrics. *Colloid Interface Sci. Commun.* **2018**, *27*, 49–59. [CrossRef]
178. Urbaniak, T.; Musiał, W. Influence of Solvent Evaporation Technique Parameters on Diameter of Submicron Lamivudine-Poly-ε-Caprolactone Conjugate Particles. *Nanomaterials* **2019**, *9*, 1240. [CrossRef]
179. Tshweu, L.; Katata, L.; Kalombo, L.; Swai, H. Nanoencapsulation of water-soluble drug, lamivudine, using a double emulsion spray-drying technique for improving HIV treatment. *J. Nanopart. Res.* **2013**, *15*, 2040. [CrossRef]
180. Reddy, P.P.; Reddy, B.M.; Prakash, J.; Latha, K.; Prashanthi, D. Formulation and characterisation of chitosan based lamivudine nanoparticles. *EJPMR* **2014**, *4*, 377–383.
181. Sanchez, C.G.; Molinski, S.V.; Gongora, R.; Sosulski, M.; Fuselier, T.; MacKinnon, S.S.; Mondal, D.; Lasky, J.A. The Antiretroviral Agent Nelfinavir Mesylate: A Potential Therapy for Systemic Sclerosis. *Arthritis Rheumatol.* **2018**, *70*, 115–126. [CrossRef] [PubMed]
182. Bowman, M.-C.; Ballard, T.E.; Ackerson, C.J.; Feldheim, D.L.; Margolis, D.M.; Melander, C. Inhibition of HIV Fusion with Multivalent Gold Nanoparticles. *J. Am. Chem. Soc.* **2008**, *130*, 6896–6897. [CrossRef]
183. Vijayakumar, S.; Ganesan, S. Gold Nanoparticles as an HIV Entry Inhibitor. *CHR* **2012**, *10*, 643–646. [CrossRef]
184. Garrido, C.; Simpson, C.A.; Dahl, N.P.; Bresee, J.; Whitehead, D.C.; Lindsey, E.A.; Harris, T.L.; Smith, C.A.; Carter, C.J.; Feldheim, D.L.; et al. Gold nanoparticles to improve HIV drug delivery. *Future Med. Chem.* **2015**, *7*, 1097–1107. [CrossRef]
185. Singh, L.; Kruger, H.G.; Maguire, G.E.M.; Govender, T.; Parboosing, R. The role of nanotechnology in the treatment of viral infections. *Ther. Adv. Infect.* **2017**, *4*, 105–131. [CrossRef]
186. Bovier, P.A. Epaxal®: A virosomal vaccine to prevent hepatitis A infection. *Exp. Rev. Vaccines* **2008**, *7*, 1141–1150. [CrossRef]
187. He, H.; Yuan, D.; Wu, Y.; Cao, Y. Pharmacokinetics and Pharmacodynamics Modeling and Simulation Systems to Support the Development and Regulation of Liposomal Drugs. *Pharmaceutics* **2019**, *11*, 110. [CrossRef]
188. Herzog, C.; Hartmann, K.; Künzi, V.; Kürsteiner, O.; Mischler, R.; Lazar, H.; Glück, R. Eleven years of Inflexal® V—A virosomal adjuvanted influenza vaccine. *Vaccine* **2009**, *27*, 4381–4387. [CrossRef]
189. Package Insert PEG-IntronTM (Peginterferon alfa-2b) Powder for Injection. Schering Corporation. Available online: https://www.accessdata.fda.gov/drugsatfda_docs/label/2001/pegsche080701LB.htm (accessed on 16 January 2020).
190. Ventola, C.L. Progress in Nanomedicine: Approved and Investigational Nanodrugs. *Pharm. Ther.* **2017**, *42*, 742–755.
191. Package Insert PEGASYS® (Peginterferon Alfa-2a) Hoffmann-La Roche Inc. Available online: https://www.accessdata.fda.gov/drugsatfda_docs/label/2002/pegihof101602LB.htm (accessed on 16 January 2020).
192. Influvac (Influenza Vaccine). Available online: https://bodyandhealth.canada.com/drug/getdrug/influvac (accessed on 16 January 2020).
193. VivaGel® BV launched in Europe. Available online: https://starpharma.com/assets/asxannouncements/190627%20VivaGel%20BV%20launched%20in%20Europe%20-%20final%20.pdf (accessed on 16 January 2020).
194. Safety. Tolerability and Immune Response to LC002, an Experimental Therapeutic Vaccine, in Adults Receiving HAART-ClinicalTrials.gov Identifier: NCT00270205. Available online: https://clinicaltrials.gov/ct2/show/NCT00270205?cond=NCT00270205&draw=2&rank=1 (accessed on 18 January 2020).
195. Bioavailability of MK-1439 Experimental Nano Formulations in Healthy Adults (MK-1439-046)-ClinicalTrials.gov Identifier: NCT02549040. Available online: https://clinicaltrials.gov/ct2/show/study/NCT02549040?term=NCT02549040&draw=2&rank=1 (accessed on 18 January 2020).
196. Study of ARB-001467 in Subjects with Chronic HBV Infection Receiving Nucleos(t)Ide Analogue Therapy-ClinicalTrials.gov Identifier: NCT02631096. Available online: https://clinicaltrials.gov/ct2/show/NCT02631096?term=NCT02631096&draw=2&rank=1 (accessed on 18 January 2020).
197. Odiba, A.; Ottah, V.; Ottah, C.; Anunobi, O.; Ukegbu, C.; Edeke, A.; Uroko, R.; Omeje, K. Therapeutic nanomedicine surmounts the limitations of pharmacotherapy. *Open Med.* **2017**, *12*. [CrossRef]
198. Parboosing, R.; Maguire, G.E.M.; Govender, P.; Kruger, H.G. Nanotechnology and the Treatment of HIV Infection. *Viruses* **2012**, *4*, 488–520. [CrossRef]
199. Li, M.; Al-Jamal, K.T.; Kostarelos, K.; Reineke, J. Physiologically Based Pharmacokinetic Modeling of Nanoparticles. *ACS Nano* **2010**, *4*, 6303–6317. [CrossRef]

200. Sanvicens, N.; Marco, M.P. Multifunctional nanoparticles—Properties and prospects for their use in human medicine. *Trends Biotechnol.* **2008**, *26*, 425–433. [CrossRef]
201. Müller, R.H.; Gohla, S.; Keck, C.M. State of the art of nanocrystals—Special features, production, nanotoxicology aspects and intracellular delivery. *Eur. J. Pharm. Biopharm.* **2011**, *78*, 1–9. [CrossRef]
202. Ochekpe, N.A.; Olorunfemi, P.O.; Ngwuluka, N.C. Nanotechnology and drug delivery part 2: Nanostructures for drug delivery. *Trop. J. Pharm. Res.* **2009**, *8*. [CrossRef]
203. Dey, P.; Bergmann, T.; Cuellar-Camacho, J.L.; Ehrmann, S.; Chowdhury, M.S.; Zhang, M.; Dahmani, I.; Haag, R.; Azab, W. Multivalent Flexible Nanogels Exhibit Broad-Spectrum Antiviral Activity by Blocking Virus Entry. *ACS Nano* **2018**, *12*, 6429–6442. [CrossRef]

pharmaceutics

MDPI

Article

pH-Sensitive Chitosan–Heparin Nanoparticles for Effective Delivery of Genetic Drugs into Epithelial Cells

Iuliia Pilipenko [1], Viktor Korzhikov-Vlakh [1], Vladimir Sharoyko [1], Nan Zhang [2], Monika Schäfer-Korting [2], Eckart Rühl [3], Christian Zoschke [2] and Tatiana Tennikova [1,*]

[1] Institute of Chemistry, Saint Petersburg State University, Peterhoff, Universitetskii pr. 26, 198504 St. Petersburg, Russia

[2] Institute of Pharmacy (Pharmacology and Toxicology), Freie Universität Berlin, Königin-Luise-Straße 2+4, 14195 Berlin, Germany

[3] Institute of Chemistry and Biochemistry (Physical Chemistry), Freie Universität Berlin, Arnimallee 22, 14195 Berlin, Germany

* Correspondence: tennikova@mail.ru; Tel.: +7-921-969-1060

check for updates

Received: 29 May 2019; Accepted: 3 July 2019; Published: 5 July 2019

Abstract: Chitosan has been extensively studied as a genetic drug delivery platform. However, its efficiency is limited by the strength of DNA and RNA binding. Expecting a reduced binding strength of cargo with chitosan, we proposed including heparin as a competing polyanion in the polyplexes. We developed chitosan–heparin nanoparticles by a one-step process for the local delivery of oligonucleotides. The size of the polyplexes was dependent on the mass ratio of polycation to polyanion. The mechanism of oligonucleotide release was pH-dependent and associated with polyplex swelling and collapse of the polysaccharide network. Inclusion of heparin enhanced the oligonucleotide release from the chitosan-based polyplexes. Furthermore, heparin reduced the toxicity of polyplexes in the cultured cells. The cell uptake of chitosan–heparin polyplexes was equal to that of chitosan polyplexes, but heparin increased the transfection efficiency of the polyplexes two-fold. The application of chitosan–heparin small interfering RNA (siRNA) targeted to vascular endothelial growth factor (VEGF) silencing of ARPE-19 cells was 25% higher. Overall, chitosan–heparin polyplexes showed a significant improvement of gene release inside the cells, transfection, and gene silencing efficiency in vitro, suggesting that this fundamental strategy can further improve the transfection efficiency with application of non-viral vectors.

Keywords: chitosan; cytotoxicity; DNA; heparin; transfection; pH-sensitive

1. Introduction

Progress in the field of gene medicine has been stimulated by basic research of novel vectors for DNA, RNA, and oligonucleotides. The use of different nucleotide-based therapeutics (e.g., siRNA, messenger RNA (mRNA), and plasmid DNA (pDNA)) is an important strategy in future treatment of diseases [1]. The cellular mechanisms of listed therapeutic agents are extremely different, with structural advantages [2]. Nevertheless, it is important to find an effective delivery system for genetic medicines to obtain high therapeutic efficiency in the cell.

Cationic biopolymers have been frequently used as a nano-carriers for siRNA and pDNA delivery because they effectively complex the anionic polynucleotides, protect them from degradation and improve their cellular uptake [3]. The colloidal vesicles of hydrophilic natural polymers, such as chitosan and its derivatives, represent a promising class of such biopolymers [4–6]. Chitosan is water-soluble and positively charged polymer with pK_a of 6.0. It forms pH-sensitive nano-sized hydrogels that may enable pH-triggered release siRNA or pDNA in the acidic medium of endosomes [7,8]. The positive charges of chitosan can be used to form colloidal nanoparticles via electrostatic interactions with polyanions and to promote their internalization into the cells. Moreover, chitosan adheres to mucosal surfaces [9] and enhances the epithelial penetration by opening tight-junctions, which is key to overcoming the skin barrier [10]. In addition, chitosan is metabolized by certain human enzymes [11] and it is considered a biocompatible polymer with low toxicity [12].

Numerous excellent reviews and regular articles on the potential of chitosan for pharmaceutical applications have been published recently [13–16]. Despite the great potential of chitosan [13], its efficiency for delivering different types of oligo- and polynucleotides strongly depends on the N/P (amino groups of polycation to phosphate groups of polyanion) ratio, molecular structure, size, and shape of the complexes. High N/P ratios in the complexes lead to high stability of the polyplexes that prevents release of the cargo within the cells and precludes translation of DNA or RNA action [15,16]. To address this issue, we incorporated heparin as a stronger polyanionic competitor for polynucleotides.

Alternative polyplexes with other polyanions, such as sodium alginate [17] and hyaluronic acid [18], have already been published as tools for improved gene delivery into epithelial cells. Sodium alginate and hyaluronic acid have a gelling behavior, high molecular weight, and broad molecular weight distribution, which can lead to aggregation and reduced stability of the particles. Heparin is biocompatible and mucoadhesive biopolymer with comparatively low molecular mass (12–14 kDa), non-gelling behavior, and stronger acidic sulfate groups. Thus, we propose that heparin could be more effective competitive polyanion for improved RNA or DNA release from the polyplexes.

The formulations based on chitosan–heparin nanoparticles were previously described [19,20]. In one study [19], chitosan–heparin particles were used for bovine serum albumin (BSA) delivery. The physico-chemical properties of such formulations, namely, size, pH-sensitivity, zeta-potential, and entrapment efficiency of BSA were described. In another study, the immobilization of VEGF-loaded chitosan–heparin nanoparticles on the scaffolds for tissue-engineering was performed [20]. This significantly increased fibroblast infiltration, extracellular matrix production, and accelerated in vivo vascularization in a mouse subcutaneous implantation model. Thus, the application of chitosan–heparin carriers as a highly biocompatible and effective delivery system represents a promising approach for genetic drug delivery into the cells.

In this work we developed pH-sensitive chitosan–heparin nanoparticles and investigated the effect of different chitosan/heparin ratios on pH, particle size and structure, zeta potential, and encapsulation efficiency of model oligonucleotide. Next, we studied the release mechanisms of model oligonucleotide and proved the biocompatibility of the drug delivery system with primary normal human keratinocytes (NHKs). Finally, we unraveled specific role of competitive polyanion heparin in the complex and showed the advances of chitosan–heparin nanocarrier to deliver plasmid DNA and VEGF silencing RNA to epithelial cells.

2. Materials and Methods

2.1. Materials

Polymers for nanoparticle preparation include chitosan (deacetylation degree 75–85%, medium molecular weight; Sigma Aldrich, Schnelldorf, Germany) and heparin (MW 12–14 kDa; AppliChem, Darmstadt, Germany). Model Cy3-labeled and non-labeled oligothymidine and oligoadenine (23 base pairs oligo-dT-dA) were purchased from Biobeagle^{TM} (Saint-Petersburg, Russia).

The reporter gene plasmid encoding green fluorescent protein (pEGFP-C2) was a generous gift from Prof. Arto Urtti (University of Eastern Finland, Kuopio, Finland). pEGFP-C2 was transformed into *Escherichia coli* XL1-Blue (Evrogen, Moscow, Russia) and purified using the Plasmid Miniprep purification kit (Evrogen, Moscow, Russia). DNA concentration was quantified by the measurement of UV absorbance at 260 nm using a Nanodrop 2000c spectrophotometer (Thermo Fischer Scientific, Vantaa, Finland). The purity of the plasmid was verified by gel electrophoresis (1% agarose gel).

The 27-base pairs (bp) double stranded (ds) RNA (sense and antisense strands) were designed as targets to the vascular endothelial growth factor gene (sense: 5′-CUUCCUACAGCAC AACAAAUGUGAAUG-3′, antisense: 3′GAAGGAUGUCGUGUUGUUUACACUUAC-5′). For visualization, Cy5-labeled (5′-modification) 27-bp VEGF siRNA were used. Cy5-labeled and non-labeled 27-bp ds VEGF siRNAs and scrambled 27-bp RNA (siC) (sense 5′-GUAAGUGUAAACAACACGA CAUCCUUC-3′, antisense: 3′-CAUUCACAUUUG UUGUGCUGUAGGAAG-5′ [21] were purchased from GenTerra (Moscow, Russia). The primers used for the target mRNA: VEGF forward primer (5′-CCCTGATGAGATCGAGTACATCTT-3′), VEGF reverse primer (5′-ACCGCCTCGGCTTGTCAC-3′), GAPDH forward primer (5′-GTCTCCTCTGACTTCAACAGCG-3′), and GAPDH reverse primer (5′-ACCACCCTGTTGCTGTAGCCAA-3′) were purchased from GenTerra (Moscow, Russia).

Primary NHKs were from therapeutically indicated circumcisions (ethical approval EA1/081/13) after parents had signed the written informed consent. Primary cells were isolated according to a standard operating procedure [22]. In brief, tissues were washed with phosphate buffered saline (PBS) pH 7.4, cut into pieces, and incubated with dispase solution overnight at 4 °C. Next, epidermis was separated from dermal pieces and collected in Trypsin–EDTA solution to isolate keratinocytes. After stopping the enzymatic reaction, keratinocytes were grown and subcultured in cell culture flasks. Cells in passage 3 were used for cell viability experiments.

Human retinal epithelial (ARPE-19) cells were obtained from American Type Culture Collection (ATCC, Manassas, VA, USA).

2.2. Particle Preparation

Chitosan–heparin nanoparticles were prepared based on spontaneous polyelectrolyte complexation in mild conditions [15]. Polyelectrolyte complex formation between polycationic chitosan and polyanionic heparin results in partial charge neutralization. Before complexation heparin was dissolved in deionized water to give solution with concentration 1 mg/mL, while chitosan was dissolved in 0.01% acetic acid solution to obtain 10 mg/mL solution and further diluted with PBS pH 8.0–8.5 or 0.1M NaOH to obtain the final concentration 0.1 mg/mL. The pH of the chitosan solutions was fixed at 6.0, except in the studies on the effects of pH on size/zeta-potential and encapsulation efficiency of oligonucleotide. Before the complexation, chitosan solution was sonicated for 60–90 s using 10% power of ultrasonic homogenizer (Bandelin Sonopuls HD 2070, Berlin, Germany). After that, chitosan–heparin nanoparticles were obtained by dropwise addition of heparin solution to chitosan solution under stirring (Vortex, Thermo Fischer Scientific, Vantaa, Finland) at room temperature. To encapsulate oligo- and polynucleotides inside chitosan–heparin nanoparticles, the anionic oligonucleotide was added to chitosan solution immediately after sonication under stirring, thereafter heparin was added to these complexes.

2.3. Particle Structure

Hydrodynamic diameter, polydispersity index (PDI) and zeta potential were determined by dynamic light scattering (DLS) using a Zetasizer Nano ZS with He-Ne laser (λ = 633 nm; Malvern Instruments, Worcester, UK) at 25 °C. Moreover, particle morphology and size were measured with scanning-transmission electron microscopy (STEM; SU8010, Hitachi, Japan) and nanoparticle tracking analysis (NTA; Nanosight NS300, Malvern Panalytical, Worcester, UK).

The concentration of particles which were used for DLS, zeta-potential, and STEM experiments was 10% from the stock particle suspension. DLS experiments were performed in PBS, pH 7.4, whereas the zeta-potentials were determined in deionized water (except the studies of pH effects).

2.4. Oligonucleotide Loading Efficiency

Oligonucleotide-loaded nanoparticles were obtained by adding 1 µg of duplex oligo-thymidine-adenine (dT-dA) or Cy3-oligo-thymidine-adenine (Cy3-dT-dA) to chitosan solution before complexation with heparin. To determine the loading efficiency of Cy3-dT-dA into chitosan–heparin nanoparticles, free oligonucleotide was separated from the loaded nanoparticles using centrifugation of 10,000× *g* at 4 °C for 20 min in filter tubes (30,000 NWML, Amicon Ultra 0.5 mL, Merck, Darmstadt, Germany). The filtrate with free Cy3-dT-dA was collected and analyzed using multimode microplate reader (Thermo Scientific Varioscan lux, Vantaa, Finland) at excitation and emission wavelengths of 550 and 570 nm, respectively. The amount of loaded oligonucleotide was calculated using a linear calibration curve. The loading efficiency was expressed as encapsulated/total ratio of oligonucleotide.

2.5. Oligonucleotide Release

In vitro release of Cy3-dT-dA from chitosan–heparin and chitosan complexes was measured for 4 h using 100 µL of test formulation diluted with 300 µL of the release media. The release media were buffer solutions with different pH: late endosome (2-(*N*-morpholino)ethanesulfonic acid buffer solution (MES) pH 6.3), lysosome (MES pH 4.5), cytosol (PBS pH 7.4) media, in Eppendorf tubes with filter (30,000 NWML, Amicon Ultra 0.5 mL, Merck, Darmstadt Germany) and shaken at 37 °C and 1000 rpm. At predetermined times, the tubes were centrifuged at 10,000× *g* for 15 min. The filtrates were collected for fluorescent measurements to estimate the amount of released oligonucleotide (λ_{ex} = 550 nm, λ_{em} = 570 nm). The release was calculated as slope × fluorescence × volume of sample.

2.6. Particle Cytotoxicity

The cytotoxicity of nanoparticles was studied using MTT (3-(4,5-dimethylthiazol-2-yl)-2,5-diphenyltetrazolium bromide) reduction assay. NHK were seeded into 96-well plates at density of 10^4 cells per well (TPP, Trasadingen, Switzerland). Keratinocyte growth medium (KGM) was prepared by mixing of keratinocyte basal medium (KBM, Lonza, Visp, Switzerland) with a KGM supplement pack, which contained 2.0 mL of BPE, 0.5 mL human epidermal growth factor protein (hEGFP), 0.5 mL of insulin, 0.5 mL of hydrocortisone, 0.5 mL of transferrin, 0.5 mL of epinephrine, and 0.5 mL of GA-1000. After 24 h the nanoparticles without any cargo and reagents for particle preparation (concentrations of 0.05% and 0.005% mass referred to chitosan amount in KGM) were added for a 24-h period. Then, 100 µL MTT solution (5 mg/mL in PBS) was added after the incubation period for 4 h. The supernatant was carefully removed, and cells were lysed in 50 µL dimethyl sulfoxide for 5 min to dissolve the formazan salt crystals. The absorbance was measured at 540 nm using a Micro Plate Reader (FLUOstar Optima, BMG Labtech, Ortenberg, Germany). For toxicity assessment, KGM was used as a reference for untreated cells (negative control), 0.005% mass. Sodium dodecyl sulfate (SDS, Sigma-Aldrich, München, Germany) in KGM was used as positive control, and 10% of distilled water in KGM was the solvent control. The mean value of solvent control (corrected for blank value) was set to 100%. As reported in the previous studies, we considered cell viability below 75% as an indicator of cytotoxicity [23].

In addition to the testing of NAD(P)H-dependent oxidoreductase activity using MTT assay, the Trypan Blue dye exclusion test was used to determine the number of viable cells present in a cell suspension. Following the same conditions, the cells were incubated with nanoparticles for a 24 h at 37 °C and 5% CO_2. Then, an equal quantity of 0.4% Trypan Blue dye was added to the cell suspension. The mixture was incubated for less than three minutes at room temperature. All cells (blue and clear) were counted on hemocytometer using a light microscope. The results were expressed as a percentage of blue non-viable cells in relation to the total number of cells.

2.7. Cellular Uptake and Transfection

Cell uptake of chitosan–heparin and chitosan nanoparticles into ARPE-19 cell line was assessed using Cy5 labeled double stranded VEGF siRNA complexed by the polymers. The transfection of ARPE-19 was performed using chitosan and chitosan–heparin nanoparticles with encapsulated pEGFP-C2.

Firstly the cells were seeded on 96-well optical-bottom plate with Coverglass base (Thermo Scientific™ Nunc™ MicroWell™) with a density of 5 × 10^4 cells/well in Dulbecco's Modified Eagle

Medium (DMEM-F12) (Biolot, Saint Petersburg, Russia)/10% fetal bovine serum (FBS) (Biowest, South America)/50 IU/mL penicillin/50 µg/mL streptomycin (Biolot, Saint Petersburg, Russia). After 8 h, the medium was removed and 90 µL serum-free DMEM-F12 medium was added to each well. In the cell transfection study, 10 µL of 0.02 mg/mL (200 ng) chitosan or chitosan–heparin nanoparticles with different concentrations of heparin (1:1; 1:2; 1:3 chitosan:heparin mass ratios) and 100 ng of pEGFP-C2 were added to each well (mass ratio chitosan:pDNA = 2:1; N/P = 4.6). To investigate cell uptake, 0.1 nmol of Cy5-siRNA complexed by chitosan or chitosan–heparin at different heparin levels (4:1; 2:1; 1:1 chitosan:heparin mass ratios; mass ratio of chitosan:siRNA was 2:1 equivalent to N/P = 4.6) was added to the wells. The cells were incubated with nanoparticles in serum-free medium for 4 h, then the medium was removed and cells were washed with 1 M NaCl in order to wash out not penetrated particles. After that, 100 µL DMEM-F12 containing 2× FBS and 2× penicillin–streptomycin was added for another 20 h of incubation. Cell uptake and transfection levels obtained with blank Cy5-siRNA and pEGFP-C2 were used as controls.

After 24 h the cells were fixed using exposure of 200 µL 3.7% of formaldehyde–methanol per well for 15 min at 37 °C. Thereafter, the cells were washed three times with PBS. Cell membranes were permeabilized with Permeabilization Buffer (0.2% Triton X-100 in PBS) for 15 min. Staining of the cell nuclei was performed using incubation with Hoechst 33,258 (1 µg/mL) for 30 min according to an established protocol [18]. Then, the cells were washed with PBS for three times, 5 min each time. In order to remove the salts, the cells were washed three times, 2 min each time, with distilled water. Then, the cell membranes were stained by 1× CellMask Green Plasma Membrane Stain (λ_{ex} = 522 nm, λ_{em} = 535 nm; Thermo Fischer Scientific, Paisley, UK), according to the manufacturer's protocol.

The cell uptake efficiency was determined by analyzing the fluorescence intensity of Cy5-siRNA (λ_{ex} = 650 nm, λ_{em} = 670 nm) using CELENA S Digital Imaging System (Logos Biosystems, GE Healthcare, South Korea) and Cytell Cell Imaging instrument (GE Healthcare, Washington, Issaquah, USA). Transfection efficiency was determined by analyzing the fluorescence intensity of GFP using a microplate reader (λ_{ex} = 488 nm, λ_{em} = 509 nm; (Thermo Scientific Varioskan lux, Vantaa, Finland) Transfection efficiency were calculated as a percentage of GFP fluorescent signal in relation to the average fluorescence of the cells treated with pEGFP-C2 alone.

2.8. Gene Silencing of VEGF

The RNA interference potency of 27 bp ds siRNA complexed with chitosan–heparin and directed against VEGF was evaluated in ARPE-19 cells.

ARPE-19 cells were seeded on 24-well plate (Costar TC-treated Multiple Well Plates, Corning, USA) at a density of 2 × 10^5 cells/well in DMEM-F12 (Biolot, Saint Petersburg, Russia)/10% FBS (Biowest, Nuaille, France)/50 IU/mL penicillin/50 µg/mL streptomycin (Biolot, Saint Petersburg, Russia) overnight prior to siRNA delivery. After that, the medium was removed and 500 µL serum-free DMEM-F12 medium was added to each well. Then, 1 nmol of the siRNA complexed with chitosan–heparin (4:1; 2:1; 1:1 chitosan:heparin mass ratios) was added to each well (mass ratio chitosan:siRNA was 2:1; N/P = 4). To investigate RNA interference in the absence of ds VEGF siRNA, chitosan–heparin complexes with scrambled siControl were added using similar conditions. After 4 h of incubation, the medium was replaced with fresh medium, and the cells were further cultured for 48 h.

The RNA interference against VEGF was evaluated by analyzing the levels of VEGF mRNA with reverse transcription polymerase chain reaction (RT-PCR). Glyceraldehyde-3-phosphate dehydrogenase (GAPDH) mRNA was used as control. The extraction of total RNA was performed using RNA extraction kit from Biosilica according to the manufacturer's protocol (Biosilica, Novosibirsk, Russia) and the concentration of RNA was determined based on the absorbance at 260 nm. Then, 60 ng of cDNA was synthesized using the MMLV RT kit (Evrogen, Moscow, Russia). Further RT-PCR analysis was conducted using 12 ng of cDNA and the relevant VEGF forward, VEGF reverse, GADPH forward, and GADPH reverse primers. The analyses were carried out using a qPCRmix-HS SYBR (Evrogen,

Moscow, Russia) according to the manufacturer's protocol. The PCR consisted of 45 amplification cycles of 95 °C for 30 s, 55 °C for 30 s, and 72 °C for 45 s.

2.9. Statistics

The data were expressed as mean (±SD). Statistical significance of differences (at least 3 measurements for each probe) was determined by one-way analysis of variance (ANOVA) with post hoc test (Bonferroni). Statistical analysis and plotting were performed using PRISM software (GraphPad Prism 5.0, La Jolla, CA, USA). $p \leq 0.05$ was considered to indicate a statistically significant difference.

3. Results and Discussion

3.1. Nanoparticle Size, Shape, and Surface Charge

Heparin should be suitable polymeric component in nanomedicines, because it has relatively low molecular weight (12–14 kDa) and it is biodegradable, biocompatible, and non-gelling. However, heparin has been rarely been used for the delivery of gene medicines. To the best of our knowledge, we are the first ones who developed self-assembling chitosan–heparin nanoparticles for cutaneous gene delivery. The ionic crosslinking of the natural polymers is an alternative to covalently cross-linked hydrogels. The electrostatic adhesion between cationic amino groups of chitosan and anionic carboxyl groups of heparin provide a strong interaction in this polyelectrolyte system [15,16,23,24].

Since natural polysaccharides usually have a broad molecular mass distribution and, in some cases, irregular composition, it was important to study the effect of various parameters on particles structure and their stability. Moreover, the size, shape and surface chemistry of nanoparticles can greatly impact cellular uptake and their delivery efficiency in vivo [25]. For determination of particle size, morphology and surface charge various methods were used in this study.

Dynamic light scattering showed that chitosan–heparin nanoparticles with mass ratios of 2:1 and 3:1 yielded the smallest particles with mean diameters of 176 nm and 192 nm, respectively (Figure 1A). The agglomeration takes place at isoelectric point (appr. mass ratio of 0.5:1). While an increased amount of chitosan decreases the particle size up to the chitosan/heparin ratio of 3:1, further addition of chitosan increases the particle size (mean size of 1088 nm at 15:1 ratio). The most compact particles with minimum size distribution were obtained at 2:1 chitosan–heparin ratio that can be related to the strongest complexation of two counterparts. The differences in mass ratio represent the relation of polymer weights. The particles formed a stable colloidal solution for at least 72 h.

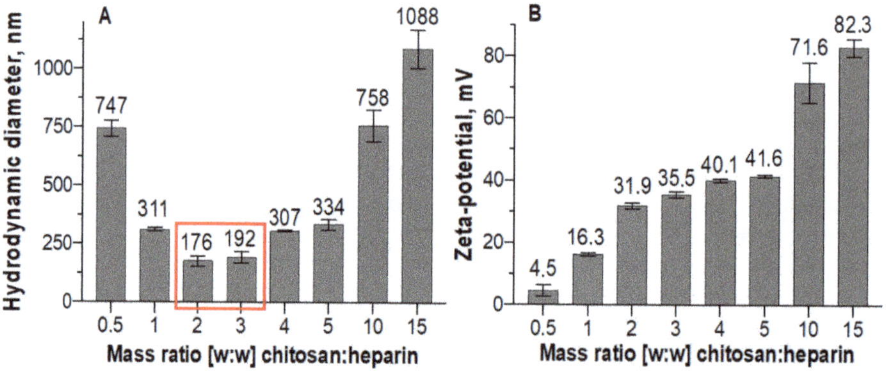

Figure 1. Size (**A**) and charge (zeta-potential) (**B**) of chitosan–heparin nanoparticles prepared with various mass ratios. The smallest sizes are shown with red rectangle. Measurements were performed in 0.01 M PBS, pH 7.4. Mean (±SD), $n = 3$.

Higher proportions of chitosan in the particles resulted in higher positive zeta-potentials (Figure 1B). Nevertheless, the surface charge of the nanoparticles increases only slightly at mass ratios from 2:1 to 5:1 and chitosan forms nanoparticles with nearly constant hydrodynamic sizes (Figure 1A) and zeta-potentials (Figure 1B) at these mass ratios. The similar effects have been shown also in hyaluronic acid/chitosan polyelectrolyte complexes [26]. Further increases to chitosan–heparin ratios of 10:1–15:1 forced the nanoparticles to form large aggregates.

We selected chitosan–heparin nanoparticles with a mass ratio of 2:1 for further experiments. These nanoparticles are expected to be colloidally stable, since they minimal size, narrow size distribution and high surface charge. The particles were loaded with negatively charged oligonucleotide dT-dA that resulted in polyplex formation (mass ratio chitosan:oligonucleotide 2:1, N/P = 4.6) between chitosan, heparin and oligonucleotide. The size of the polyplexes was 145 ± 27 nm and mean PDI was 0.21. This PDI demonstrates a narrow size distribution compared to many other nanoparticles [25]. The zeta potential decreased to 24.9 ± 0.6 mV after complexation with the oligonucleotide.

The various chitosan–heparin nanoparticles were studied with scanning-transmission electron microscopy and nanoparticle tracking analysis to get further information about their size and particle structure. Chitosan–heparin nanoparticles were uniformly spherical at both pH 6.0 (Figure 2A) and pH 8.0 (Figure 2B). The particles had a mean diameter of about 50 nm at pH 6.0 and slightly acidic environment caused swelling of individual particles (Figure 2A). In contrast, at pH 8.0 the nanoparticles collapsed and aggregated (Figure 2B), indicating pH-sensitive properties of chitosan–heparin nanoparticles. The different sizes obtained by DLS and STEM are due to the technical differences of both methods and this has been reported earlier [27].

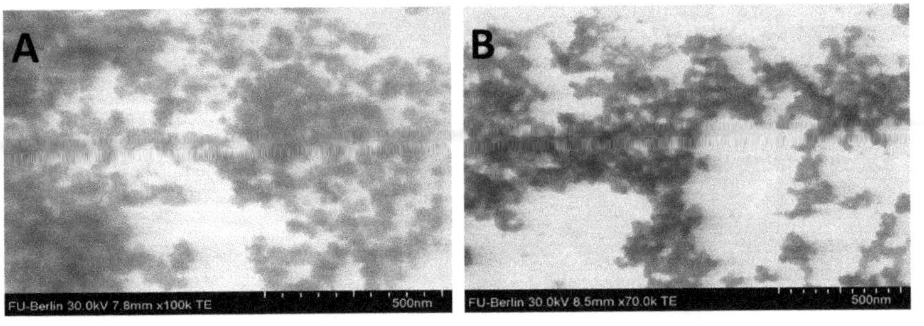

Figure 2. Structure of a chitosan–heparin nanoparticle with mass ratio 2:1 prepared at pH 6.0 (**A**) and pH 8.0 (**B**).

NTA is a method for the determination of a hydrodynamic diameter and size distribution profile of small particles in liquid suspension. NTA tracks the Brownian motion of individual particles or their aggregates at size range of 10–1000 nm and calculates their size taking into account the medium viscosity and particle shape [28].

NTA determination resulted in mean particle size of about 98 nm, smaller diameter than the one obtained with DLS (Figures 1A and 3A). Moreover, small nanoparticles (about 20–30 nm) were clearly visible with the Nanosight system, but harder to capture due to the presence of larger particles that may lead to overexposure when increasing the focus. With increased amounts of chitosan (mass ratio 5:1) (Figure 3B), large conglomerates consisting of small nanoparticles were regularly seen. At increasing chitosan concentration, the nanoparticles formed an ionic network structure in a liquid suspension resulting in size of approximately 1000 nm. These conglomerates are hardly capable of penetrating into the cells or mediating transgene transfection.

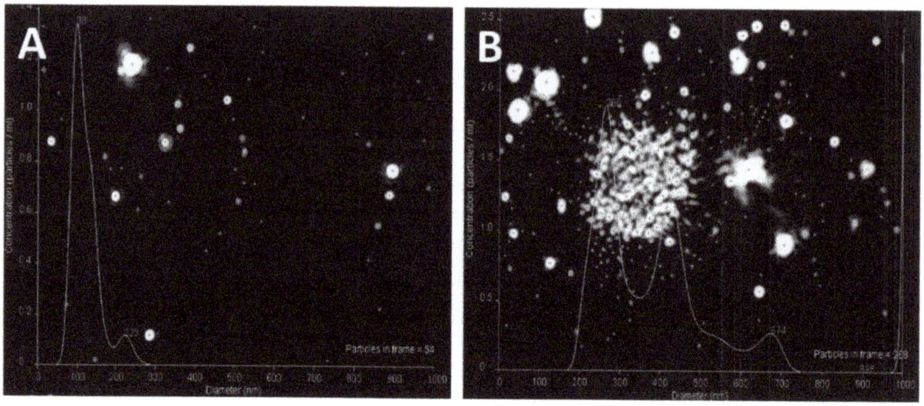

Figure 3. Nanoparticle tracking analysis images of chitosan–heparin nanoparticles with mass ratios of 2:1 (**A**) and 5:1 (**B**) diluted in 0.01 M PBS 7.4.

3.2. *Oligonucleotide Entrapment Efficiency*

In order to deliver genetic materials into the target cells, DNA or siRNA can be bound to polyelectrolyte chitosan–heparin nanoparticles by ionic forces. We used a fluorescently labeled oligonucleotide as an siRNA model. Due to the pH-sensitive nature of chitosan, defined by its pK_a 6.0, we examined the impact of the pH value on the entrapment efficiency of oligonucleotide into chitosan–heparin nanoparticles (Figure 4A), as well as their size and zeta-potential (Figure 4B).

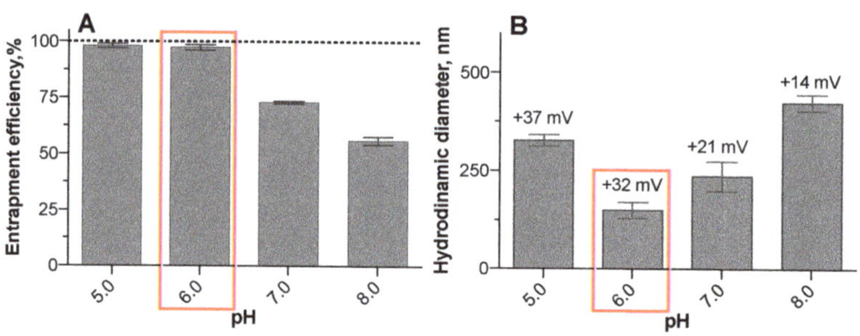

Figure 4. Effect of pH of the medium on entrapment efficiency (**A**), hydrodynamic diameter and zeta-potential (the value above the column) (**B**). Particles with smallest hydrodynamic diameter incubated under pH 6.0 are enclosed in red rectangle. Measurements were performed in various buffer solutions (MES 5.0; MES 6.0; PBS 7.0; PBS 8.0). Mean (±SD), $n = 3$.

Chitosan–heparin complexes encapsulate the oligonucleotide at various degrees, depending on pH of the medium. Almost 100% of labeled oligonucleotide was entrapped at acidic conditions (Figure 4A). The amino groups of chitosan were protonated to the greatest extent in the acidic media resulting in maximal binding efficiency. Thus, the optimal pH for oligonucleotide entrapment has to be around pH 6.0 or below.

The investigations of particle size and zeta-potential at different pH values showed increasing particle size at weakly alkaline conditions due to the deprotonation of amino groups of chitosan that resulted in the aggregation of nanoparticles (Figure 4B). At acidic media with (pH = 5.0) the particle size also increased due to the protonation of amino groups of chitosan followed by the swelling of chitosan–heparin nanoparticles. We assume that protonation/deprotonation processes cause the decreased zeta-potential when pH is shifting from slightly acidic to weakly alkaline media (Figure 4B). Compact and stable nanoparticle complexes were observed at pH 6.0 and below.

Consequently, chitosan–heparin nanoparticles must be prepared at controlled pH conditions to obtain their minimal size.

3.3. *Oligonucleotide Release*

The impact of added heparin on oligonucleotide release was investigated in media with different pH values, mimicking cytosol and bloodstream (pH 7.5), early endosomes (pH 6.3), or lysosomes (pH 4.5) (Figure 5). Major part of oligonucleotide was released from chitosan and chitosan–heparin nanoparticles in 4 h. The release rate increased especially at pH 4.5, and heparin accelerated oligonucleotide release from the polyplexes at this pH (Figure 5). This might be related to the altered swelling properties of chitosan–heparin nanoparticles. Since heparin is stronger polyanion than oligonucleotide, the displacement of cargo can accelerate release rate. Another reason may be related to breaking of chitosan–heparin nanoparticles into smaller pieces at pH 4.5, which would speed up the release after 2 h of incubation. At endosomal and cytosolic pH, the release was slower in both systems and the curves had similar slopes. At pH 6.3 the release kinetics of both systems were similar (Figure 5). It might be explained by stability factors of nanoparticle colloidal solutions. The increase of pH to 7.4 did not result in a major difference in oligonucleotide release as compared to pH 6.3. Overall chitosan–heparin systems protect the oligonucleotide cargo from degradation and presents relatively fast and complete release at slightly acidic medium, an important factor in facilitation of transfection.

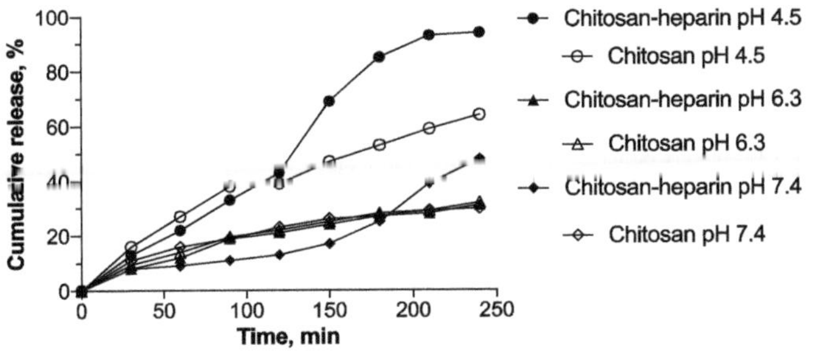

Figure 5. Effect of heparin on oligonucleotide release. Chitosan and chitosan–heparin (mass ratio 2:1) polyplexes with encapsulated Cy3-dT-dA were incubated at pH 4.5 (MES), pH 6.3 (MES), pH 7.4 (PBS) for 240 min.

The kinetic patterns were approximated to elucidate the release mechanism [29] (Table 1). The oligonucleotide release from chitosan and chitosan–heparin nanoparticles was best correlated to the Higuchi model (correlation coefficients >0.95). Thus, we assumed that diffusion is an important release mechanism, based on further confirmation by a good correlation with the Baker–Lonsdale model. The release exponent $n = 1.0$–1.2, obtained from a Korsmeyer–Peppas model for chitosan nanoparticles, proved erosion of the chitosan chain to define the cargo release. In contrast, the release exponent $n = 0.5$–0.86 for chitosan–heparin nanoparticles corresponded to non-Fickian diffusion, and, thus proved a combination of both diffusion and erosion to control release rates. The correlation coefficient value r was around 0.95 for both release from chitosan and chitosan–heparin, best correlated with zero-order model, i.e., drug release at constant rate. The comparison of correlation coefficient values for zero-order ($r \approx 0.95$) and first-order ($r \approx 0.8$) models led us to consider that the release is best correlated with zero-order kinetics. The release coefficients (k) obtained by Hixson-Crowell model confirms the process defined by diffusion rather than dissolution of the carrier.

Table 1. Release kinetics of oligonucleotides from chitosan or chitosan–heparin nanoparticles, with *r* as correlation coefficient value, *k* as the release constant, and *n* as the diffusion or release exponent.

Nanoparticle	pH	Model					
		Zero Order	First Order	Higuchi	Hixson-Crowell	Korsmeyer–Peppas	Baker–Lonsdale
Chitosan	4.5	r = 0.9498 k = 13.48	r = 0.8150 k = −0.51	r = 0.9445 k = 28.52	r = 0.8741 k = −0.84	r = 0.9394 n = 1.26	r = 0.9430 k = 8.41
	6.3	r = 0.9537 k = 16.13	r = 0.7897 k = −0.41	r = 0.9704 k = 34.47	r = 0.8662 k = −0.78	r = 0.9688 n = 1.04	r = 0.9583 k = 14.61
	7.4	r = 0.9811 k = 33.80	r = 0.9037 k = −0.47	r = 0.9400 k = 70.43	r = 0.9478 k = −1.09	r = 0.9489 n = 1.12	r = 0.8687 k = 78.14
Chitosan–heparin	4.5	r = 0.9644 k = 6.52	r = 0.7834 k = −0.33	r = 0.9943 k = 14.01	r = 0.8651 k = −0.57	r = 0.9795 n = 0.86	r = 0.9841 k = 2.42
	6.3	r = 0.9666 k = 6.80	r = 0.8193 k = −0.24	r = 0.9535 k = 19.06	r = 0.8517 k = −0.45	r = 0.9506 n = 0.59	r = 0.9639 k = 5.91
	7.4	r = 0.9377 k = 14.40	r = 0.7855 k = −0.26	r = 0.9839 k = 31.32	r = 0.8483 k = −0.56	r = 0.9768 n = 0.66	r = 0.9836 k = 16.87

3.4. Biocompatibility

After showing the stability of nanoparticles for 48 h in the test medium by DLS, we assessed their cytotoxicity in cultured cells. The hydrodynamic sizes of chitosan and chitosan–heparin nanoparticles (mass ratio 1:1 and 2:1) without cargo after incubation in DMEM during 48 h were not higher than 240 nm, suggesting colloidal stability in the medium. Therefore, we investigated cellular viability with MTT and Trypan Blue assays and primary human keratinocytes. The cells were exposed to chitosan or chitosan–heparin nanoparticles for 24 h and chitosan and heparin bulk materials were used as references.

In general, chitosan–heparin nanoparticles induce minor cytotoxicity in human keratinocytes, as the viability declined to 74% in the worst case (Figure 6A). About 18% of cells lost membrane integrity (Figure 6B) for chitosan–heparin nanoparticles at concentration of 0.005%. Nevertheless, higher nanoparticle concentrations only slightly increased the cytotoxicity. This effect remained marginal with minimal viability of 75% and maximum 22% average dead cells. Different chitosan–heparin ratios did not affect cellular toxicity, but the combination of heparin and chitosan improved the safety of the nanoparticles as compared to chitosan alone.

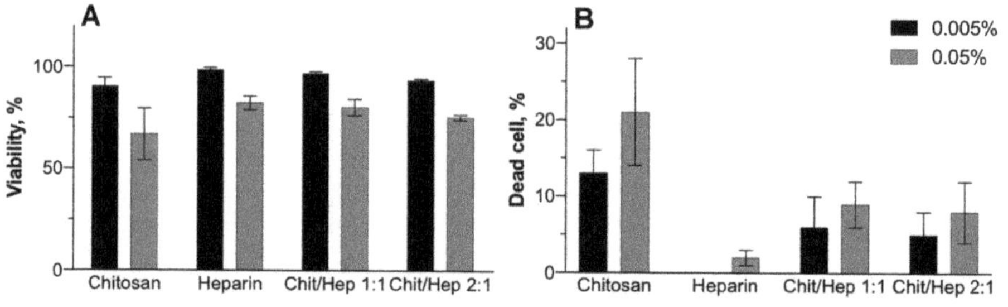

Figure 6. Nanoparticle cytotoxicity in normal human keratinocytes. MTT (**A**) and Trypan Blue Assay (**B**) after 24 h of incubation. Mean (±SD), n = 3.

3.5. Transfection Efficiency

Cellular DNA-transfection with chitosan based polyplexes has been described previously, and some factors affecting the efficiency (e.g., molecular weight of chitosan, chitosan/DNA ratio, particle size, zeta-potential) have been investigated [7,8,18]. Following endocytic cellular uptake, release of DNA from endo-lysosomal compartment to the cytosol and nucleus was described. It was based on protonation of amino groups in chitosan, consequent particle swelling, and bursting of endosomal membrane [7]. There are a few reports about inclusion of polyanions into chitosan-based polyplexes, namely hyaluronic acid [26] or sodium alginate [30]. The transfection of epithelial cells with those polyplexes with pDNA or siRNA resulted in better cell penetration and release of the cargo as compared to chitosan alone [26]. However, the effects of heparin in the chitosan-based polyplexes have not been studied earlier.

Human retinal epithelial cell line was transfected with the reporter gene that encodes green fluorescence protein using chitosan and chitosan–heparin nanoparticles. The fluorescence signal was used as an indicator of transfection efficiency (Figure 7).

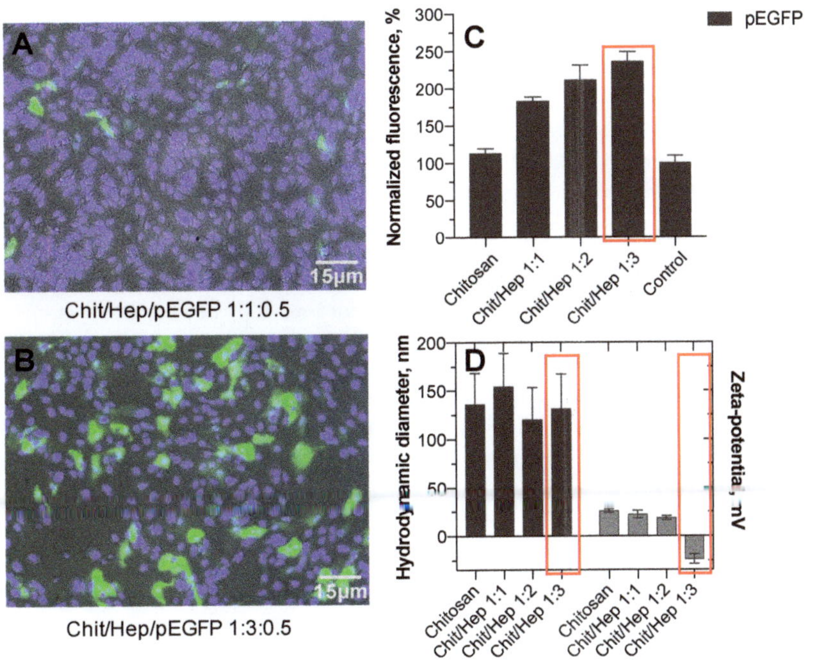

Figure 7. Transfection of ARPE-19 cells by chitosan–heparin-pEGFP polyplexes at mass ratio 1:1:0.5 (**A**) and 1:3:0.5 (**B**). Normalized fluorescence of ARPE-19 cells after transfection with pEGFP chitosan–heparin polyplexes (**C**). Transfected cells using pEGFP alone were used as negative control and the transfection data were normalized to this negative control. Hydrodynamic diameter (DLS) and surface zeta-potential of polyplexes (**D**). The particle size (black bars) and zeta potential (grey bars) of the polyplexes. The formulation with the best transfection efficiency is marked with red rectangles. Mean (±SD), $n = 3$.

The data demonstrates that the cells were transfected at much higher efficiency with chitosan–heparin–pEGFP polyplexes as compared to the control transfection with pEGFP-C2 alone. Furthermore, chitosan–heparin-based polyplexes exhibited higher transfection efficiency than chitosan nanoparticles with pEGFP.

Transfection with chitosan and chitosan–heparin based nanoparticles depends on the cellular endocytosis of polyplexes. DNA release from the polyplexes is believed to proceed through proton sponge effect due to the swelling of chitosan–heparin complexes in the endo-lysosomes [31]. Chitosan is known to be a strong polycation with high charge density. Thus, the inclusion of heparin, (polyanionic glycosaminoglycan) on the nanoparticles was proposed to enhance DNA release and transfection of cells due to interactions between heparin and chitosan, which may lead to DNA displacement and its release to the cytoplasm.

Polyplexes were also characterized in terms of particle size, polydispersity index and zeta-potential. Excess of heparin on the nanoparticle surface at a chitosan:heparin mass ratio of 1:3 resulted in negative surface charge (zeta-potential −22 ± 4 mV). However, there were no significant size differences among polyplexes at different chitosan:heparin mass ratios (Figure 7D). Hence, the polyplexes with negative zeta potentials were capable to bind DNA and transfer it into the cells.

The lowest rate of transfection was obtained with pEGFP-C2 alone. Chitosan–pEGFP formulation increased the tranfection levels only by 17% (Figure 7C). In contrast, GFP expression after chitosan/heparin polyplex transfection was almost two times higher than that of chitosan polyplexes (Figure 7C). In addition, significant differences were found among polyplexes with different chitosan/heparin mass ratios, with chitosan/heparin polyplexes at a mass ratio 1:3 showing the highest transfection levels (Figure 7C).

3.6. *Gene Silencing of VEGF*

Vascular endothelial growth factor is an endothelial cell-specific mitogen and an angiogenesis inducer in vivo. Its activity has been linked to tumor growth and formation of metastases [32]. VEGF has been implicated in the disruption of retinal pigment epithelium barrier function and accumulation of subretinal fluid from the leaky neo-vessels. Therefore, it is important to find a powerful technologies for VEGF inhibition [33].

RNA interference technologies are currently widely used in functional genomic studies [34–36]. Also, small interfering RNA is being developed as therapeutics against cancers and other indications [37]. However, the drug delivery systems for siRNA have not been fully developed for in vivo use. We studied the use of chitosan–heparin nanoparticles and the role of heparin amount in these polyplexes for the delivery of anti-VEGF siRNA into ARPE-19 cells.

We investigated the RNA interference efficiency of anti-VEGF siRNA [21] in ARPE-19 cells that express VEGF constitutively. The VEGF mRNA expression in the cells was analyzed by RT-PCR. Chitosan–heparin polyplexes at mass ratios 1:1 and 2:1 demonstrated two times stronger inhibitory effects than chitosan–heparin polyplexes at a 4:1 mass ratio (Figure 8).

Fluorescence microscopy images of ARPE-19 cells treated with Cy5-dsRNA–chitosan–heparin polyplexes showed effective cell penetration of Cy5-labeled dsRNA. In the case of 1:1 and 2:1 chitosan–heparin mass ratios, most siRNAs were located in the cytoplasm (Figure 8). At a 4:1 mass ratio, chitosan–heparin particles have larger particle size due to the aggregate formation. Hence, the particle penetration into the cells was partially limited (Figure 8).

Overall, addition of heparin to chitosan-based polyplexes enhanced siRNA and DNA release and cytosolic delivery, thereby improving gene silencing and transgene expression in the cells.

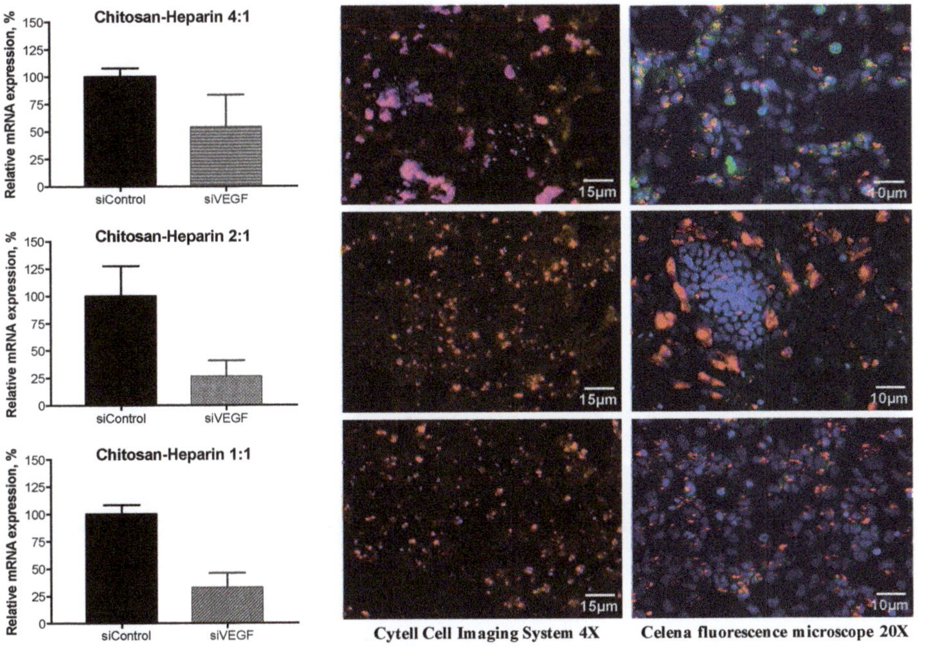

Figure 8. RNA interference by anti-VEGF small interfering RNA that was delivered in chitosan–heparin nanoparticles at different mass ratios (1:1; 2:1; 4:1) in the ARPE-19 cell line. LEFT. Relative VEGF mRNA expression data from RT-PCR assays are shown (left panel). GADPH mRNA was measured as an intrinsic control. Mean (±SD), $n = 3$. RIGHT. Fluorescence microscopy images of ARPE-19 cells that were transfected with Cy5-dsRNA (red color) complexed by chitosan–heparin at different mass ratios (1:1; 2:1; 4:1). The cell nuclei were stained by Hoechst 33,258 (blue color) and the plasma membranes were stained using CellMask Green Plasma Membrane Stain (yellow and green color).

4. Conclusions

In present work, biocompatible nanoparticles based on chitosan–heparin complexes were developed. The particles were prepared in mild conditions using a drop-wise addition method. The model oligonucleotide was associated with cationic chitosan–heparin complexes via ionic interactions that led to the formation of chitosan–heparin-oligonucleotide polyplexes. The pH-sensitive behavior of nanoparticles was demonstrated. DLS, NTA, and STEM methods informed that the mean particle sizes of the particles were 100–200 nm depending on the heparin amount. The increasing chitosan concentrations in the polyplexes led to the particle aggregation and increased particle size (400–1000 nm). The in vitro release studies and mathematical modelling of the release kinetics demonstrated increased release rate of oligonucleotide at slightly acidic conditions (pH 4.5) due to swelling and diffusion of cargo. In vitro release from chitosan–heparin polyplexes was faster than from chitosan polyplexes, possibly due to the effect of heparin as competitive polyanion.

The cell viability tests demonstrated lack of cytotoxicity for chitosan–heparin. Addition of heparin to the polyplexes did not affect cellular uptake, but increased the DNA-transfection efficiency in the ARPE cells. Likewise, VEGF silencing in the ARPE-19 cells was enhanced when heparin was used in the chitosan polyplexes. These improved effects were facilitated by the enhanced release of siRNA in the cells. This approach could be used to increase the rate of cargo release from other nanocarriers. The chitosan–heparin nanocarriers also have potential for the delivery of genetic drugs into epithelial tissues because they may open the epithelial tight junctions and they have mucoadhesive properties.

Author Contributions: Conceptualization, T.T., V.K.-V., I.P.; methodology, V.K.-V. and I.P.; validation, I.P., V.K.-V., N.Z., C.Z.; formal analysis, M.S.-K. and E.R.; investigation, I.P.; resources, T.T., M.S.-K., E.R.; data curation, V.K.-V., C.Z., V.S.; writing-original draft preparation, I.P., V.K.-V., T.T., C.Z., N.Z., V.S.; writing-review and editing, V.K.-V., T.T., C.Z., V.S.; visualization, I.P.; supervision, V.K.-V.; project administration, T.T.

Funding: The financial support was provided by Government of Russian Federation (Megagrant # 14.W03.31.0025, "Biohybrid technologies for modern medicine"). The authors also thank the German Academic Exchange Service DAAD and the German–Russian Interdisciplinary Science Center for the funding of internships (Project №B-2017b-11).

Acknowledgments: The authors are grateful to Professor Dr. Arto Urtti, University of Eastern Finland for his valuable comments to the article. The authors are grateful to Ms. Anke Schindler, Freie Universität Berlin, for STEM experiments.

Conflicts of Interest: The authors declare no conflict of interest. The funders had no role in the design of the study; in the collection, analyses, or interpretation of data; in the writing of the manuscript, or in the decision to publish the results.

References

1. Chakraborty, C.; Sharma, A.R.; Sharma, G.; Doss, C.G.P.; Lee, S.-S. Therapeutic miRNA and siRNA: Moving from Bench to Clinic as Next Generation Medicine. *Mol. Ther. Nucleic Acids* **2017**, *8*, 132–143. [CrossRef] [PubMed]
2. Song, J.; Lee, M.; Kim, T.; Na, J.; Jung, Y.; Jung, G.Y.; Park, N. A RNA producing DNA hydrogel as a platform for a high performance RNA interference system. *Nat. Commun.* **2018**, *9*, 4331. [CrossRef] [PubMed]
3. Jeong, J.H.; Kim, S.W.; Park, T.G. Molecular design of functional polymers for gene therapy. *Prog. Polym. Sci.* **2007**, *32*, 1239–1274. [CrossRef]
4. Tobío, M.; Gref, R.; Sánchez, A.; Langer, R.; Alonso, M.J. Stealth PLA-PEG nanoparticles as protein carriers for nasal administration. *Pharm. Res.* **1998**, *15*, 270–275. [CrossRef] [PubMed]
5. Germershaus, O.; Mao, S.; Sitterberg, J.; Bakowsky, U.; Kissel, T. Gene delivery using chitosan, trimethyl chitosan or polyethylenglycol-graft-trimethyl chitosan block copolymers: Establishment of structure-activity relationships in vitro. *J. Control. Release* **2008**, *125*, 145–154. [CrossRef] [PubMed]
6. Sato, T.; Ishii, T.; Okahata, Y. In vitro gene delivery mediated by chitosan. Effect of pH, serum, and molecular mass of chitosan on the transfection efficiency. *Biomaterials* **2001**, *22*, 2075–2080. [CrossRef]
7. Huang, M.; Fong, C.W.; Khor, E.; Lim, L.Y. Transfection efficiency of chitosan vectors: Effect of polymer molecular weight and degree of deacetylation. *J. Control. Release* **2005**, *106*, 391–406. [CrossRef] [PubMed]
8. Lehr, C.M.; Bouwstra, J.A.; Schacht, E.H.; Junginger, H.E. In vitro evaluation of mucoadhesive properties of chitosan and some other natural polymers. *Int. J. Pharm.* **1992**, *78*, 43–48. [CrossRef]
9. Schipper, N.G.M.; Olsson, S.; Hoogstraate, J.A.; DeBoer, A.G.; Vårum, K.M.; Artursson, P. Chitosans as absorption enhancers for poorly absorbable drugs 2: Mechanism of absorption enhancement. *Pharm. Res.* **1997**, *14*, 923–929. [CrossRef]
10. Muzzarelli, R.A.A. Natural chelating polymers; alginic acid, chitin, and chitosan. *Dermatol. Mon.* **1973**, *158*, 465–473.
11. Koga, D.; Mitsutomi, M.; Kono, M.; Matsumiya, M. Biochemistry of chitinases. *Chitin Chitinases* **1999**, *87*, 111–123. [CrossRef]
12. Zhang, Z.; Tsai, P.C.; Ramezanli, T.; Michniak-Kohn, B.B. Polymeric nanoparticles-based topical delivery systems for the treatment of dermatological diseases. *Wiley Interdiscip. Rev. Nanomed. Nanobiotechnol.* **2013**, *5*, 205–218. [CrossRef] [PubMed]
13. Grenha, A. Chitosan nanoparticles: A survey of preparation methods. *J. Drug Target.* **2012**, *20*, 291–300. [CrossRef] [PubMed]
14. Hembram, K.C.; Prabha, S.; Chandra, R.; Ahmed, B.; Nimesh, S. Advances in preparation and characterization of chitosan nanoparticles for therapeutics. *Artif. CellsNanomed. Biotechnol.* **2016**, *44*, 305–314. [CrossRef] [PubMed]
15. Duceppe, N.; Tabrizian, M. Advances in using chitosan-based nanoparticles for in vitro and in vivo drug and gene delivery. *Expert Opin. Drug Deliv.* **2010**, *7*, 1191–1207. [CrossRef] [PubMed]

16. Köping-Höggård, M.; Vårum, K.M.; Issa, M.; Danielsen, S.; Christensen, B.E.; Stokke, B.T.; Artursson, P. Improved chitosan-mediated gene delivery based on easily dissociated chitosan polyplexes of highly defined chitosan oligomers. *Gene Ther.* **2004**, *11*, 1441–1452. [CrossRef] [PubMed]
17. Douglas, K.L.; Piccirillo, C.A.; Tabrizian, M. Effects of alginate inclusion on the vector properties of chitosan-based nanoparticles. *J. Control. Release* **2006**, *115*, 354–361. [CrossRef]
18. De La Fuente, M.; Seijo, B.; Alonso, M.J. Novel Hyaluronic Acid-Chitosan Nanoparticles for Ocular Gene Therapy. *Investig. Opthalmol. Vis. Sci.* **2008**, *49*, 2016–2024. [CrossRef]
19. Liu, Z.; Jiao, Y.; Liu, F.; Zhang, Z. Heparin/chitosan nanoparticle carriers prepared by polyelectrolyte complexation. *J. Biomed. Mater. Res. Part A* **2007**, *83*, 806–812. [CrossRef]
20. Tan, Q.; Tang, H.; Hu, J.; Hu, Y.; Zhou, X.; Tao, Y.; Wu, Z.; Hu, Y.; Zhou, X.; Tao, R. Controlled release of chitosan/heparin nanoparticle-delivered VEGF enhances regeneration of decellularized tissue-engineered scaffolds. *Int. J. Nanomed.* **2011**, *6*, 929–942. [CrossRef]
21. Kubo, T.; Mihara, K.; Seyama, T.; Yanagihara, K.; Takei, Y.; Sato, Y. Lipid-Conjugated 27-Nucleotide Double-Stranded RNAs with Dicer-Substrate Potency Enhance RNAi-Mediated Gene Silencing. *Mol. Pharm.* **2012**, *9*, 1374–1383. [CrossRef] [PubMed]
22. Zoschke, C.; Kral, V.; Gonska, H.; Kapfer, C.; Schafer-Korting, M. Production of Organotypic NMSC Models. 2017. Available online: http://www.bb3r.de/en/forschungsplattform/Standardarbeitsvorschriften/index.html (accessed on 27 May 2019).
23. Balansin Rigon, R.; Kaessmeyer, S.; Wolff, C.; Hausmann, C.; Zhang, N.; Sochorová, M.; Zoschke, C. Ultrastructural and Molecular Analysis of Ribose-Induced Glycated Reconstructed Human Skin. *Int. J. Mol. Sci.* **2018**, *19*, 3521. [CrossRef] [PubMed]
24. Abcam. *Ab228550 Protocol Hoechst 33258 Staining*; Abcam: Cambridge, UK, 2018.
25. Banerjee, A.; Qi, J.; Gogoi, R.; Wong, J.; Mitragotri, S. Role of Nanoparticle Size, Shape and Surface Chemistry in Oral Drug Delivery. *J. Control. Release* **2016**, *238*, 176–185. [CrossRef] [PubMed]
26. Almalik, A.; Donno, R.; Cadman, C.J.; Cellesi, F.; Day, P.J.; Tirelli, N. Hyaluronic acid-coated chitosan nanoparticles: Molecular weight-dependent effects on morphology and hyaluronic acid presentation. *J. Control. Release* **2013**, *172*, 1142–1150. [CrossRef] [PubMed]
27. Hoo, C.M.; Starostin, N.; West, P.; Mecartney, M.L. A comparison of atomic force microscopy (AFM) and dynamic light scattering (DLS) methods to characterize nanoparticle size distributions. *J. Nanopart. Res.* **2008**, *10*, 89–96. [CrossRef]
28. Gross, J.; Sayle, S.; Karow, A.R.; Bakowsky, U.; Garidel, P. Nanoparticle tracking analysis of particle size and concentration detection in suspensions of polymer and protein samples: Influence of experimental and data evaluation parameters. *Eur. J. Pharm. Biopharm.* **2016**, *104*, 30–41. [CrossRef] [PubMed]
29. Costa, P.; Lobo, J.M.S. Modeling and comparison of dissolution profiles. *Eur. J. Pharm. Sci.* **2001**, *13*, 123–133. [CrossRef]
30. Rafiee, A.; Alimohammadian, M.H.; Gazori, T.; Riazi-Rad, F.; Fatemi, S.M.R.; Parizadeh, A.; Haririan, I.; Havaskary, M. Comparison of chitosan, alginate and chitosan/alginate nanoparticles with respect to their size, stability, toxicity and transfection. *Asian Pac. J. Trop. Dis.* **2014**, *4*, 372–377. [CrossRef]
31. Freeman, E.C.; Weiland, L.M.; Meng, W.S. Modeling the proton sponge hypothesis: Examining proton sponge effectiveness for enhancing intracellular gene delivery through multiscale modeling. *J. Biomater. Sci. Polym. Ed.* **2013**, *24*, 398–416. [CrossRef]
32. Kim, K.J.; Li, B.; Winer, J.; Armanini, M.; Gillett, N.; Phillips, H.S.; Ferrara, N. Inhibition of vascular endothelial growth factor-induced angiogenesis suppresses tumour growth in vivo. *Nature* **1993**, *362*, 841–844. [CrossRef]
33. Ablonczy, Z.; Crosson, C.E. VEGF modulation of retinal pigment epithelium resistance. *Exp. Eye Res.* **2007**, *85*, 762–771. [CrossRef] [PubMed]
34. Fire, A.; Xu, S.; Montgomery, M.K.; Kostas, S.A.; Driver, S.E.; Mello, C.C. Potent and specific genetic interference by double-stranded RNA in *Caenorhabditis elegans*. *Nature* **1998**, *391*, 806–811. [CrossRef] [PubMed]
35. Elbashir, S.M.; Harborth, J.; Lendeckel, W.; Yalcin, A.; Weber, K.; Tuschl, T. Duplexes of 21-nucleotide RNAs mediate RNA interference in cultured mammalian cells. *Nature* **2001**, *411*, 494–498. [CrossRef] [PubMed]

36. Carpenter, A.E.; Sabatini, D.M. Systematic genome-wide screens of gene function. *Nat. Rev. Genet.* **2004**, *5*, 11–22. [CrossRef] [PubMed]
37. Ryther, R.C.C.; Flynt, A.S.; Phillips, J.A.; Patton, J.G. SiRNA therapeutics: Big potential from small RNAs. *Gene Ther.* **2005**, *12*, 5–11. [CrossRef]

pharmaceutics

MDPI

Article

Skin Irritation Testing beyond Tissue Viability: Fucoxanthin Effects on Inflammation, Homeostasis, and Metabolism

Renata Spagolla Napoleão Tavares [1], Silvya Stuchi Maria-Engler [2], Pio Colepicolo [3], Hosana Maria Debonsi [1], Monika Schäfer-Korting [4], Uwe Marx [5], Lorena Rigo Gaspar [1] and Christian Zoschke [4,*]

1 School of Pharmaceutical Sciences of Ribeirão Preto, University of São Paulo, Av. do Café s/n, Monte Alegre, Ribeirão Preto, SP 14040-903, Brazil; renatasnt@usp.br (R.S.N.T.); hosana@fcfrp.usp.br (H.M.D.); lorena@fcfrp.usp.br (L.R.G.)
2 Clinical and Toxicological Analyses Department, School of Pharmaceutical Sciences, University of São Paulo, Av. Prof. Lineu Prestes, 748, Cidade Universitária, São Paulo, SP 05508-000, Brazil; silvya@usp.br
3 Institute of Chemistry, University of São Paulo, Av. Prof. Lineu Prestes, 748, Cidade Universitária, São Paulo, SP 05508-000, Brazil; piocolep@iq.usp.br
4 Institute of Pharmacy (Pharmacology & Toxicology), Freie Universität Berlin, Königin Luise Str 2+4, 14195 Berlin, Germany; monika.schaefer-korting@fu-berlin.de
5 TissUse GmbH, Oudenarder Str. 16, 13347 Berlin, Germany; uwe.marx@tissuse.com
* Correspondence: christian.zoschke@fu-berlin.de; Tel.: +49-30-838-56189

Received: 20 November 2019; Accepted: 31 January 2020; Published: 5 February 2020

check for updates

Abstract: UV light catalyzes the ozone formation from air pollutants, like nitrogen oxides. Since ozone reacts with cutaneous sebum lipids to peroxides and, thus, promotes inflammation, tumorigenesis, and aging, even broad-spectrum sunscreens cannot properly protect skin. Meanwhile, xanthophylls like fucoxanthin, proved their antioxidant and cytoprotective functions, but the safety of their topical application in human cell-based models remains unknown. Aiming for a more detailed insight into the cutaneous fucoxanthin toxicity, we assessed the tissue viability according to OECD test guideline no. 439 as well as changes in inflammation (IL-1α, IL-6, IL-8), homeostasis (EGFR, HSPB1) and metabolism (NAT1). First, we proved the suitability of our 24-well-based reconstructed human skin for irritation testing. Next, we dissolved 0.5% fucoxanthin either in alkyl benzoate or in ethanol and applied both solutions onto the tissue surface. None of the solutions decreased RHS viability below 50%. In contrast, fucoxanthin ameliorated the detrimental effects of ethanol and reduced the gene expression of pro-inflammatory interleukins 6 and 8, while increasing NAT1 gene expression. In conclusion, we developed an organ-on-a-chip compatible RHS, being suitable for skin irritation testing beyond tissue viability assessment. Fucoxanthin proved to be non-irritant in RHS and already showed first skin protective effects following topical application.

Keywords: antioxidants; epidermal growth factor receptor; interleukins; irritation; metabolism response; N-acetyltransferase; small heat and shock protein beta 1

---

1. Introduction

Epidemiological data and clinical presentation provide conclusive evidence for UV radiation as the major cause of skin aging, cancer, and inflammation [1,2]. UVB radiation promote the dimerization of pyrimidine DNA bases to cyclobutane dimers and subsequently C to T base transitions. Abundant C to T base transitions and CC to TT tandem mutations are referred to as a UVB signature or fingerprint. UVA radiation increases numbers of reactive oxygen species, which oxidize DNA bases to 8-hydroxyguanine, and cause G-to-T base transversion [3]. Thereby, the production of reactive oxygen species has two major causes. First, UV radiation directly oxidizes the major sebum component squalene and, thus, drives the expression of pro-inflammatory cytokines [4]. Second, UVA reacts with nitrogen oxides and volatile organic compounds, both abundant air pollutants in urban areas, to ozone [5]. Ozone itself does not penetrate the skin, but the ozone-mediated peroxidation of unsaturated lipids on the skin's surface induces oxidative stress and inflammatory responses in deeper skin layers [6].

Prevention is considered as the prime strategy to reduce the number of skin cancer patients because treatment remains insufficiently effective in terms of cure rates and recurrence [7]. Yet, the widely recommended sunscreen application also provides insufficient protection against direct or indirect sunlight effects. Frequent underdosage [8] and abuse of sunscreens for intentional prolonged sun exposure [9] contribute to their low efficacy. Moreover, current sunscreens currently do not absorb VIS and IR light. This filter gap allows VIS and IR radiation penetrating deep skin layers and to produce 50% of the total amount of reactive oxygen species in skin [10]. Due to minor amounts of endogenous antioxidants in deeper skin layers, reactive oxygen species are hardly eliminated in the dermis [11] as well as in aged skin [12]. Taken together, even optimal application of currently existing sunscreen will not provide full protection from the effects of sunlight and air pollution, emphasizing the need for effective antioxidants in skin care.

Antioxidants address this issue by lowering the amount of reactive oxygen species. Administered either orally in nutraceuticals or topically in cosmetic products, antioxidants comprise carotenoids, squalene, and vitamins, to name but a few. Although nutraceuticals received considerable interest, studies on bioavailability, efficacy, and mechanism of action are scarce [13]. Furthermore, their uncontrolled use might cause secondary effects and interactions with medicinal products. Focusing on carotenoids, β-carotene and fucoxanthin are known for their antioxidant and cytoprotective functions and exemplarily represent the two carotenoid classes, carotenes and xanthophylls. Fucoxanthin was first isolated from the marine brown seaweeds and differs from other carotenoids due to an unusual allenic carbon (C-7′),5,6-monoepoxide, two hydroxyl groups, a carbonyl group, and an acetyl group in the terminal ring of fucoxanthin [14]. This particular chemical structure causes similar or higher antioxidant activities in comparison to those of α-tocopherol as well as suppressive effects on superoxide anion and nitric oxide generation [15,16]. Despite of the high antioxidant activity, oral fucoxanthin application does not result in efficient cutaneous concentrations. Fucoxanthin and its metabolites were primarily found in adipose tissue, liver, lungs, kidney, heart, and the spleen of mice [17]. Thus, a topical application of the lipophilic compound should achieve higher fucoxanthin doses at the target site.

The poor water solubility of fucoxanthin (log$P$ 7.8) poses a challenge for the formulation of topically-used fucoxanthin products. Alkyl benzoate and ethanol are frequently used as solubilizers for highly lipophilic substances [18], and especially ethanol enhances the skin penetration [19].

Herein, we investigated the toxicological effects of fucoxanthin, dissolved either in alkyl benzoate or ethanol, on inflammation (interleukin-1α, 6, 8), homeostasis (epidermal growth factor receptor, small heat and shock protein beta 1), and metabolism (N-acetyltransferase 1) as well as on tissue viability in reconstructed human skin from primary human cells. To evaluate fucoxanthin effects, we selected pro-inflammatory cytokines IL-1α, IL-6, and IL-8, as well as EGFR, to study the beginning of re-epithelialization and HSPB1 to monitor the protective functions under stress conditions as well as NAT1, since fucoxanthin is totally deacetylated in the intestinal lumen.

2. Materials and Methods

2.1. Reconstructed Human Skin

The experimental procedures conformed to the principles of the Declaration of Helsinki and were approved by the ethics committees of Charité—Universitätsmedizin Berlin (EA1/081/13). Informed written consent was obtained from all the donors or their parent or legal guardian. The reconstructed human skin (RHS) was cultured in 24-well plates with primary human keratinocytes and fibroblasts (passage 3, pooled from three donors) from foreskin with ethical committee approval (EA1/081/13), and after parents had signed the written informed consent. We made the dermal compartment on day 01 by pouring 0.5 mL collagen I (Biochrom, Berlin, Germany) with 1.14×10^5 normal human dermal fibroblasts into the insert (0.4 μm pore size; Millicell, Merck, Darmstadt, Germany). We seeded 3.7×10^5 normal human keratinocytes onto the dermal compartment on day 2 and raised the constructs to the air-liquid interface on day 3. The culture medium was changed three times a week for seven days.

2.2. Test Substance Application

After the RHS were fully differentiated on day 10, we placed them into new 24-well plates containing 0.5 mL of fresh medium and performed the test according to OECD test guideline no. 439 [20]. In brief, 10 μL of the following test substances were applied for 15 min: phosphate-buffered saline (PBS, Sigma-Aldrich, München, Germany), 5% (*w*/*v*) sodium dodecyl sulfate (SDS, CAS-no: 151-21-3; Carl Roth, Berlin, Germany), 0.5% (*w*/*v*) all-*trans*-fucoxanthin (≥95% pure, CAS-no: 3351-86-8; Sigma-Aldrich), C12-15 alkyl benzoate (Crodamol™ AB, CAS-no 68411-27-8, Croda, Brazil), and ethanol (99.5%, CAS-no: 64-17-5, Merck, Germany). Subsequently, the constructs were washed 12 times with 0.5 mL PBS, dried with a sterile cotton swab, and placed into new 24-well plates with fresh medium for 42 h at 37 °C, 5% CO_2.

2.3. Viability Assay

Constructs were incubated with the test substances for 15 min followed by a 42 h post-incubation period. RHS viability was determined by measuring the metabolic activity of the constructs after exposure and post-incubation using a colorimetric test according to OECD test guideline no. 439 [20]. The reduction of mitochondrial dehydrogenase activity was assessed via the decreased formazan production following incubation with MTT (3-(4,5-dimethylthiazol-2-yl)-2,5-diphenyltetrazolium bromide, Sigma-Aldrich). The formazan production was measured at 570 nm (FLUOstar OPTIMA, BMG Labtech, Ortenberg, Germany). We performed tests for interference of chemicals with MTT endpoint and correction in accordance to MatTek's "In Vitro EpiDerm™ Skin Irritation Test" ([21], steps 1–4). The formazan readings were corrected by the fucoxanthin-related optical densities as well as by the optical densities due to direct MTT reduction and compared to those of negative control RHS [20]. Data are presented as the relative viability according to Equation (1):

$$Relative\ viability\ (\%) = 100 \times [OD\ (test\ substance) / mean\ OD\ (negative\ control)] \quad (1)$$

2.4. Morphology and Immunofluorescence

Each RHS was snap frozen, sectioned into 8 μm slices (Leica CM 1510S, Wetzlar, Germany), and analyzed by hematoxylin-eosin or immunofluorescence staining. Antibodies against the following proteins were purchased from Abcam (Cambridge, UK): filaggrin (1:1000; ab81468), involucrin (1:500; ab111781), and from Dianova (Hamburg, Germany): keratin-10 (1:200, cat-no. AF0197-01). Pictures were taken with a fluorescence microscope (BZ-8000, Keyence, Neu-Isenburg, Germany) and analyzed by ImageJ software 1.52a [22].

2.5. Gene Expression

Real-time qPCR endpoint analysis was performed according to established procedures [23]. Briefly, the epidermal and dermal compartments of the RHS were mechanically separated. RNA from the epidermis was isolated using the NucleoSpin RNA II Kit (Macherey-Nagel, Düren, Germany). The TaqMan® Reverse Transcription Kit (Thermo Fisher Scientific Inc., Waltham, MA, USA) synthesized cDNA by reverse transcription of 100 ng total RNA. Quantitative PCR was performed using the QuantStudio 5 Real-Time PCR System (Thermo Fisher Scientific) and the SensiFAST SYBR Lo-ROX Kit (Bioline, Luckenwalde, Germany) according to the manufacturer's instructions. The primers were designed with UCSC Genome Browser [24] as described in Table 1 and were synthesized by EuroPrime (Invitrogen, Berlin, Germany). Gene expressions were normalized to the housekeeping gene ACTB.

Table 1. Primer sequences for PCR studies.

Gene Name	Used Primer Sequence	
	Forward	Reverse
ACTB	CCACCATGTACCCTGGCATT	GCTTGCTGATCCACATCTGCT
EGFR	GCCGACAGCTATGAGATGGAG	TGGAGGTGCAGTTTTTGAAGTG
HSPB1	GACCCCACCCAAGTTTCCTC	TCGGATTTTGCAGCTTCTGG
IL-1α	GTGACTGCCCAAGATGAAGACC	TGCCAAGCACACCCAGTAGTC
IL-6	CTGGATCAGGACTTTTGTACTCATCT	CCAATCTGGATTCAATGAGGAGACT
IL-8	GTGGAGAAGTTTTTGAAGAGGGC	TCTGGCAACCCTACAACAGAC
NAT1	ATCCGAGCTGTTCCCTTTGAG	AACATACCCTCCCAACATCGTG

2.6. Statistical Analysis

Data are presented as the mean + SD obtained from three independent experiments. Due to the explorative data analysis, a level of $p \leq 0.05$ was considered to indicate a statistically significant difference. One-way ANOVA and subsequent Tukey post hoc tests were performed with GraphPad Prism 5.0 (La Jolla, CA, USA).

3. Results

3.1. Fucoxanthin Effects on RHS Morphology

The 10-day culture of reconstructed human skin (RHS) resulted in a stratified epidermis with well-expressed *stratum basale, spinosum, granulosum,* and *corneum* (Figure 1a). Keratinocyte differentiation induced cell flattening and the expression of keratin-10 in suprabasal layers (Figure 1b). Keratin-10 and 14 was expressed throughout all epidermal layers (Figure 1b). Moreover, involucrin found in the *stratum corneum* showed the formation of a cornified envelope (Figure 1c). Although we observed also parakeratosis—an almost regular feature of skin models—and only small amounts of filaggrin, the RHS morphology suggested a sufficient skin barrier formation.

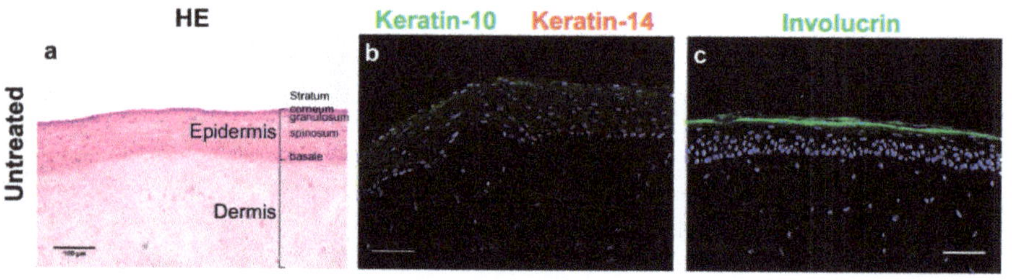

Figure 1. Morphology and protein expression in reconstructed human skin. (**a**) Hematoxylin-eosin staining showing all layers of human skin. (**b**) Immunolocalization of keratin-10 (green) in suprabasal epidermal layers and keratin-14 (red) in all epidermal layers. (**c**) Immunolocalization of involucrin (green), indicating terminal keratinocyte differentiation. Images are representative of three batches; nuclei in blue (DAPI); scale bars = 100 µm.

Neither fucoxanthin nor the vehicle alkyl benzoate disturbed tissue morphology (Figure 2a,b). Ethanol caused a detachment of the stratum corneum from the viable epidermis as well as slight damages in the coherence of the viable epidermis (Figure 2c). The filaggrin expression in RHS was slightly increased following exposure to fucoxanthin and ethanol (Figure 2d,f). Involucrin was homogenously distributed in the stratum corneum following all substance exposures (Figure 2g–i).

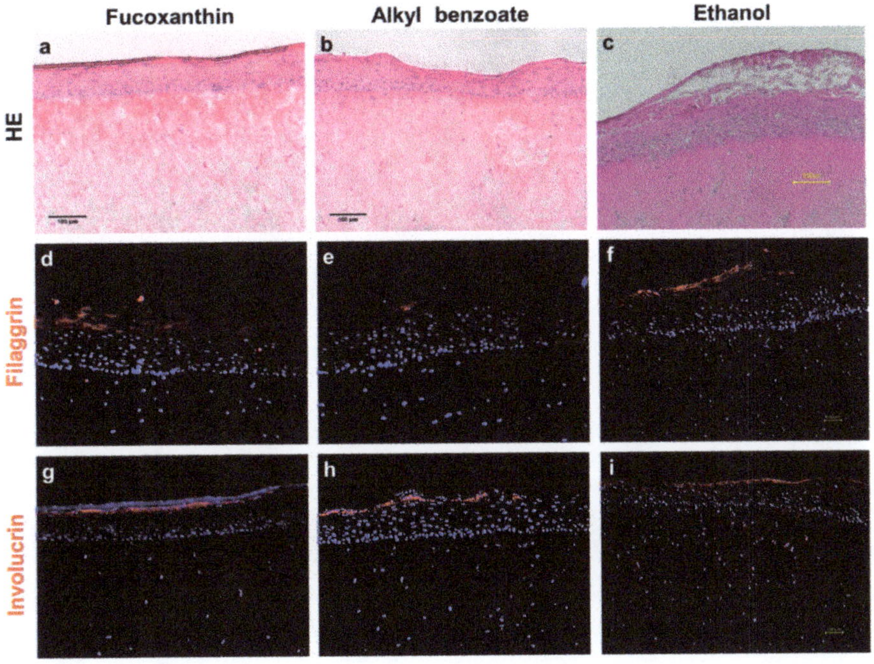

Figure 2. Morphology and protein expression in reconstructed human skin following fucoxanthin exposure versus the vehicles alone. (**a–c**) Hematoxylin-eosin staining with slightly disturbed morphology in c. (**d–f**) Immunolocalization of filaggrin with enhanced expression in d and f. (**g–i**) Similar immunolocalization of involucrin in all samples. Images are representative of at least three batches; nuclei in blue (DAPI); scale bars = 100 µm.

3.2. Fucoxanthin Effects on RHS Viability

Before assessing the RHS viability following substance exposure, we evaluated the interference of fucoxanthin with an MTT test. Fucoxanthin solutions as well as RHS treated with fucoxanthin were stained red, resulting in absorbance at 450 nm after isopropanol extraction although no MTT was added to these control tissues. Next, we investigated the direct reduction of MTT by fucoxanthin in solution as well as in freeze-killed tissues. Although we observed only a minor effect, we subtracted the absorbance due to fucoxanthin's color and due to a direct interaction with MTT from all absorbance values of fucoxanthin-treated RHS in the viability tests. Phosphate-buffered saline (PBS) and sodium dodecyl sulfate (SDS) served as negative and positive controls, respectively, as recommended by the OECD.

The viability of RHS following SDS exposure and 42 h post-treatment incubation period decreased to 2.6 ± 2.4%, correctly identifying SDS as skin irritant (Figure 3). Moreover, the standard deviation between tissue replicates fell far below 18%, and, thus, met the acceptability criteria of OECD test guideline no. 439.

Fucoxanthin showed a significantly higher viability than the positive control (Figure 3). Since the values exceeded also the threshold of 50%, fucoxanthin was not irritant to RHS. When testing the solvent controls, we observed a marked decrease in viability to 52.8 ± 9.0%. This tissue damage was ameliorated in the ethanolic test solution of fucoxanthin, as seen by a relative viability of 75.7%.

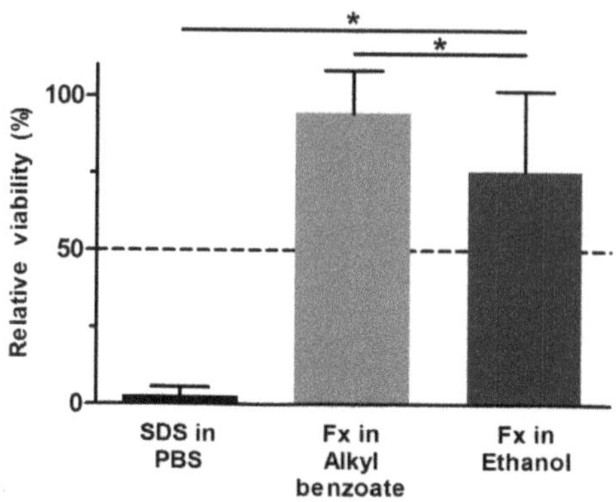

Figure 3. Fucoxanthin effects on the viability of reconstructed human skin. Test substances below the 50% threshold (dashed line) are predicted to be skin irritant. PBS—phosphate buffered saline (negative control); SDS—sodium dodecyl sulfate (positive control); Fx—fucoxanthin. Alkyl benzoate and ethanol were exposed for 15 min to the RHS and subsequently washed off. Mean + SD, $n \geq 3$, * $p \leq 0.05$ compared to SDS.

3.3. *Fucoxanthin Effects on RHS Gene Expression*

Next, we evaluated the gene expression following substance exposure and 42 h post-treatment incubation period to get a more detailed insight into the toxicity of fucoxanthin (Figure 4). SDS markedly increased the gene expression of N-acetyltransferase 1 (NAT1) as well as of pro-inflammatory genes, like interleukin (IL)-1α, 6, and 8, compared to the levels in PBS-treated RHS. Moreover, the gene expression of epidermal growth factor receptor (EGFR) and small heat and shock protein beta 1 (HSPB1) were slightly elevated.

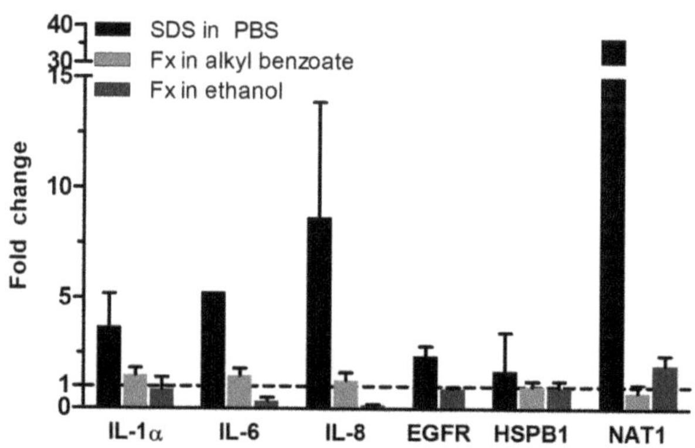

Figure 4. Fucoxanthin effects on the gene expression of reconstructed human skin. Fold change in gene expression was calculated in relation to the gene expression in reconstructed human skin exposed to the respective solvent (dashed line), SDS to PBS and Fx to alkyl benzoate or to ethanol. SDS—sodium dodecyl sulfate; PBS—phosphate-buffered saline; Fx—fucoxanthin; IL—interleukin; EGFR—epidermal growth factor receptor; HSPB1—small heat and shock protein beta 1; NAT1—N-acetyltransferase 1. Mean + SD, $n = 3$.

Fucoxanthin exposure increased none of these gene expressions. When applied in alkyl benzoate solutions, we detected almost no change in gene expression. The ethanolic solution of fucoxanthin decreased the gene expression of IL-6 and 8 compared to the gene expression in ethanol-treated RHS to

33% or 15% compared to the solvent control samples. NAT1 gene expression was doubled compared to ethanol-treated RHS.

4. Discussion

Intrigued by the marked antioxidant efficacy of fucoxanthin [15,16], we investigated the potential toxicity of fucoxanthin in two different solvents, which are frequently used in dermatological products. Our results proved fucoxanthin to be non-irritant and suggested using alkyl benzoate as a solvent for the fucoxanthin (Figure 3). Absolute ethanol reduced the viability of RHS, but was necessary to dissolve the lipophilic fucoxanthin. Both viability testing and gene expression analysis revealed the ameliorating effects of fucoxanthin on the ethanol-induced inflammation in RHS (Figures 3 and 4). Moreover, we proved the applicability of our novel 24-well-sized RHS, being fully compatible with organ-on-a-chip applications (Figures 1 and 2). Using RHS extends the approach of OECD test guideline no. 439 and becomes mandatory when investigating biochemical pathways due to the known epidermal-dermal cross-talk in normal and diseased skin models [25,26].

The strong clinical need for skin protection is emphasized by the extraordinary increase of skin cancer patients. In Brazil, non-melanoma skin cancer accounts for 30% of diagnosed cancers [27], while in Australia, more people have been diagnosed with skin cancer than all other cancers combined [28]. The underlying biochemical mechanisms clearly provided the correlation between the cumulative exposure to UV radiation and skin cancer [3] and the contribution of VIS and IR radiation and air pollutants to cutaneous carcinogenesis has been described [6,10] as well. Aging accelerates a vicious circle of cumulative damage to the skin, reduced amounts of antioxidants among other age-associated conditions, like xerosis, impaired skin barrier, and wound healing, promoting the penetration of more pollutants into the skin [29]. The correlation of an additional 10 μg/m^3 NO_2 in the air with 25% more pigment spots on female cheeks [30] shows the cosmetic, and may indicate a medical, need for the prevention from extrinsic factors. However, protection from UV, or even from the entire spectrum of solar radiation cannot prevent from its detrimental effects, since nitric oxide, volatile organic compounds, and particulate matter also contributes to skin aging, inflammation, and cancer [5]. Thus, the strategy against skin aging must include a multitude of different approaches, including cleansing products to reduce the particle load on skin, agents that strengthen the skin barrier function, products that protect from sunlight, anti-inflammatory agents, and antioxidants [31].

Antioxidants from natural sources, like fucoxanthin, address these clinical and cosmetic needs due to the antioxidant properties. Herein, we exclusively investigated the effects of the commercially available fucoxanthin and not multi-compound algae extracts. Nevertheless, fucoxanthin needs to accumulate in sufficient amounts at the target site. The lipophilicity (log*P* 7.8) of fucoxanthin, as well as the accumulation of fucoxanthin and its degradation products in murine adipose tissue, liver, lungs, kidney, heart, and spleen [17], questions the efficacy of orally-administered fucoxanthin in skin. Thus, we applied the fucoxanthin solution topically to RHS and selected 0.5% as concentration, being in the range used for antioxidants in cosmetic formulations (0.01–1%). We dissolved the highly lipophilic fucoxanthin in alkyl benzoate, being devoid of genotoxic properties and frequently used in cosmetics as a solvent, emollient, preservative, and plasticizer [32]. Moreover, alkyl benzoate is already used to solubilize UV filters in sunscreens [18]. For comparison, we also included an ethanolic solution based on the recommendations for poorly water-soluble substances in skin irritation testing [33]. Moreover, the high ethanol concentration was required to dissolve fucoxanthin and allowed us to investigate the potentially ameliorating effects of this antioxidant.

Even the small amount of 10 μL ethanol solution per tissue, equal to 17 μL/cm^2, disturbed the RHS morphology (Figure 2c), viability (Figure 3), and altered the gene expression (Figure 4). This is well in accordance with previous observations, where high concentrations were necessary for drug dissolution, as well [34]; more than 80% of cell death occurred even following the exposure to 50% ethanol in skin models [35]. The mechanism of tissue damage by ethanol is related to oxidative stress and well-known from the oral cavity [36]. Ethanol directly reacts to hydroxyethyl radicals and subsequently contributes to the formation of other ROS species. Chronic ethanol exposure causes CYP induction, mitochondrial damage, reduced antioxidant defense mechanisms, and thereby amplifies ROS-related tissue damage. Meanwhile, ethanol is also most effective in increasing the skin absorption of lipophilic drugs, like butenafine (log*P* 6.6) [37]. Assuming that ethanol also effectively increased the penetration of fucoxanthin (log*P* 7.8), fucoxanthin can counteract the pro-oxidative effects of ethanol. Thus, both ethanol and fucoxanthin toxicity data in our study are in accordance with previously-published results on polyphenols in grapes and red wine, which are suggested to protect from ethanol damage [38]. However, the efficacy of fucoxanthin will depend on the fucoxanthin concentrations within the tissue, as observed for other antioxidants.

The change in IL-6 and 8 gene expressions following test substance exposure correlates to the expression of IL-1α, HSPB1, and EGFR. IL-1α activates the p38 MAPK pathway to increase the expression of HSPB1, which causes anti-apoptotic effects, possesses chaperone-like activity and refolds denatured proteins, and is cytoprotective against heat shock [35]. EGFR also increases the expression of IL-8 [39] and plays an essential role in re-epithelialization by increasing keratinocyte proliferation and cell migration in wounded skin [40]. The constant gene expressions of EGFR and HSPB1 substantiate the absence of damage from fucoxanthin, but does not explain the reduction in IL-6 and IL-8 gene expression.

HSPB1 gene expression remained stable, even following the exposure to SDS. Although HSPB1 exerts protective functions under stress conditions [41], previous results with increased HSPB1 expressions were obtained 24 h after SDS treatment by Western blotting. Thus, the 42 h period of incubation was recommended for skin irritation testing [20], yet probably too long to detect maximum increase in HSPB1 gene expression.

Finally, we investigated the expression of NAT1 in RHS, since fucoxanthin is totally deacetylated in the intestinal lumen (for review see [14]). We observed an upregulation of NAT1 following the application of fucoxanthin in ethanol (Figure 4), indicating an activation of cutaneous metabolism, being relevant when applying drugs or cosmetic actives to the skin.

Instead of using murine models [42], we developed a full-thickness RHS to assess the potential toxic effects of fucoxanthin. Accumulating insights into the predictive power of animal-based tests in toxicology [43] emphasize the need for human cell-based models. Nevertheless, the skin model in this study consisted of primary juvenile cells and did not consider age-related changes in skin function. Future studies should investigate the efficacy of fucoxanthin in aged RHS [44], thereby taking age-related differences between young, middle-aged, and senior patients into account. Moreover, emerging markets for cosmetics demand the use of human cell-based test systems. For example, Brazil banned cosmetic products from being tested in animal models in 2019, but the Brazilian legislation gap in the use of biological material of human origin, which until recently impeded access to commercially-available skin models [45]. Thus, in-house or open-source protocols are urgently needed, but also need to be validated for their intended use. In the present study, we presented a fast (10-day culture) RHS protocol at reasonable price (3.7×10^5 keratinocytes per construct), which is fully compatible to organ-on-a-chip applications. Multi-organ-chips provide the opportunity to study substance effects in an interconnected and perfused in vitro model, aiming to include the influence of liver function into substance evaluation [46].

5. Conclusions

Taken together, our study clearly demonstrated the non-irritancy of fucoxanthin. Fucoxanthin ameliorated detrimental effects of ethanol on tissue viability and inflammatory response. We observed metabolism activation by NAT1 upregulation, but no change in HSPB1 regulation. Finally, we proved the applicability of our novel, organ-on-a-chip compatible RHS for the evaluation of substance effects.

Author Contributions: Conceptualization: R.S.N.T., H.M.D., L.R.G., and C.Z.; investigation: R.S.N.T., S.S.M.-E., U.M., and C.Z.; writing—original draft preparation: R.S.N.T., M.S.-K., L.R.G., and C.Z.; writing—review and editing: R.S.N.T., L.R.G., and C.Z.; visualization: R.S.N.T. and C.Z.; supervision: L.R.G. and C.Z.; project administration: L.R.G. and C.Z.; funding acquisition: R.S.N.T., P.C., H.M.D., M.S.K., and L.R.G. All authors have read and agreed to the published version of the manuscript.

Funding: This research was funded by The São Paulo Research Foundation, FAPESP (grant numbers 2014/50928-2 and 2016/06931-4), by CNPq (grant number 465687/2014-8), and by the Bundesministerium für Bildung und Forschung (Berlin-Brandenburg Research Platform BB3R, grant number 031A262A). R.S.N.T. gratefully acknowledges a doctoral scholarship of DAAD/CAPES/CNPq 15° 2017 (grant number 290281/2017-2). We acknowledge support by the German Research Foundation and the Open Access Publication Fund of the Freie Universität Berlin.

Acknowledgments: The authors highly appreciated the excellent technical assistance of Carola Kapfer.

Conflicts of Interest: The authors declare no conflict of interest. Uwe Marx is employee of the TissUse GmbH. Neither the company nor the funding source had a role in the design of the study; in the collection, analyses, or interpretation of data; in the writing of the manuscript; or in the decision to publish the results.

References

1. Flament, F.; Bazin, R.; Laquieze, S.; Rubert, V.; Simonpietri, E.; Piot, B. Effect of the sun on visible clinical signs of aging in caucasian skin. *Clin. Cosmet. Investig. Dermatol.* **2013**, *6*, 221–232. [CrossRef]
2. Xu, F.; Yan, S.; Wu, M.; Li, F.; Xu, X.; Song, W.; Zhao, J.; Xu, J.; Kan, H. Ambient ozone pollution as a risk factor for skin disorders. *Br. J. Dermatol.* **2011**, *165*, 224–225. [CrossRef]
3. Nishisgori, C. Current concept of photocarcinogenesis. *Photochem. Photobiol. Sci.* **2015**, *14*, 1713–1721. [CrossRef] [PubMed]
4. Awad, F.; Assrawi, E.; Louvrier, C.; Jumeau, C.; Giurgea, I.; Amselem, S.; Karabina, S.A. Photoaging and skin cancer: Is the inflammasome the missing link? *Mech. Ageing Dev.* **2018**, *172*, 131–137. [CrossRef]
5. Burke, K.E. Mechanisms of aging and development—A new understanding of environmental damage to the skin and prevention with topical antioxidants. *Mech. Ageing Dev.* **2018**, *172*, 123–130. [CrossRef] [PubMed]
6. Valacchi, G.; Muresan, X.M.; Sticozzi, C.; Belmonte, G.; Pecorelli, A.; Cervellati, F.; Demaude, J.; Krol, Y.; Oresajo, C. Ozone-induced damage in 3d-skin model is prevented by topical vitamin c and vitamin e compound mixtures application. *J. Dermatol. Sci.* **2016**, *82*, 209–212. [CrossRef] [PubMed]
7. Ulrich, C.; Jürgensen, J.S.; Degen, A.; Hackethal, M.; Ulrich, M.; Patel, M.J.; Eberle, J.; Terhorst, D.; Sterry, W.; Stockfleth, E. Prevention of non-melanoma skin cancer in organ transplant patients by regular use of a sunscreen: A 24 months, prospective, case–control study. *Br. J. Dermatol.* **2009**, *161*, 78–84. [CrossRef] [PubMed]
8. Heerfordt, I.M.; Torsnes, L.R.; Philipsen, P.A.; Wulf, H.C. Sunscreen use optimized by two consecutive applications. *PLoS ONE* **2018**, *13*, e0193916. [CrossRef]
9. Autier, P. Sunscreen abuse for intentional sun exposure. *Br. J. Dermatol.* **2009**, *161* (Suppl. 3), 40–45. [CrossRef]
10. Lohan, S.B.; Muller, R.; Albrecht, S.; Mink, K.; Tscherch, K.; Ismaeel, F.; Lademann, J.; Rohn, S.; Meinke, M.C. Free radicals induced by sunlight in different spectral regions—In vivo versus ex vivo study. *Exp. Dermatol.* **2016**, *25*, 380–385. [CrossRef]
11. Thiele, J.J.; Schroeter, C.; Hsieh, S.N.; Podda, M.; Packer, L. The antioxidant network of the stratum corneum. *Curr. Probl. Dermatol.* **2001**, *29*, 26–42. [CrossRef] [PubMed]
12. Rhie, G.; Shin, M.H.; Seo, J.Y.; Choi, W.W.; Cho, K.H.; Kim, K.H.; Park, K.C.; Eun, H.C.; Chung, J.H. Aging- and photoaging-dependent changes of enzymic and nonenzymic antioxidants in the epidermis and dermis of human skin in vivo. *J. Investig. Dermatol.* **2001**, *117*, 1212–1217. [CrossRef] [PubMed]
13. Perez-Sanchez, A.; Barrajon-Catalan, E.; Herranz-Lopez, M.; Micol, V. Nutraceuticals for skin care: A comprehensive review of human clinical studies. *Nutrients* **2018**, *10*, 403. [CrossRef] [PubMed]
14. Peng, J.; Yuan, J.-P.; Wu, C.-F.; Wang, J.-H. Fucoxanthin, a marine carotenoid present in brown seaweeds and diatoms: Metabolism and bioactivities relevant to human health. *Mar. Drugs* **2011**, *9*, 1806–1828. [CrossRef] [PubMed]
15. Sachindra, N.M.; Sato, E.; Maeda, H.; Hosokawa, M.; Niwano, Y.; Kohno, M.; Miyashita, K. Radical scavenging and singlet oxygen quenching activity of marine carotenoid fucoxanthin and its metabolites. *J. Agric. Food Chem.* **2007**, *55*, 8516–8522. [CrossRef] [PubMed]

16. Murakami, A.; Nakashima, M.; Koshiba, T.; Maoka, T.; Nishino, H.; Yano, M.; Sumida, T.; Kim, O.K.; Koshimizu, K.; Ohigashi, H. Modifying effects of carotenoids on superoxide and nitric oxide generation from stimulated leukocytes. *Cancer Lett.* **2000**, *149*, 115–123. [CrossRef]
17. Hashimoto, T.; Ozaki, Y.; Taminato, M.; Das, S.K.; Mizuno, M.; Yoshimura, K.; Maoka, T.; Kanazawa, K. The distribution and accumulation of fucoxanthin and its metabolites after oral administration in mice. *Br. J. Nutr.* **2009**, *102*, 242–248. [CrossRef]
18. Gaspar, L.R.; Tharmann, J.; Maia Campos, P.M.; Liebsch, M. Skin phototoxicity of cosmetic formulations containing photounstable and photostable uv-filters and vitamin a palmitate. *Toxicol. In Vitro* **2013**, *27*, 418–425. [CrossRef]
19. Lane, M.E. Skin penetration enhancers. *Int. J. Pharm.* **2013**, *447*, 12–21. [CrossRef]
20. OECD 439. Test no. 439: In Vitro Skin Irritation: Reconstructed Human Epidermis Test Method. In *Oecd Guidelines for the Testing of Chemicals, Section 4: Health Effects*. 2019. Available online: https://doi.org/10.1787/9789264242845-en (accessed on 12 January 2020).
21. Kandárová, H.; Hayden, P.; Klausner, M.; Kubilus, J.; Sheasgreen, J. An in vitro skin irritation test (sit) using the epiderm reconstructed human epidermal (rhe) model. *JoVE* **2009**, e1366. [CrossRef]
22. Rueden, C.T.; Schindelin, J.; Hiner, M.C.; DeZonia, B.E.; Walter, A.E.; Arena, E.T.; Eliceiri, K.W. Imagej2: Imagej for the next generation of scientific image data. *BMC Bioinform.* **2017**, *18*, 529. [CrossRef] [PubMed]
23. Hübner, J.; Raschke, M.; Rutschle, I.; Grassle, S.; Hasenberg, T.; Schirrmann, K.; Lorenz, A.; Schnurre, S.; Lauster, R.; Maschmeyer, I.; et al. Simultaneous evaluation of anti-egfr-induced tumour and adverse skin effects in a microfluidic human 3d co-culture model. *Sci. Rep.* **2018**, *8*, 15010. [CrossRef] [PubMed]
24. Kent, W.J.; Sugnet, C.W.; Furey, T.S.; Roskin, K.M.; Pringle, T.H.; Zahler, A.M.; Haussler, D. The human genome browser at ucsc. *Genome Res.* **2002**, *12*, 996–1006. [CrossRef]
25. Maas-Szabowski, N.; Stark, H.J.; Fusenig, N.E. Keratinocyte growth regulation in defined organotypic cultures through il-1-induced keratinocyte growth factor expression in resting fibroblasts. *J. Investig. Dermatol.* **2000**, *114*, 1075–1084. [CrossRef]
26. Berroth, A.; Kühnl, J.; Kurschat, N.; Schwarz, A.; Stäb, F.; Schwarz, T.; Wenck, H.; Fölster-Holst, R.; Neufang, G. Role of fibroblasts in the pathogenesis of atopic dermatitis. *J. Allergy Clin. Immun.* **2013**, *131*, 1547–1554.e1546. [CrossRef]
27. The Brazilian National Cancer Institute. Non-Melanoma Skin Cancer. In *Cancer Types*; 2008. Available online: https://www.inca.gov.br/tipos-de-cancer/cancer-de-pele-nao-melanoma (accessed on 12 January 2020).
28. International Agency for Research on Cancer. *World Cancer Report 2014*; World Health Organization: Geneva, Switzerland, 2014; Available online: http://apps.who.int/bookorders/anglais/detart1.jsp?codlan=1&codcol=80&codcch=275 (accessed on 12 January 2020).
29. Parrado, C.; Mercado-Saenz, S.; Perez-Davo, A.; Gilaberte, Y.; Gonzalez, S.; Juarranz, A. Environmental stressors on skin aging. Mechanistic insights. *Front. Pharmacol.* **2019**, *10*, 759. [CrossRef]
30. Hüls, A.; Vierkotter, A.; Gao, W.; Kramer, U.; Yang, Y.; Ding, A.; Stolz, S.; Matsui, M.; Kan, H.; Wang, S.; et al. Traffic-related air pollution contributes to development of facial lentigines: Further epidemiological evidence from caucasians and asians. *J. Investig. Dermatol.* **2016**, *136*, 1053–1056. [CrossRef]
31. Mistry, N. Guidelines for formulating anti-pollution products. *Cosmetics* **2017**, *4*, 57. [CrossRef]
32. Becker, L.C.; Bergfeld, W.F.; Belsito, D.V.; Hill, R.A.; Klaassen, C.D.; Liebler, D.; Marks, J.G., Jr.; Shank, R.C.; Slaga, T.J.; Snyder, P.W.; et al. Safety assessment of alkyl benzoates as used in cosmetics. *Int. J. Toxicol.* **2012**, *31*, 342s–372s. [CrossRef]
33. Kandárová, H.; Bendova, H.; Letasiova, S.; Coleman, K.P.; De Jong, W.H.; Jirova, D. Evaluation of the medical devices benchmark materials in the controlled human patch testing and in the rhe in vitro skin irritation protocol. *Toxicol. In Vitro* **2018**, *50*, 433–438. [CrossRef]
34. Wanjiku, B.; Yamamoto, K.; Klossek, A.; Schumacher, F.; Pischon, H.; Mundhenk, L.; Rancan, F.; Judd, M.M.; Ahmed, M.; Zoschke, C.; et al. Qualifying X-ray and stimulated raman spectromicroscopy for mapping cutaneous drug penetration. *Anal. Chem.* **2019**, *91*, 7208–7214. [CrossRef] [PubMed]
35. Li, L.N.; Margolis, L.B.; Hoffman, R.M. Skin toxicity determined in vitro by three-dimensional, native-state histoculture. *Proc. Natl. Acad. Sci. USA* **1991**, *88*, 1908–1912. [CrossRef] [PubMed]
36. Zukowski, P.; Maciejczyk, M.; Waszkiel, D. Sources of free radicals and oxidative stress in the oral cavity. *Arch. Oral Biol.* **2018**, *92*, 8–17. [CrossRef] [PubMed]

37. Zhang, A.; Jung, E.C.; Zhu, H.; Zou, Y.; Hui, X.; Maibach, H. Vehicle effects on human stratum corneum absorption and skin penetration. *Toxicol. Ind. Health* **2017**, *33*, 416–425. [CrossRef] [PubMed]
38. Sun, A.Y.; Ingelman-Sundberg, M.; Neve, E.; Matsumoto, H.; Nishitani, Y.; Minowa, Y.; Fukui, Y.; Bailey, S.M.; Patel, V.B.; Cunningham, C.C.; et al. Ethanol and oxidative stress. *Alcohol. Clin. Exp. Res.* **2001**, *25*, 237s–243s. [CrossRef]
39. Frankart, A.; Coquette, A.; Schroeder, K.R.; Poumay, Y. Studies of cell signaling in a reconstructed human epidermis exposed to sensitizers: Il-8 synthesis and release depend on egfr activation. *Arch. Dermatol. Res.* **2012**, *304*, 289–303. [CrossRef]
40. Barrientos, S.; Stojadinovic, O.; Golinko, M.S.; Brem, H.; Tomic-Canic, M. Perspective article: Growth factors and cytokines in wound healing. *Wound Rep. Reg.* **2008**, *16*, 585–601. [CrossRef]
41. Verschuure, P.; Tatard, C.; Boelens, W.C.; Grongnet, J.F.; David, J.C. Expression of small heat shock proteins hspb2, hspb8, hsp20 and cvhsp in different tissues of the perinatal developing pig. *Eur. J. Cell Biol.* **2003**, *82*, 523–530. [CrossRef]
42. Rodríguez-Luna, A.; Ávila-Román, J.; González-Rodríguez, M.L.; Cózar, M.J.; Rabasco, A.M.; Motilva, V.; Talero, E. Fucoxanthin-containing cream prevents epidermal hyperplasia and uvb-induced skin erythema in mice. *Mar. Drugs* **2018**, *16*, 378. [CrossRef]
43. Bailey, J.; Thew, M.; Balls, M. An analysis of the use of animal models in predicting human toxicology and drug safety. *Altern. Lab. Anim.* **2014**, *42*, 181–199. [CrossRef]
44. Hausmann, C.; Vogt, A.; Kerscher, M.; Ghoreschi, K.; Schäfer-Korting, M.; Zoschke, C. Optimizing skin pharmacotherapy for older patients: The future is at hand but are we ready for it? *Drug Discov. Today* **2020**. [CrossRef] [PubMed]
45. De Vecchi, R.; Dakic, V.; Mattos, G.; Rigaudeau, A.-S.; Oliveira, V.; Garcia, C.; Alépée, N.; Bouez, C. Implementation, availability and regulatory status of an OECD accepted reconstructed human epidermis model in brazil. *Vigilância Sanitária Debate* **2018**, *6*, 64. [CrossRef]
46. Maschmeyer, I.; Hasenberg, T.; Jaenicke, A.; Lindner, M.; Lorenz, A.K.; Zech, J.; Garbe, L.A.; Sonntag, F.; Hayden, P.; Ayehunie, S.; et al. Chip-based human liver-intestine and liver-skin co-cultures—A first step toward systemic repeated dose substance testing in vitro. *Eur. J. Pharm. Biopharm.* **2015**, *95*, 77–87. [CrossRef] [PubMed]

pharmaceutics

MDPI

Article

Amphiphilic Polypeptides for VEGF siRNA Delivery into Retinal Epithelial Cells

Olga Osipova [1], Vladimir Sharoyko [1], Natalia Zashikhina [1,2], Natalya Zakharova [2], Tatiana Tennikova [1], Arto Urtti [1,3,4] and Evgenia Korzhikova-Vlakh [1,2,*]

1 Institute of Chemistry, Saint Petersburg State University, Universitetskii pr. 26, 198584 Saint Petersburg, Russia; olya_osipova_06_01@mail.ru (O.O.); sharoyko@gmail.com (V.S.); nzashihina@bk.ru (N.Z.); tennikova@mail.ru (T.T.); arto.urtti@uef.fi (A.U.)

2 Institute of Macromolecular Compounds, Russian Academy of Sciences, Bolshoy pr. 31, 199004 Saint-Petersburg, Russia; na_zar@inbox.ru

3 School of Pharmacy, University of Eastern Finland, Yliopistonranta 1 C, 70210 Kuopio, Finland

4 Faculty of Pharmacy, University of Helsinki, Viikinkaari 5 E, 00790 Helsinki, Finland

* Correspondence: vlakh@mail.ru; Tel.: +7-812-323-0461

Received: 27 November 2019; Accepted: 31 December 2019; Published: 2 January 2020

check for updates

Abstract: Polyethyleneimine, poly-L-lysine, chitosan and some others cationic polymers have been thoroughly studied as nucleic acid delivery systems in gene therapy. However, the drug release from these systems proceeds at a very low rate due to extremely high binding between a carrier and gene material. To reduce these interactions and to enhance drug release, we developed a set of amphiphilic polypeptides containing positively and negatively charged amino acids as well as a hydrophobic one. The copolymers obtained were characterized by size-exclusion chromatography, static light scattering, HPLC amino acid analysis and ^{1}HNMR spectroscopy. All copolymers formed particles due to a self-assembly in aqueous media. Depending on polypeptide composition, the formation of particles with hydrodynamic diameters from 180 to 900 nm was observed. Stability of polymer particles, loading and release efficiency were carefully studied. Cellular uptake of the particles was efficient and their cytotoxicity was negligible. The application of polymer carriers, containing siRNA, to vascular endothelial growth factor (VEGF-A165) silencing of ARPE-19 cells was successful. The gene silencing was confirmed by suppression of both messenger RNA and protein expression.

Keywords: amphiphilic polypeptides; self-assembly; nanoparticles; siRNA delivery; VEGF; gene silencing

1. Introduction

Diabetic retinopathy (DR) and age-related macular degeneration (AMD) are among the most common ocular diseases that lead to the impairment or loss of vision [1,2]. Proliferative neovascular forms of DR and AMD are characterized by the formation of new vessels from the existing ones, because the proliferation and migration of endothelial cells are stimulated by over-expression of vascular endothelial growth factor (VEGF) [3]. Inhibition of VEGF via intraocular injections of anti-VEGF drugs are the corner-stone approach in the treatment of AMD and diabetic macular edema that is associated with DR. Currently, a common clinical therapy is based on application of the monoclonal antibodies (mABs) that specifically bind to VEGF, and neutralize its function. Such recombinant monoclonal antibodies as bevacizumab and its Fab-fragment ranibizumab bind with high affinity to the site present in all VEGF isoforms [4,5]. Besides antibodies, there is also the aptamer pegaptanib (28-nucleotide RNA aptamer) that was registered for the treatment of neovascular AMD fifteen years ago [4,5]. In spite of the existing clinical therapeutics, some drawbacks still exist in the treatment of neovascular AMD. Monthly or bimonthly injections into the vitreous of the eye constitute a major burden to the patients and healthcare. Furthermore, the intravitreal injections may lead to some complications, such as retinal detachment, intraocular pressure increase, ocular infection, and hyperemia.

In contrast to the mABs and aptamers, small-interfering RNAs (siRNAs) do not block protein function but play a key role in the post-transcriptional gene silencing [6]. siRNAs have a great potential as drugs, including the treatment of ocular diseases. For instance, phase II clinical trials of ocular siRNA treatments were reported [7]. These trials by the Quark and Pfizer companies aim at developing treatments for wet AMD, glaucoma and some other ocular diseases.

To inhibit intracellular VEGF synthesis, usually picomole quantities of siRNA are necessary [8]. Moreover, siRNAs provide a long therapeutic effect with a single administration and thereby may reduce the risks and burden associated with intravitreal anti-VEGF therapy. Another benefit of siRNA is the simple synthesis and possibility to up-scale the production. However, the successful application of siRNAs is limited by the necessity to overcome several biological barriers [9]. The first one is the penetration across the cell membrane. The physicochemical properties of siRNAs, such as hydrophilicity, relatively large size and net negative charge, complicate their entrance to the cell. After cellular internalization, siRNA must escape the endosomes and reach the cytosol, where it must be involved into formation of the RNA-inducing silencing protein complex (RISC) and recognize the target mRNA. Furthermore, siRNA is subject to enzymatic degradation, but its stability can be improved with chemical modifications (e.g., phosphorothioate) and formulation techniques.

To overcome the obstacles, different siRNA delivery systems have been investigated. The carriers for siRNA delivery should meet some requirements. They must be biocompatible, biodegradable and they must have suitable combination of chemical-physical properties, including charge, size, stability, and possibility to surface functionalization. Currently, the encapsulation of the siRNAs into nanoparticles is based on lipids [10,11] or polymers [12,13], or their complexation with cell-penetrating peptides [14,15] or cationic-polymers [16–19]. Such technologies have been summarized in some recent reviews [20,21]. Self-assembled into micelles or liposomes, the cationic lipids can entrap siRNA and facilitate its cellular entry by endocytosis. However, a poor stability of such systems has induced a wave of studies on their covalent and non-covalent modification with PEG or its copolymers [22,23] The preparation of complex delivery systems, for example, consisting of siRNA-cationic lipid complex encapsulated into folate-PEG-co-(polylactic acid-co-polyglycolic acid)-polyketal nanoparticles [12] or based on poly(ε-caprolactone)-co-PEG-co-poly(L-histidine) [13] were recently reported.

Protonable cationic polymers are among the leading candidates for nucleic acid binding and intracellular delivery. They are characterized with high buffering capacity at endosomal pH range (about 5.5–7.4) and it is hypothesized that they can mediate endosomal escape by acting as 'proton sponges'. Polyethyleneimine (PEI), poly-L-lysine, chitosan and its derivatives are the most widely studied cationic polymers for nucleic acid delivery [17,19,24]. Despite high binding and cellular transfection efficiency of many polymers, they are not biodegradable (e.g., PEI, chitosan) or form stable polyplexes with limited release of nucleic acids (e.g., poly-L-lysine). To facilitate the release of nucleic acid from polyplex, recently, Pilipenko et al. suggested the introduction of heparin as concurrent polyanion to chitosan [25]. Chitosan-heparin polyplexes with different ratios of positive and negative polysaccharides were prepared and their efficiency for delivery of DNA and siRNA into human keratinocytes was estimated.

Following that idea, in present work, we developed a set of random amphiphilic polypeptides consisting of positively and negatively charged amino acids as well as a hydrophobic one. L-Lysine and L-glutamic acid were selected as oppositely charged amino acids to control siRNA binding and release. L-Phenylalanine or L-isoleucine provided polypeptide self-assembly and possibly facilitated delivery of siRNA across the cell membrane. The ratio between the three amino acids was varied to find the composition of polypeptides that provides optimal particle size, zeta potential, encapsulation efficiency, cytotoxicity and VEGF-A165 silencing in retinal pigment epithelial cells.

2. Materials and Methods

2.1. Materials

ε-Z-L-Lysine, γ-benzyl-L-glutamate, L-phenylalanine, L-isoleucine, triphosgene, α-pinene, trifluoroacetic acid (TFA), trifluoromethanesulfonic acid (TFMSA), and other reagents for NCA and polymer synthesis were ordered in Sigma-Aldrich (Darmstadt, Germany). All organic solvents, i.e., *N,N*-dimethylformamide (DMF), dimethyl sulfoxide (DMSO) 1,4-dioxane, petroleum ether, ethyl acetate, and some others were purchased from Vecton Ltd. (St. Petersburg, Russia), and purified before use. The buffer solutions were prepared by dissolving salts of ACS reagent grade purchased from Vecton Ltd. (St. Petersburg, Russia) in deionized water and afterwards the solutions were filtered through a 0.45-μm membrane Milex, Millipore Merck (Darmstadt, Germany). For purification of synthesized polymers, Spectra/Pore® dialysis bags (MWCO: 1000, Rancho Dominguez, CA, USA) were used. Filter tubes (30,000) were ordered in Merck and in Sartorius (Göttingen, Germany).

SEC column calibration was performed with the use of poly(methyl methacrylate) (PMMA) standards (M_w = 17,000–250,000; Đ ≤ 1.14) purchased from Supelco (Bellefonte, PA, USA). Poly-L-lysine (M_w = 15,000–30,000) was a product of Sigma-Aldrich.

Model non-labeled and Cy3-labeled 23-base pairs duplex of oligothymidine and oligoadenine (oligo-dT-dA) were purchased from Biobeagle™ (St. Petersburg, Russia). Dulbecco's modified Eagle's medium (DMEM), penicillin, streptomycin and fetal bovine serum (FBS) were obtained from BioloT (St. Petersburg, Russia).

The 27-base pairs (bp) double stranded RNA were selected as target to VEGF-A165 gene. siRNA sequence was the following: Sense—5′-CUUCCUACAGCACAACAAAUGUGAAUG-3′, antisense: 3′-GAAGGAUGUCGUGUUGUUUACACUUAC-5′. The same siRNA labelled with Cy5 was utilized for the visualization experiments. Cy5-labeled and non-labeled 27-bp VEGF siRNAs and scrambled 27-bp RNA for control (siControl) (sense 5′-GUAAGUGUAAACAACACGACAUCCUUC-3′, antisense: 3′-CAUUCACAUUUGUUGUGCUGUAGGAAG-5′ [21] were purchased from GenTerra (Moscow, Russia). The primers used for the target mRNA: VEGF forward and reverse primers, GAPDH forward and reverse primers were obtained from GenTerra.

Human retinal pigment epithelial (ARPE-19) cell line was a product of American Type Culture Collection (ATCC, Manassas, VA, USA), while HEK-293 (human embryonic kidney) and BEAS-2B (human bronchial epithelium) cell lines were obtained from the German Collection of Microorganisms and Cell Culture (DSMZ, Braunschweig, Germany).

2.2. Methods

2.2.1. Synthesis and Characterization of Polypeptides

Polypeptides were synthesized by ring-opening polymerization (ROP) of α-amino acid *N*-carboxyanhydrides (NCA) as random structures. NCA monomers of Lys(Z), Glu(OBzl), Phe and Ile were prepared as described elsewhere [26]. In all cases, anhydrous dioxane was used as a solvent. Acquired NCAs were purified by recrystallization twice from anhydrous ethyl acetate/*n*-hexane. Yields: Lys(Z) NCA—87%, Glu(OBzl) NCA—89%, Phe NCA—74%, Ile NCA—69%.

The structure and purity of NCAs were testified by ^{1}H-NMR at 25 °C in $CDCl_3$. The spectra were recorded using a 400 MHz Avance instrument (Bruker, Karlsruhe, Germany). Lys(Z) NCA: δ 7.43–7.28 (m, 5H), 6.97 (s, 1H), 5.12 (s, 2H), 4.97 (s, 1H), 4.32–4.23 (t, J = 5.2, 1H) (s, 1H), 3.29–3.14 (m, 2H), 2.03–1.90 (m, 1H), 1.90–1.75 (m, 1H), 1.73–1.28 (m, 4H); Glu(OBzl) NCA: 2.05–2.39 (m, 2H), 2.63 (t, 2H), 4.39 (t, 1H), 5.17 (s, 2H), 6.40 (br. s., 1H), 7.39 (m, 5H); Phe NCA: 2.94–3.35 (m, 2H), 4.55 (m, 1H), 6.12 (s, 1H), 7.19–7.41 (m, 5H); Ile NCA: 0.836 (t, 3H), 0.871 (d, 3H), 1.236 (dq, 2H), 1.941 (qtd, 1H), 4.28 (d, 1H).

Two series of polypeptides, e.g., P(Lys(Z)-*co*-Glu(OBzl)-*co*-Phe) and P(Lys(Z)-*co*-Glu(OBzl)-*co*-Ile), were synthesized using *n*-hexylamine as initiator. The NCA/initiator molar ratio was 100. The following initial ratio of NKAs were used for the synthesis: (1) Lys(Z)/Glu(OBzl)/Phe = 60/20/20 (KEF1), 40/40/20 (KEF2) and 20/60/20 (KEF3); (2) Lys(Z)/Glu(OBzl)/Ile = 60/10/30 (KEI1), 50/20/30 (KEI2) and 40/30/30 (KEI3). The polymerization was performed in anhydrous 1,4-dioxane, preparing 4 wt% solution of NCAs. Then α-pinene and triphosgene were added. The reaction was carried out at 55 °C over 4 h. The product was precipitated with an excess of diethyl ether, then the precipitate was filtrated, washed with diethyl ether and then dried.

Molecular-weight characteristics (weight average and number average molecular weights, M_w and M_n, respectively, as well as dispersity, Đ) of polymers were evaluated by SEC. Shimadzu LC-20 Prominence system supplied with refractometric RID 10-A detector (Kyoto, Japan) and 7.5 mm × 300 mm Agilent PLgel MIXED-D column (Chrom Tech, Apple Valley, MN, USA) were applied for SEC analysis. DMF with 0.1 M LiBr was used as a mobile phase. The analysis was performed at 40 °C under 1.0 mL/min of the mobile phase flow rate. SEC LC Solutions software (Shimadzu, Kyoto, Japan) was used for calculations of M_w, M_n and Đ regarding to the calibration curve plotted for PMMA standards.

Additionally, the molecular weights and hydrodynamic radius $R_{h\text{-}D}$ for macromolecules were measured by static and dynamic light scattering methods in solutions in DMSO at 21.0 °C. Light scattering was studied on a Photocor Complex unit (Photocor Instruments, Moscow, Russia); a Photocor-DL diode laser served as a light source (power of 5–30 mW, wavelength λ = 659.1 nm). The unit was calibrated by benzene ($R_V = 2.32 \times 10^{-5}$ cm^{-1}). The correlation function of the scattered light intensity was obtained with the use of a Photocor-PC2 correlator with 288 channels and was processed with DynalS software. In these solutions, the asymmetry of light scattering was absent; thus, M_w of copolymers was determined by the Debay method. The refractive index increments were measured on a Refractometer RA-620 (KEM, Kyoto, Japan).

The Bzl- and Z-protective groups were removed by TFMSA/TFA mixture at a ratio 1/10 using known procedure [27]. After deprotection, the product was dissolved in DMF, placed into a dialysis bag (MWCO 1000), and dialyzed against water for 36 h.

The amino acid compositions of the polymers were determined using HPLC amino acid analysis after total hydrolysis of polypeptides up to free amino acids. The hydrolysis of 1 mg of a polypeptide dissolved in 1 mL of 6 M HCl with 0.0001% phenol was carried out in a sealed ampoule for 48 h at 110 °C. The solvent was evaporated several times with water to eliminate HCl and to reach neutral pH. The hydrolysates were analyzed using LC-20 Shimadzu system with refractometric RID-20A detector (all from Shimadzu, Kyoto, Japan) equipped with 4.6 × 125 mm Shodex IC YS-50 column, 5 μm beads (Showa Denko, Kyoto, Japan). The isocratic elution mode was applied and 3 mM H_3PO_4 solution was used as eluent. The mobile phase flow rate was 1.0 mL/min.

2.2.2. Preparation and Characterization of Polypeptide Particles

Polymer nanoparticles were obtained by phase inversion during dialysis from DMSO to water followed by freeze drying for 2 days and finally dispersing for 30 s under sonication using 10% power of ultrasonic homogenizer (Bandelin Sonopuls HD2070, Berlin, Germany) at necessary concentration (0.20–1.50 mg/mL) in water or buffer of choice.

Average hydrodynamic diameter of particles (D_H) and polydispersity index (PDI) were established at 25 °C by dynamic light scattering (DLS) Zetasizer Nano ZS equipped with a He–Ne laser beam at λ = 633 nm and a scattering angle of 173° (Malvern Instruments, Worcester, UK). DLS experiments were performed in deionized water, DMEM and PBS, pH 7.4, whereas the zeta-potentials were determined in deionized water. To study the pH effect, ζ-potential was measured in deonized water containing 10^{-3} M NaCl and adjusted with 0.1 M HCl/NaOH to pH 3–12.

The morphology was investigated by transmission electronic microscopy (TEM) using a JEM-2100 microscope (Jeol, Tokyo, Japan) operated at an acceleration voltage of 160 kV. The samples were prepared by dropping 3 μL of particles' suspension (0.3 mg/mL) on copper grid covered with carbon and further staining with 3 *w/w*% uranyl acetate solution for 1 min. The grids were washed gently with pure water and dried for 30 min before the measurement.

Critical micelle concentration (CMC) was determined by conductometry. The measurements of conductivity were performed at 25 °C in the range of concentrations 0.5–35 μg/mL with the use of SevenCompact Cond meter S230 conductometer (Mettler Toledo, Columbus, OH, USA). CMC was determined as the intersection point of two linear sections of the conductivity vs. polymer concentration plots.

2.2.3. Encapsulation and Release of RNA, Duplex Oligo-dT-dA and siRNA

Due to the presence of ε-amino groups of lysine in polypeptides obtained, negative oligo- and polynucleotides can easily bind to them due to the ionic interactions. Firstly, the suspension of polypeptide particles with concentration of 1 mg/mL was prepared and then diluted to a desired concentration. Then, the solution of oligonucleotide (0.1 nmol/μL) or siRNA (0.05 nmol/μL) was added quickly to suspension under stirring (Vortex, Thermo Fischer Scientific, Vantaa, Finland).

To determine the loading efficiency of polypeptide nanoparticles regarding to oligonucleotides we used Cy3-oligo-dT-dA. After encapsulation, non-bound Cy3-oligo-dT-dA was separated from nanoparticles by centrifugation (10,000× *g*) at 4 °C for 20 min in Amicon Ultra filter tubes with MWCO 30,000 (Merck). The filtrate containing free Cy3-oligo-dT-dA was collected and then analyzed using a Thermo Scientific Varioscan microplate reader at excitation and emission wavelengths of 550 and 570 nm, respectively. The amount of free Cy3-oligo-dT-dA was determined using a linear calibration plot and then the amount of loaded Cy3-oligo-dT-dA was calculated as a difference between initial and non-bound oligonucleotide amounts. The entrapment efficacy *(EE)* was calculated using the following equation:

$$EE = (m1 - m2)/m1 \times 100\% \quad (1)$$

where $m1$ is initial amount of Cy3-oligo-dT-dA, but $m2$ is the amount of non-bound Cy3-oligo-dT-dA.

In vitro release of Cy3-oligo-dT-dA from nanoparticles was investigated over 5 days. 100 μL of test formulation was diluted with 300 μL of the release medium (0.01 M Na-phosphate buffer, pH 7.4, cell culture medium DMEM or DMEM-F12 with 10% (*v/v*) fetal calf serum, FCS) and placed in Eppendorf filter tubes, which were incubated in a thermoshaker at 37 °C and stirring for 300 rpm.

After a certain period, the tubes were centrifuged at 10,000× *g* for 10 min. The filtrates that contained free Cy3-oligo-dT-dA were collected and fluorescence of Cy3 was analyzed using a fluorometer (λ_{ex} = 550 nm, λ_{em} = 570 nm). The amount of free Cy3-oligo-dT-dA was calculated using a linear calibration plot.

2.2.4. Cytotoxicity of Particles

ARPE-19, HEK-293 and BEAS-2B were used for cytotoxicity evaluation. 10^4 cells per well were cultured in a 96-well plate (200 μL/well) in DMEM-F12 containing 10% (*v/v*) fetal calf serum and 1% (*v/v*) penicillin/streptomycin (72 h, 37 °C, a humidified atmosphere of 5% CO_2). After that, the cells were cultivated in the culture medium containing test empty polypeptide particles at the concentrations from 3 to 125 μg/mL for 24 and 72 h. The viability was evaluated using MTT-assay. For that, culture medium was aspirated and 100 μL/well of MTT solution (1.0 mg/mL in DMEM-F12) was added. The plate was incubated for 2 h. Finally, the solution was removed and 100 μL of DMSO was added to each well. After 10 min of gentle shaking, the solution absorbance was measured at 570 nm with the use of Fluoroscan Ascent microplate reader (Thermo Fisher Scientific Inc., Waltham, MA, USA). The relative cell viability (%) was calculated as following:

$$Cell\ viability = (Asample - Ablanck)/\ (Acontrol - Ablanck) \times 100\% \quad (2)$$

2.2.5. Cellular Uptake

To visualize cell penetration of nanoparticles loaded with siRNA we used polypeptide particles loaded with Cy5 labeled VEGF siRNA. Approximately 10,000 ARPE-19 cells were seeded per well in a 96-well plate (Thermo Scientific^{TM} Nunc^{TM} MicroWell^{TM}) in DMEM-F12. After 12 h, the medium was removed by aspiration and 90 µL serum-free DMEM-F12 medium with penicillin and streptomycin was added to each well. Peptide nanoparticles loaded with siRNA were added to cells (concentration of Cy-5 labeled siRNA was 1 nmol/mL). The cells were incubated with particles at 37 °C in serum-free medium for 4 h. Then the cells were washed with 1 M sodium chloride to remove remaining extra particles from cell surface, and the wells were filled with DMEM-F12 and the incubation continued for 44 h more. In each well the cells were fixed using 200 µL of 3.7% solution of formaldehyde in methanol. After that, the cells were washed three times with PBS.

We used Permeabilization Buffer (0.2% Triton X-100 in PBS) to permeabilize cell membranes. Cell nuclei were stained with Hoechst 33,258 (1 µg/mL) for 30 min according to the protocol published elsewhere [28]. After that, the cells were washed for three times with PBS and two times with water in order to remove crystals of salt.

The cell membranes were stained by CellMask Green Plasma Membrane Stain (Thermo Fischer Scientific, Paisley, UK), according to the manufacturer's protocol. The study of cellular uptake was performed using CELENA S Digital Imaging System (Logos Biosystems, GE Healthcare, Anyang, Korea) and Cytell Cell Imaging instrument (GE Healthcare, Issaquah, WA, USA) by analysis of the fluorescence intensity of Cy5-siRNA.

To confirm particle penetration into the cells, we performed an additional experiment. For this purpose, we labeled polypeptide particles with Cy5 and used Cy3-oligo-dT-dA for encapsulation. 10 µL of 0.1 mg/mL solution of Cy5 in DMSO was added to 500 µL of 0.1 mg/mL suspension of nanoparticles in 0.01 M Na-phosphate buffer, pH 7.4, and stained for 1 h at 37 °C. Non-bound Cy5 was separated from nanoparticles by centrifugation (10,000× g) using filter tubes with MWCO 10,000 (Merck, Darmstadt, Germany) 7–8 times until the filtrate was free from Cy5. The encapsulation of Cy3-dT-dA and the cell experiments were the same as described above.

2.2.6. VEGF Gene Silencing

The efficacy of siRNA, loaded in peptide nanoparticles, against VEGF was evaluated in ARPE-19 cells. Approximately 4×10^4 ARPE-19 cells were seeded per well in a 24-well plate (Thermo Scientific^{TM} Nunc^{TM} MicroWell^{TM}) in DMEM-F12/10% FBS (Biowest, Nuaille, France)/50 IU/mL penicillin/50 µg/mL streptomycin (BioloT) overnight. After that, the medium was removed by aspiration and 500 µL serum-free DMEM-F12 medium was added to each well. Then, 0.05 nmol of the siRNA loaded in 33.5 µg of polypeptide particles was added to each well (mass ratio particles:siRNA was equal to 4:1). The siRNA concentration was 100 nM and it was selected from data published elsewhere for ARPE-19 cells where the concentration range from 10 to 1000 nM was investigated [29,30]. The cells were incubated with particles at 37 °C for 4 h. Then the medium was replaced with fresh one and the incubation continued additionally for 44 h. To investigate RNA interference in the absence of VEGF we used the nanoparticles with scrambled siControl. The conditions were the same as for target siRNA.

2.2.7. Total RNA Isolation, Reverse Transcription and Quantitative Real-Time PCR Analysis

The efficiency of VEGF silencing was estimated by measuring VEGF mRNA expression quantitatively using real-time reverse transcription polymerase chain reaction (RT-PCR). Glyceraldehyde-3-phosphate dehydrogenase (GAPDH) mRNA expression level was used as a housekeeping gene. After lysis of ARPE-19 cells, the total RNA was isolated using an RNA extraction kit from Biosilica (Novosibirsk, Russia) according to the manufacturer's instructions and the RNA concentration was measured by the absorbance at 260 nm wavelength using an UV NanoDrop spectrometer (Thermo Scientific, Rockford, IL, USA). The ratio of absorbance at 260 nm and 280 nm for noncontaminated RNA samples was in the range of >1.8 and <2.2 [31]. The integrity of total RNA was determined by native 1% agarose gel electrophoresis [31]. Intact 28S and 18S rRNA were detected on a gel as sharp bands with intensity ratio ~2:1 and no signs of genomic DNA contaminations were observed. Then, 60 ng of cDNA was synthesized using the MMLV RT kit (Evrogen, Moscow, Russia). After that, 12 ng of cDNA was applied for RT-PCR with the respective VEGF forward, VEGF reverse, GADPH forward, and GADPH reverse primers. RT-PCR was performed using a qPCRmix-HS SYBR (Evrogen) according to the manufacturer's instruction. The PCR performed for 40 amplification cycles of 95 °C for 30 s, 55 °C for 30 s, and 72 °C for 45 s. The amount of VEGF mRNA was determined relative to the amount of housekeeping gene glyceraldehyde 3-phosphate dehydrogenase (GAPDH) mRNA in the same sample by the equation $X_0/R_0 = 2C_tR{-}C_tX$, where X_0 is the original amount of VEGF mRNA, R_0 is the original amount of GAPDH mRNA, C_tR is the C_t (cycle threshold) value for GAPDH, and C_tX is the C_t value for VEGF.

2.2.8. Western Blotting

To confirm the results of VEGF-silencing on protein level, western-blotting was performed as well. ARPE-19 cells were seeded on 6-well plate at a density of 8×10^5 cells/well in DMEM-F12. After 12 h, the medium was aspirated, and 2 mL serum-free DMEM-F12 medium with penicillin and streptomycin was added. Then, 0.8 nmol of the siRNA loaded in polypeptide nanoparticles was added (mass ratio polymer:siRNA was 4:1). The nanoparticles with scrambled siRNA sequence were used as a control. For analysis of VEGF level suppression in a presence of nanoparticles loaded with siRNA, ARPE-19 cells were cultured for 5 days (instead of 2 days as for mRNA study) due to the slow turnover of the protein [13,32]. Afterwards, the cells were lysed with radioimmunoprecipitation assay (RIPA) buffer. Total protein in the lysate was quantified by Bradford assay using the albumin calibration curve as a standard. The equal amount of protein (20 μg) was separated on a 12% SDS-PAGE gel and electroblotted onto polyvinylidenedifluoride (PVDF) membrane (0.2 μm, Thermo Scientific, Rockford, IL, USA) at 400 mA for 20 min. The PVDF membrane was blocked with 5% (*w*/*v*) nonfat dried milk diluted in TBS buffer with 0.1% (*v*/*v*) Tween 20 for 1 h at room temperature. Recombinant humanized monoclonal anti-(human VEGF) antibody (1:20,000; Avastin, Roche, Welwyn Garden City, UK) and monoclonal mouse anti-(human β-actin) antibody (1:1000) (Santa Cruz Biotechnology, Heidelberg, Germany) were used as primary antibodies and horseradish peroxidase-linked rabbit anti-human IgG Fc antibody (1:10,000; Thermo Fisher Scientific) and rabbit anti-goat IgG (H + L) antibody (1:10,000; Thermo Fisher Scientific) were used as secondary antibodies. Blots were developed with enhanced chemiluminescence Amersham Hyperfilm ECL reagent and detection was done using Amersham Imager 600 Instrument (GE Healthcare, Little Chalfont, UK).

2.2.9. Statistical Analysis

To analyze the statistical significance among the groups, one-way analysis of variants (ANOVA) in Excel with the XLSTAT was used. Data were expressed as mean ± SD ($n = 3$). $p \leq 0.05$ was counted as a statistically significant.

3. Results and Discussion

3.1. *Synthesis and Characterization of Polypeptides*

Two series of polypeptides with different hydrophobic amino acid contents were synthesized by varying the initial ratio of amino acids in the synthesis. Polypeptides had a random primary structure that was generated during copolymerization of NCAs of L-Lys(Z), L-Glu(OBzl) and L-Phe/L-Ile via ROP technique [33]. Weight average molecular weight (M_w), number average molecular weight (M_n) and dispersity ($Đ$) of protected copolymers were evaluated using size-exclusion chromatography (SEC) in DMF with refractometric detection regarding poly(methyl methacrylate) standards (Table 1). Additionally, for some samples M_w values were also determined by static light scattering (SLS) (Table 1 and Supplementary Materials Table S1) to testify the applicability of used SEC for such kind of polymers. M_w values measured by both methods were found to be close. The polydispersity of the samples was in the range of 1.10–1.33 that corresponds to quite narrow molecular weight distribution.

Table 1. Molecular-weight characteristics of protected polypeptides determined by SEC and SLS.

Sample	Polymer Characteristics			
	SEC			SLS
	M_n	M_w	Đ	M_w
	P(Lys(Z)$_n$-co-Glu(OBzl)$_m$-co-Phe$_k$)			
KEF1	20,170	23,600	1.17	23,000
KEF2	9400	17,000	1.80	17,500
KEF3	10,000	13,200	1.32	17,400
	P(Lys(Z)$_n$-co-Glu(OBzl)$_m$-co-Ile$_k$)			
KEI1	21,350	28,400	1.33	18,800
KEI2	16,090	21,240	1.32	17,100
KEI3	17,560	22,300	1.27	-

The amphiphilic copolymers were prepared via deprotection of ε-amino groups of lysine and γ-carboxylic groups of glutamic acid. After deprotection, the copolymers acquired the tendency to self-assembly in aqueous media. The completeness of removal of the protective groups was verified by ^{1}HNMR spectroscopy (Supplementary Materials Figure S1).

HPLC analysis of hydrolyzed polypeptide samples allowed for the determination of the ratio between amino acids (Table 2). The samples KEF1, KEF2, and KEI1-KEI3 were enriched with lysine whereas KEF3 was enriched with glutamic acid. Additionally, to evaluate polymer composition, the samples were analyzed by ^{1}H-NMR spectroscopy. ^{1}HNMR spectra and information on correlations of signals can be found in Supplementary Materials (Figures S2 and S3). For both Phe- and Ile containing copolymers, the content of hydrophilic (Lys + Glu) and hydrophobic (Phe or Ile) amino acids (mol%) established by HPLC amino acid analysis was in an agreement with data obtained by ^{1}HNMR spectroscopy. Due to the overlap of signals related to the carboxybenzyl and benzyl groups of Lys(Z) and Glu(OBzl), the separate calculation of Lys and Glu content could not be possible by ^{1}HNMR method.

Table 2. Composition of amphiphilic amino acid polypeptides.

Sample	Determined Polymer Composition (mol%)				
	HPLC			^{1}HNMR	
	Lys	*Glu*	*Phe/Ile*	*Lys + Glu*	*Phe/Ile*
	P(Lys$_n$-co-Glu$_m$-co-Phe$_k$)				
KEF1	55	25	20	76	24
KEF2	75	9	16	75	25
KEF3	21	54	25	64	36
	P(Lys$_n$-co-Glu$_m$-co-Ile$_k$)				
KEI1	66	16	18	85	15
KEI2	57	31	12	88	12
KEI3	54	36	10	-	-

3.2. Preparation and Characterization of Polypeptide Particles

The preparation of nanoparticles (NPs) based on the synthesized polypeptides was carried out by gradient solvent inversion approach favoring to slow polymer self-assembly and followed by freeze-drying. Before application, the necessary amount of sample was redispersed in water or

0.01 M PBS via short-term ultrasonic exposure (30 s). Usually, the main force for the self-assembly of amphiphilic polymers is the interactions between hydrophobic regions of macromolecules in organic-aqueous or aqueous media to reduce the area of contact with water. At the same time, the hydrophilic fragments are exposed into aqueous medium because of their good solvation with water. In our amphiphilic macromolecules, except hydrophobic interactions, the electrostatic forces play a role due to the presence in polypeptides positively charged Lys and negatively charged Glu residues. Such complex interactions between polypeptide chains should provide high stability of self-assembled nanoparticles.

Size, surface charge and stability are among the most important parameters of nanoparticles, affecting cellular uptake and cytotoxicity. Various methods were used in this study to characterize the polypeptide nanoparticles obtained. Such parameters as hydrodynamic diameter (D_H), polydispersity index (PDI) and electrokinetic potential (ζ-potential) of obtained nanoparticles were determined by dynamic light scattering (DLS) and electrophoretic light scattering, respectively (Table 3). The smallest particles with the average hydrodynamic diameter about 200 nm were formed when Lys/Glu ratio ranged from 1.9 to 2.8 (samples KEF1, KEI1 and KEI2). The same samples were also characterized with the lowest PDI. As expected from the copolymer's composition, all samples, except KEF3, had positive ζ-potential, while KEF3 sample, enriched with glutamic acid (Table 2), was charged negatively. We chose the samples KEF1, KEI1 and KEI2 for further studies.

Table 3. Characteristics of polypeptide nanoparticles obtained (DLS, in 0.01 PBS, pH 7.4).

Sample	Particle Characteristics		
	D_H, nm	PDI	ζ-Potential, mV
	P(Lys$_n$-co-Glu$_m$-co-Phe$_k$)		
KEF1	200 ± 8	0.22	+12 ± 2
KEF2	550 ± 14	0.31	+18 ± 5
KEF3	900 ± 38	0.39	−6 ± 3
	P(Lys$_n$-co-Glu$_m$-co-Ile$_k$)		
KEI1	232 ± 11	0.14	+49 ± 2
KEI2	180 ± 19	0.19	+45 ± 3
KEI3	440 ± 31	0.30	+31 ± 5

The critical micelle concentration (CMC) were determined for the selected polymers by conductometry method [34]. With increasing polymer concentration, the formation of thermodynamically stable aggregates starts and it is can be detected as a change in solution conductivity. The CMC values were calculated by a linear fitting of the two data subsets and calculating the concentration at their intersection (Supplementary Materials Figure S4). CMC values for KEF1, KEI1 and KEI2 were found to be close to each other and equal to 8.9, 10.6 and 6.2 mg/L, respectively. The values determined were close [35] or even lower [36] than these established for other amphiphilic random copolymers.

Polypeptide nanoparticles were studied with nanoparticle tracking analysis (NTA) and transmission electron microscopy (TEM) to receive more information about their size, shape and structure. As an example, Figure 1 illustrates TEM and NTA images obtained for sample KEI2. According to TEM investigations, the nanoparticles of chosen sample represented spheres with a mean diameter of about 160 ± 50 nm. NTA allows for the determination of hydrodynamic diameter and size distribution profile of small particles that are hard to capture by DLS. According to NTA, there were three modes of nanoparticles with D_H around 130, 180 and 240 nm. In turn, the DLS measurements gave the average value for the same sample equal to 180 nm with no detection of particles of different size (Supplementary Materials Figure S5). Similar results were obtained for two other selected samples (KEF1 and KEI1).

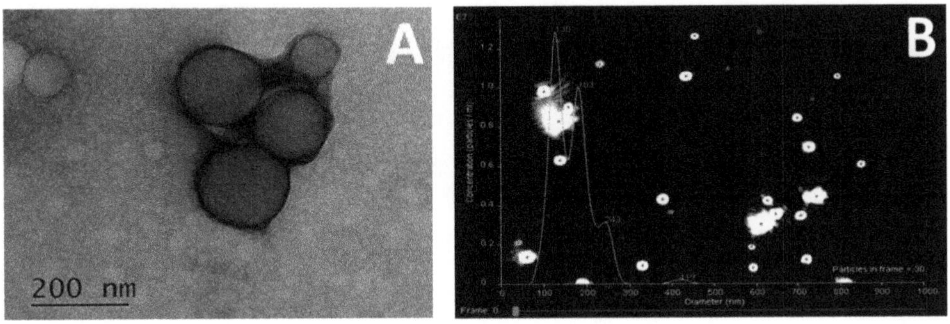

Figure 1. Size and structure of polypeptide nanoparticles (sample KEI2), investigated by TEM (**A**) and NTA (**B**).

The pH of solutions did not affect the particle size until it reached 10.5 (Figure 2A). At this point, ε-amino groups of lysine are deprotonated and became neutrally charged. As a result, an aggregation takes place. Such behavior was observed previously for random copolymers of lysine and phenylalanine [37]. In that case, the 10-fold increase in D_H was detected. For terpolymers containing two oppositely charged amino acids (lysine and glutamic acid), the hydrodynamic diameter of the particles also increased but not so drastic: Only two-fold growth of D_H was registered. Moreover, further pH elevation to 11 and 12 followed with charge switching to negative, −17 and −36 mV, respectively. The negative charge is explained by deprotonating of γ-carboxylic groups of glutamic acid, while amino group of lysine was uncharged. Moreover, the hydrodynamic diameter of nanoparticles at pH 12 returned to the initial value.

The stability of nanoparticles was studied in PBS and DMEM over 7 days. All samples proved to be colloidally stable, keeping their size and surface charge constant (Figure 2B).

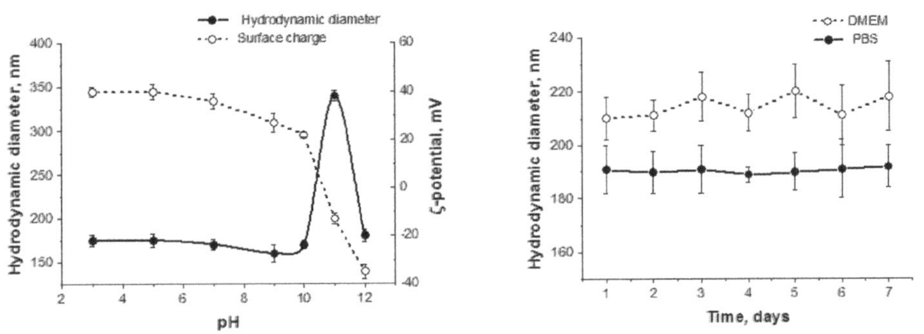

(**A**) (**B**)

Figure 2. pH sensitivity (**A**) and stability (**B**) of polypeptide nanoparticles (sample KEI2).

3.3. Entrapment and Release of Duplex Oligo-dT-dA

Entrapment efficacy and release kinetics are crucial characteristics for developed delivery system. At first, it was important to understand how the loading of oligonucleotide influences the physico-chemical characteristics of nanoparticles. For this purpose, duplex oligo-dT-dA was used as the cheaper and stable model of siRNA. The positively charged nanoparticles (sample KEI2) were loaded with negatively charged oligo-dT-dA due to polyelectrolyte interactions. The hydrodynamic diameter of the loaded particles as well as the surface charge depended on polymer:oligonucleotide ratio, while mean PDI was constant and equal to 0.18±0.01. The particle's hydrodynamic diameter decreased gradually and the surface charge lowered steadily with the increase of encapsulated duplex oligo-dT-dA (Figure 3). A significant aggregation was observed when the polymer:oligonucleotide ratio reached 2:1 and the surface charge dropped to +8 mV. Thus, it can be deduced that polymer:oligonucleotide ratio 4:1 provided the smallest particles with a noticeable surface charge.

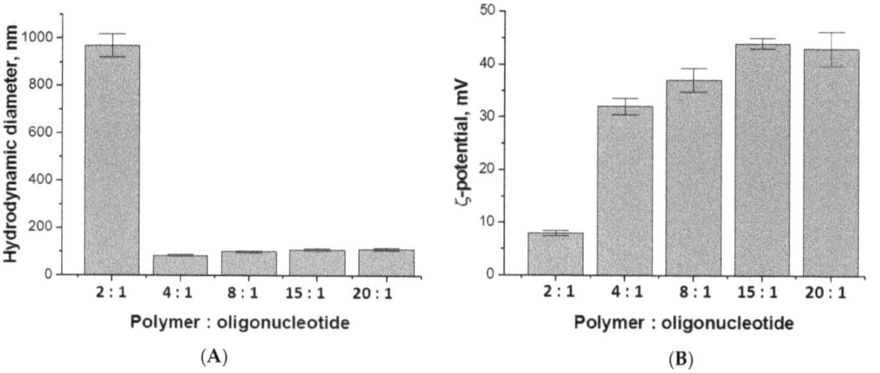

Figure 3. Hydrodynamic diameter (**A**) and surface charge (**B**) of polypeptide nanoparticles (sample KEI2), loaded with duplex oligo-dT-dA.

To compare the entrapment efficacy and to study the oligonucleotide release kinetics, we used the same oligonucleotide but labeled with Cy3 fluorescent dye. Since the particle size was stable until the polymer:oligo-dT-dA ratio reached 2:1, the oligonucleotide encapsulation was carried out in the range of this ratio from 4:1 to 20:1. Since pH does not affect particle size and morphology (except at pH 10.5) the encapsulation was performed in pure water. Almost 100% of labeled oligonucleotide was entrapped into polypeptide nanoparticles, regardless polymer:oligonucleotide ratio (Figure 4A). Therefore, the ratio 4:1 was appeared to be suitable for further experiments.

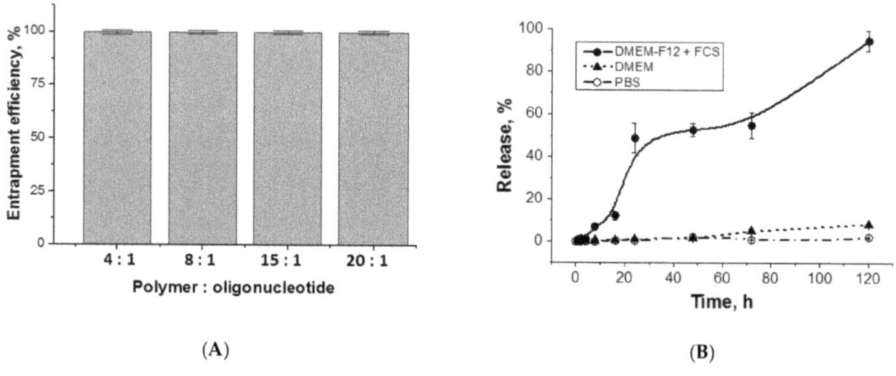

Figure 4. Entrapment efficacy (**A**) and release (**B**) of duplex oligo-dT-dA (sample KEI2) from the complex (polymer:oligo-dT-dA ratio equal to 4:1).

The release of duplex oligo-dT-dA was studied for the system obtained at polymer:oligo-dT-dA ratio = 4:1. A simple buffer system (PBS, pH 7.4), a complex medium containing amino acids, vitamins and inorganic salts (DMEM) and DMEM-F12 containing 10% of fetal calf serum (DMEM-F12 + FCS) were used as the media for experiments. About 50% of oligonucleotide duplex was released over 24 h in DMEM-F12 + FCS (Figure 4B). After 5 days of incubation in DMEM-F12 + FCS at 37 °C, the release reached 95%. However, oligonucleotide duplex did not escape polymer particles in PBS during 5 days that means good storage stability of formulation. At the same time, the release in DMEM without serum was negligible. The release in DMEM-F12 + FCS medium is explained by the presence of different negatively charged components like proteins and vitamins at relatively high concentrations that could interact with polypeptide and, therefore, gradually replace oligonucleotide. Since there are a lot of proteins and other molecules with negative charge in the cytoplasm, we suppose that this mechanism of substitution will work also inside a cell to release siRNA.

Additionally, we prepared the polyplex of poly-L-lysine and duplex oligo-dT-dA in the same ratio (4:1) and studied the release of oligonucleotide in DMEM and DMEM-F12 + FCS over 5 days (37 °C). It was established that the oligo-dT-dA release in DMEM did not exceed 2.5%. The presence of FCS in medium slightly facilitated the release of target compound, however, even in this case it did not exceed 8% (Supplementary Materials Figure S6). Thus, the combination of positively and negatively charged amino acids, e.g., lysine and glutamic acid, indeed facilitate the release of oligonucleotides in comparison with poly-L-lysine.

3.4. *Cytotoxicity*

Since the stability of nanoparticles for 7 days in the test medium was proven with DLS, we proceeded the investigation of their cytotoxicity in cultured cells (Figure 5 and Supplementary Materials Figure S7). For this purpose, MTT assay was performed using three cell lines, e.g., HEK-293, BEAS-2B and ARPE-19 cells. The cells were incubated in the presence of polypeptide nanoparticles (KEF1, KEI1, KEI2) for 72 h. It was established that the samples KEF1 and KEI2 were not toxic for HEK-293 and BEAS-2B cells in the concentration range of 4–125 µg/mL. For the sample KEI1, modest cytotoxicity was observed at concentration 125 µg/mL (viability 65 ± 1%). In the case of ARPE-19, cell viability was higher than 75% for KEI1 and KEI2 samples in the whole tested concentration range. The sample KEF1 demonstrated evident cytotoxicity at the concentration of 125 µg/mL.

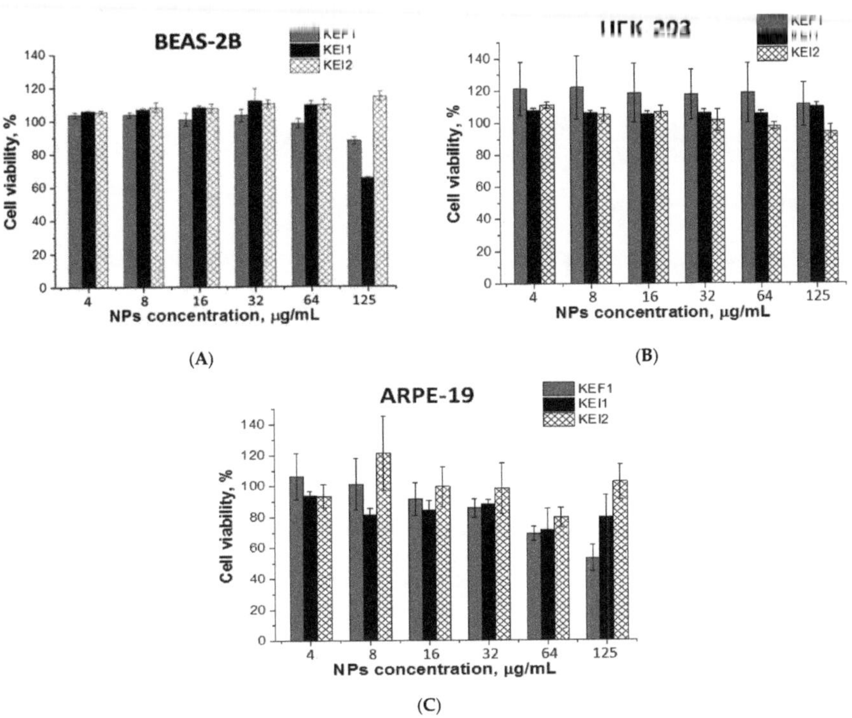

Figure 5. Cytotoxicity of the nanoparticles in BEAS-2B (**A**), HEK-293 (**B**) and ARPE-19 (**C**) cells (72 h).

3.5. VEGF Gene Silencing

To investigate VEGF gene silencing, KEF1, KEI1 and KEI2 nanoparticles were selected as the delivery systems for anti-VEGF siRNA. We used ARPE-19 cells that express VEGF constitutively. The experiment was carried out using the polymer:siRNA mass ratio of 4:1; the concentration of polymer particles used for experiment was 67 μg/mL. RNA interference of anti-VEGF siRNA was analyzed by RT-PCR.

All investigated samples demonstrated strong inhibitory effect on mRNA expression (Figure 6). The minimal inhibitory effect of 60% was reached, using KEF1 as a delivery system, while KEI1 and KEI2 inhibited VEGF gene expression more effectively ($p < 0.01$) (Figure 6). In particular, VEGF silencing reached with the use of KEI1 and KEI2 siRNA delivery systems corresponded to 82 and 71%. For comparison, 40–70% GFP gene knockdown was reported by Zhu et al. for the application of poly(dimethylaminoethyl methacrylate)-b-polycaprolactone-b-poly(dimethylaminoethyl methacrylate) systems for delivery of siRNA [38]. Wei et al. reported the application of chitosan-based systems for delivery of mTERT siRNA. The reached gene silencing was equal to 56% [39]. As to VEGF knockdown, the delivery systems based on poly(ε-caprolactone)- polyethyleneglycol-poly(L-histidine) provided up to 55% VEGF gene silencing in MCF-7 and HUVEC cells (siRNA concentration and incubation time with cells were 100 nM and 5 days, respectively) [13]. Thus, the polypeptide systems, developed in present work, can be considered as efficient for intracellular siRNA delivery and VEGF gene knockdown.

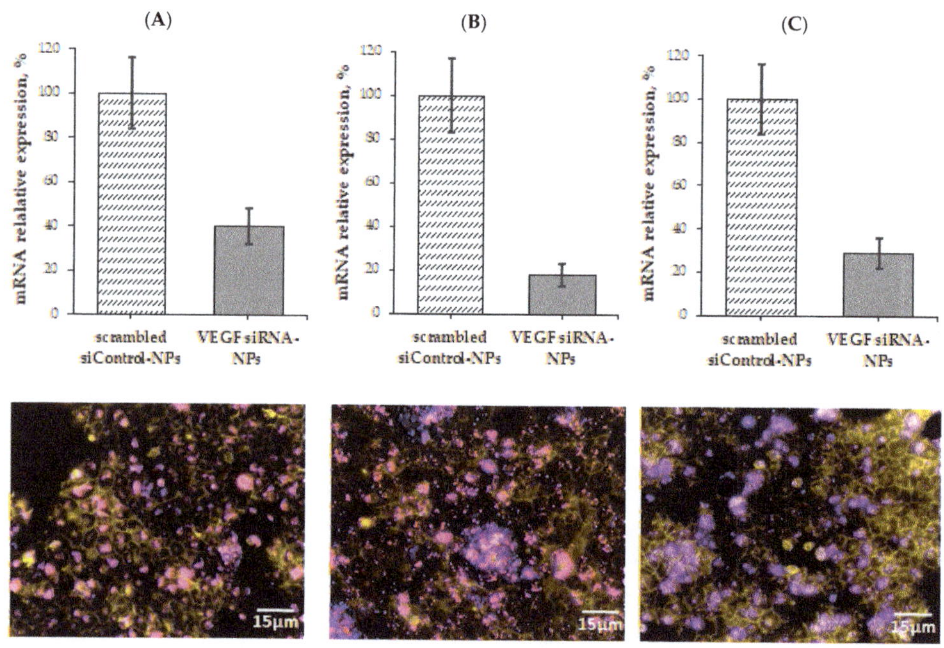

Figure 6. mRNA interference by anti-VEGF siRNA delivered by KEF1 (**A**), KEI1 (**B**) and KEI2 (**C**) nanoparticles (polymer:siRNA ratio = 4:1) in the ARPE-19 cells (mean ± SD, $n = 3$). Cell membranes were stained with Cell Mask Green Plasma Membrane dye (green-yellow color), cell nuclei were stained with Hoechst 33,258 (blue color). Cy3 labelled siRNA was depicted in pink color.

Fluorescence microscopy images of ARPE-19 cells treated with Cy5-siRNA–polypeptide particles showed effective cell penetration of Cy5-labeled siRNA. Most siRNAs were located in the cytoplasm (Figure 6). According to the fluorescent images, the delivery of Cy5-labled siRNA with KEI1 nanoparticles followed with much more pronounced aggregation of NPs. Taking this into account as well as gene silencing results and the data on cytotoxicity, KEI2 sample was selected for further western blotting analysis.

To verify the suppression of VEGF expression due to the gene silencing the western blotting was carried out. For this, KEI2 nanoparticles loaded with siRNA was used as sample whereas scrambled siRNA loaded in the same kind of NPs was applied as negative control. Additionally, empty NPs were also applied as control to reveal if they can influence on the process under study. According to the results obtained, the expression of VEGF under siRNA delivery system application was suppressed approximately by 60% ($p < 0.03$) (Figure 7). The results obtained are in a good agreement with data on RNA interference evaluated by RT-PCR where the same sample provided gene silencing by 70%. No suppressing effect was detected for both empty NPs and scrambled siControl-NPs.

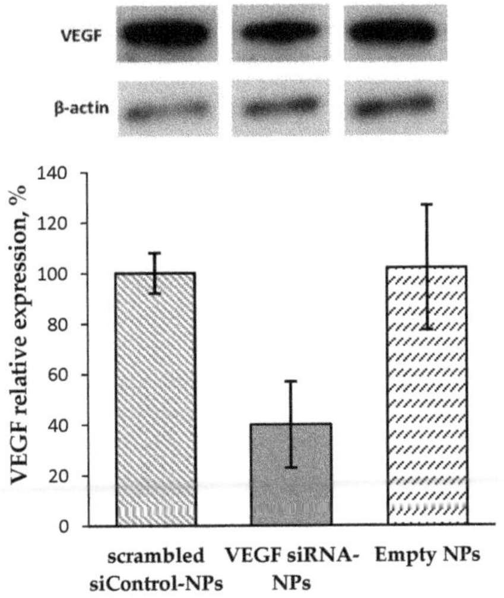

Figure 7. VEGF protein expression in the ARPE-19 cells determined by western blotting (KEI2 NPs).

4. Conclusions

In this study, polymers based on lysine, glutamic acid and phenylalanine/isoleucine were developed for the delivery of oligonucleotides. The polymers self-assembled into spherical particles with average hydrodynamic diameters of 180–900 nm that were colloidally stable in PBS and cell culture medium (DMEM). The model duplex of oligonucleotides (oligo-dT-dA) was encapsulated in the polymer nanoparticles up to 4:1 mass ratio (polymer:oligo-dT-dA) without any size change in nanoparticles. The interaction between oligonucleotide duplex and the polymer was strong and no oligonucleotide release in PBS was detected, but the presence of various negatively charged components like proteins, vitamins and amino acids in DMEM-F12 + FCS enhanced the release and 56% of oligonucleotide was freed over 48 h. The cell viability test demonstrated the absence of cytotoxicity for the sample KEI2 for three cell lines whereas for some other samples slight toxic effect was detected. The efficient VEGF gene silencing was established at the application of developed polypeptide nanoparticles as siRNA delivery systems.

Supplementary Materials: The following are available online at http://www.mdpi.com/1999-4923/12/1/39/s1, Figure S1: ^{1}HNMR spectrum of deprotected sample KEI1, Figure S2: ^{1}HNMR spectrum of protected sample KEF1, Figure S3: ^{1}HNMR spectrum of protected sample KEI2, Figure S4: Dependences of conductivity on polymer concentration: (a) sample KEF1 and (b) sample KEI1, Figure S5: DLS analysis of polymer nanoparticles prepared from P(Lys-co-Glu-co-Ile), sample KEI2, Figure S6: Release of duplex oligo-dT-dA from the complex with poly-L-lysine:oligonucleotide (ratio 4:1), Figure S7: Nanoparticle's cytotoxicity in BEAS-2B and HEK-293 (a–c) and ARPE-19 cells (d) (24 h): (a) sample KEF1, (b) sample KEI1, (c) sample KEI2 and (d) sample KEI2, Table S1: Data of static (SLS) and dynamic (DLS) light scattering of polymer solutions (DMSO).

Author Contributions: Conceptualization—T.T., A.U. and E.K.-V.; polymer synthesis—O.O.; polymer characterization—O.O., N.Z. and N.Z.; particles preparation and characterization—O.O., biological experiments—V.S., O.O. and N.Z., formal analysis—O.O., N.Z., V.S. and E.K.-V.; data curation—O.O., V.S. and E.K.-V.; visualization—O.O., N.Z., V.S. and E.K.-V.; funding acquisition—A.U.; scientific supervision—E.K.-V.; project administration—A.U.; writing—Original draft preparation, O.O. and E.K.-V.; writing—Review and editing, T.T., E.K.-V. and A.U. All authors have read and agreed to the published version of the manuscript.

Funding: The financial support was provided by Government of Russian Federation (Megagrant #14.W03.31.0025).

Acknowledgments: Chemical Analysis and Materials Research Centre, and Resource Center for Magnetic Resonance (Saint-Petersburg State University) are acknowledged for amino acid HPLC analysis and NMR-spectroscopy, respectively.

Conflicts of Interest: The authors declare no conflict of interest.

References

1. Cheung, N.; Wong, I.Y.; Wong, T.Y. Ocular anti-VEGF therapy for diabetic retinopathy: Overview of clinical efficacy and evolving applications. *Diabetes Care* **2014**, *37*, 900–905. [CrossRef]
2. Amoaku, W.M.; Chakravarthy, U.; Gale, R.; Gavin, M.; Ghanchi, F.; Gibson, J.; Harding, S.; Johnston, R.L.; Kelly, S.; Lotery, A.; et al. Defining response to anti-VEGF therapies in neovascular AMD. *Eye* **2015**, *29*, 721–731. [CrossRef] [PubMed]
3. Ferrara, N. VEGF and Intraocular Neovascularization: From Discovery to Therapy. *Transl. Vis. Sci. Technol.* **2016**, *5*, 10. [CrossRef] [PubMed]
4. Tolentino, M. Systemic and Ocular Safety of Intravitreal Anti-VEGF Therapies for Ocular Neovascular Disease. *Surv. Ophthalmol.* **2011**, *56*, 95–113. [CrossRef] [PubMed]
5. Amadio, M.; Govoni, S.; Pascale, A. Targeting VEGF in eye neovascularization: What's new?: A comprehensive review on current therapies and oligonucleotide-based interventions under development. *Pharmacol. Res.* **2016**, *103*, 253–269. [CrossRef]
6. Carthew, R.W.; Sontheimer, E.J. Origins and Mechanisms of miRNAs and siRNAs. *Cell* **2009**, *136*, 642–655. [CrossRef]
7. Bordet, T.; Behar-Cohen, F. Ocular gene therapies in clinical practice: Viral vectors and nonviral alternative. *Drug Discov. Today* **2019**, *24*, 1685–1693. [CrossRef]
8. Cavallaro, G.; Sardo, C.; Craparo, E.F.; Porsio, B.; Giammona, G. Polymeric nanoparticles for siRNA delivery: Production and applications. *Int. J. Pharm.* **2017**, *525*, 313–333. [CrossRef]
9. Dominska, M.; Dykxhoorn, D.M. Breaking down the barriers: siRNA delivery and endosome escape. *J. Cell Sci.* **2010**, *123*, 1183–1189. [CrossRef]
10. Semple, S.C.; Akinc, A.; Chen, J.; Sandhu, A.P.; Mui, B.L.; Cho, C.K.; Sah, D.W.Y.; Stebbing, D.; Crosley, E.J.; Yaworski, E.; et al. Rational design of cationic lipids for siRNA delivery. *Nat. Biotechnol.* **2010**, *28*, 172–176. [CrossRef]
11. Schroeder, A.; Levins, C.G.; Cortez, C.; Langer, R.; Anderson, D.G. Lipid-based nanotherapeutics for siRNA delivery. *J. Intern. Med.* **2010**, *267*, 9–21. [CrossRef] [PubMed]
12. Sun, X.; Dong, S.; Li, X.; Yu, K.; Sun, F.; Lee, R.J.; Li, Y.; Teng, L. Delivery of siRNA using folate receptor-targeted pH-sensitive polymeric nanoparticles for rheumatoid arthritis therapy. *Nanomed. Nanotechnol. Biol. Med.* **2019**, *20*, 102017. [CrossRef] [PubMed]
13. Yang, Y.; Meng, Y.; Ye, J.; Xia, X.; Wang, H.; Li, L.; Dong, W.; Jin, D.; Liu, Y. Sequential delivery of VEGF siRNA and paclitaxel for PVN destruction, anti-angiogenesis, and tumor cell apoptosis procedurally via a multi-functional polymer micelle. *J. Control. Release* **2018**, *287*, 103–120. [CrossRef]
14. Egorova, A.; Petrosyan, M.; Maretina, M.; Balashova, N.; Polyanskih, L.; Baranov, V.; Kiselev, A. Anti-angiogenic treatment of endometriosis via anti-VEGFA siRNA delivery by means of peptide-based carrier in a rat subcutaneous model. *Gene Ther.* **2018**, *25*, 548–555. [CrossRef] [PubMed]
15. Egorova, A.; Shubina, A.; Sokolov, D.; Selkov, S.; Baranov, V.; Kiselev, A. CXCR4-targeted modular peptide carriers for efficient anti-VEGF siRNA delivery. *Int. J. Pharm.* **2016**, *515*, 431–440. [CrossRef]
16. Imperiale, J.C.; Acosta, G.B.; Sosnik, A. Polymer-based carriers for ophthalmic drug delivery. *J. Control. Release* **2018**, *285*, 106–141. [CrossRef]

17. Helmschrodt, C.; Höbel, S.; Schöniger, S.; Bauer, A.; Bonicelli, J.; Gringmuth, M.; Fietz, S.A.; Aigner, A.; Richter, A.; Richter, F. Polyethylenimine Nanoparticle-Mediated siRNA Delivery to Reduce α-Synuclein Expression in a Model of Parkinson's Disease. *Mol. Ther. Nucleic Acids* **2017**, *9*, 57–68. [CrossRef]
18. Ragelle, H.; Vanvarenberg, K.; Vandermeulen, G.; Préat, V. Chitosan Nanoparticles for SiRNA Delivery In Vitro. In *SiRNA Delivery Methods*; Humana Press: New York, NY, USA, 2016; pp. 143–150.
19. Kodama, Y.; Kuramoto, H.; Mieda, Y.; Muro, T.; Nakagawa, H.; Kurosaki, T.; Sakaguchi, M.; Nakamura, T.; Kitahara, T.; Sasaki, H. Application of biodegradable dendrigraft poly-l-lysine to a small interfering RNA delivery system. *J. Drug Target.* **2017**, *25*, 49–57. [CrossRef]
20. Hennig, R.; Goepferich, A. Nanoparticles for the treatment of ocular neovascularizations. *Eur. J. Pharm. Biopharm.* **2015**, *95*, 294–306. [CrossRef]
21. Tatiparti, K.; Sau, S.; Kashaw, S.K.; Iyer, A.K. siRNA Delivery Strategies: A Comprehensive Review of Recent Developments. *Nanomaterials* **2017**, *7*, 77. [CrossRef]
22. Auguste, D.; Furman, K.; Wong, A.; Fuller, J.; Armes, S.; Deming, T.; Langer, R. Triggered release of siRNA from poly(ethylene glycol)-protected, pH-dependent liposomes. *J. Control. Release* **2008**, *130*, 266–274. [CrossRef] [PubMed]
23. Xia, Y.; Tian, J.; Chen, X. Effect of surface properties on liposomal siRNA delivery. *Biomaterials* **2016**, *79*, 56–68. [CrossRef] [PubMed]
24. Raik, S.V.; Andranovitš, S.; Petrova, V.A.; Xu, Y.; Lam, J.K.-W.; Morris, G.A.; Brodskaia, A.V.; Casettari, L.; Kritchenkov, A.S.; Skorik, Y.A. Comparative Study of Diethylaminoethyl-Chitosan and Methylglycol-Chitosan as Potential Non-Viral Vectors for Gene Therapy. *Polymers* **2018**, *10*, 442. [CrossRef] [PubMed]
25. Pilipenko, I.; Korzhikov-Vlakh, V.; Sharoyko, V.; Zhang, N.; Schäfer-Korting, M.; Rühl, E.; Zoschke, C.; Tennikova, T. pH-Sensitive Chitosan–Heparin Nanoparticles for Effective Delivery of Genetic Drugs into Epithelial Cells. *Pharmaceutics* **2019**, *11*, 317. [CrossRef]
26. Wilder, R.; Mobashery, S. The use of triphosgene in preparation of *N*-carboxy-alpha-amino acid anhydrides. *J. Org. Chem.* **1992**, *57*, 2755–2756. [CrossRef]
27. Schmuck, C.; Hernandez-Folgado, L. Synthesis of a new artificial host for the binding of dipeptides in water. *Org. Biomol. Chem.* **2007**, *5*, 2390–2394. [CrossRef]
28. Abcam. ab228550 Hoechst 33258 Staining Dye Solution, 2018. Available online: https://www.abcam.com/ps/products/228/ab228550/documents/ab228550%20Hoechst%2033258%20Staining%20Dye%20Solution%20v1a%20(website).pdf (accessed on 13 November 2019).
29. Chen, C.W.; Yeh, M.K.; Shiau, C.Y.; Chiang, C.H.; Lu, D.W. Efficient downregulation of VEGF in Retinal pigment epithelial cells by integrin ligand-labeled liposome-mediated siRNA delivery. *Int. J. Nanomed.* **2013**, *8*, 2613–2627.
30. Murata, M.; Takanami, T.; Shimizu, S.; Kubota, Y.; Horiuchi, S.; Habano, W.; Ma, J.X.; Sato, S. Inhibition of ocular angiogenesis by diced small interfering RNAs (siRNAs) specific to vascular endothelial growth factor (VEGF). *Curr. Eye Res.* **2006**, *31*, 171–180. [CrossRef]
31. Sambrook, J.; Russell, D.W. *Molecular Cloning: A Laboratory Manual*, 3rd ed.; Cold Spring Harbor: New York, NY, USA, 2001.
32. Kim, B.; Tang, Q.; Biswas, P.S.; Xu, J.; Schiffelers, R.M.; Xie, F.Y.; Ansari, A.M.; Scaria, P.V.; Woodle, M.C.; Lu, P.; et al. Inhibition of ocular angiogenesis by siRNA targeting vascular endothelial growth factor pathway genes: Therapeutic strategy for herpetic stromal keratitis. *Am. J. Pathol.* **2004**, *165*, 2177–2185. [CrossRef]
33. Cheng, J.; Deming, T.J. Synthesis of Polypeptides by Ring-Opening Polymerization of α-Amino Acid N-Carboxyanhydrides. In *Peptide-Based Materials*; Springer: Berlin, Germany, 2011; pp. 1–26.
34. Nesměrák, K.; Němcová, I. Determination of Critical Micelle Concentration by Electrochemical Means. *Anal. Lett.* **2006**, *39*, 1023–1040. [CrossRef]
35. Imai, S.; Hirai, Y.; Nagao, C.; Sawamoto, M.; Terashima, T. Programmed Self-Assembly Systems of Amphiphilic Random Copolymers into Size-Controlled and Thermoresponsive Micelles in Water. *Macromolecules* **2018**, *51*, 398–409. [CrossRef]
36. Pegg, J.C.; Czajka, A.; Hill, C.; James, C.; Peach, J.; Rogers, S.E.; Eastoe, J. Alternative Route to Nanoscale Aggregates with a pH-Responsive Random Copolymer. *Langmuir* **2017**, *33*, 2628–2638. [CrossRef] [PubMed]
37. Zashikhina, N.; Sharoyko, V.; Antipchik, M.; Tarasenko, I.; Anufrikov, Y.; Lavrentieva, A.; Tennikova, T.; Korzhikova-Vlakh, E. Novel Formulations of C-Peptide with Long-Acting Therapeutic Potential for Treatment of Diabetic Complications. *Pharmaceutics* **2019**, *11*, 27. [CrossRef] [PubMed]

38. Zhu, C.; Jung, S.; Luo, S.; Meng, F.; Zhu, X.; Park, T.G.; Zhong, Z. Co-delivery of siRNA and paclitaxel into cancer cells by biodegradable cationic micelles based on PDMAEMA-PCL-PDMAEMA triblock copolymers. *Biomaterials* **2010**, *31*, 2408–2416. [CrossRef]
39. Wei, W.; Lv, P.P.; Chen, X.M.; Yue, Z.G.; Fu, Q.; Liu, S.Y.; Yue, H.; Ma, G.H. Codelivery of mTERT siRNA and paclitaxel by chitosan-based nanoparticles promoted synergistic tumor suppression. *Biomaterials* **2013**, *34*, 3912–3923. [CrossRef]

pharmaceutics

MDPI

Article

Improved In Vitro Model for Intranasal Mucosal Drug Delivery: Primary Olfactory and Respiratory Epithelial Cells Compared with the Permanent Nasal Cell Line RPMI 2650

Simone Ladel [1,2,†], Patrick Schlossbauer [1,†], Johannes Flamm [1], Harald Luksch [3], Boris Mizaikoff [2] and Katharina Schindowski [1,*]

1 Institute of Applied Biotechnology, University of Applied Science Biberach, Hubertus-Liebrecht Straße 35, 88400 Biberach, Germany

2 Institute of Analytical and Bioanalytical Chemistry, University of Ulm, Albert-Einstein-Allee 11, 89081 Ulm, Germany

3 School of Life Sciences, Technical University of Munich, Liesel-Beckmann-Straße 4, 85354 Freising-Weihenstephan, Germany

* Correspondence: schindowski@hochschule-bc.de; Tel.: +49-7351-582-498

† These authors contributed equally to this work.

Received: 24 June 2019; Accepted: 20 July 2019; Published: 1 August 2019

check for updates

Abstract: Background: The epithelial layer of the nasal mucosa is the first barrier for drug permeation during intranasal drug delivery. With increasing interest for intranasal pathways, adequate in vitro models are required. Here, porcine olfactory (OEPC) and respiratory (REPC) primary cells were characterised against the nasal tumour cell line RPMI 2650. Methods: Culture conditions for primary cells from porcine nasal mucosa were optimized and the cells characterised via light microscope, RT-PCR and immunofluorescence. Epithelial barrier function was analysed via transepithelial electrical resistance (TEER), and FITC-dextran was used as model substance for transepithelial permeation. Beating cilia necessary for mucociliary clearance were studied by immunoreactivity against acetylated tubulin. Results: OEPC and REPC barrier models differ in TEER, transepithelial permeation and MUC5AC levels. In contrast, RPMI 2650 displayed lower levels of MUC5AC, cilia markers and TEER, and higher FITC-dextran flux rates. Conclusion: To screen pharmaceutical formulations for intranasal delivery in vitro, translational mucosal models are needed. Here, a novel and comprehensive characterisation of OEPC and REPC against RPMI 2650 is presented. The established primary models display an appropriate model for nasal mucosa with secreted MUC5AC, beating cilia and a functional epithelial barrier, which is suitable for long-term evaluation of sustained release dosage forms.

Keywords: barrier model; nose-to-brain; primary cells; RPMI 2650; olfactory epithelium; respiratory epithelium

---

1. Introduction

Drug delivery to the brain for the treatment of central nervous system (CNS)-related diseases became of great interest and one of the most challenging research areas in the last decade [1,2]. The brain has a unique barrier to restrict the entry of neurotoxic substances into the CNS: the blood–brain barrier (BBB). This endothelial barrier, with a low rate of pinocytosis and strong tight junctions, is a major hurdle in CNS drug delivery. Over the last years, the need for alternative drug delivery strategies became more and more obvious [3,4]. One of these strategies is the application of drugs via the intranasal route [5]. In contrast to invasive methods like intraparenchymal, intracerebroventricular, or intrathecal injections/infusions, the noninvasive intranasal approach can be used for chronically given drugs potentially resulting in a higher patient compliance. Further advantages of the intranasal application are the large surface area for absorption, rapid uptake and the avoidance of the hepatic first-pass elimination. Disadvantages of this form of application are the limitation to potent drugs (only small volumes can be applied), the mucociliary clearance, and the physico-chemical features of the mucus layer (low pH, enzymatic degradation) [6,7]. In general, there are three relevant consecutive steps for nose-to-brain drug delivery (N2B): The first step is overcoming the epithelial barrier, the second step is the transport from the mucosa to sites of brain entry and the last step is the transport to other sites in the CNS [7]. In this work, for simplification only the first step is modelled. The nasal mucosa can be divided into four epithelial types: the squamous epithelium, the respiratory epithelium, the transitional epithelium and the olfactory epithelium. Of these epithelial types, the respiratory and the olfactory epithelium are most likely to be sites of substance uptake [8]. The detailed composition of those epithelial types have been extensively studied [9–12]. Briefly, the olfactory mucosa covers ~10% of the nasal cavity surface in man. It is characterised by a pseudostratified columnar epithelium that is located in the upper dorsal part of the nasal cavity. Olfactory sensory neurons are located in the epithelial layer, with neurites spanning from the nasal cavity into the olfactory bulb in the brain. In addition to these neurons, three further cell types are part of the olfactory epithelial layer: structuring sustentacular cells that function as epithelial supporting cells, tubular type duct cells from Bowman's glands and basal cells (progenitor cells) [13,14]. Approximately 90% of the human nasal mucosa is covered with the respiratory epithelium. This pseudostratified columnar secretory epithelium consist of goblet cells, ciliated cells, intermediate cells and basal cells [15].

There are two pathways to overcome the epithelial barrier of the respiratory and the olfactory epithelium: the intracellular pathway and the extracellular pathway. Intracellularly transported substances can either be endocytosed into olfactory sensory neurons and subsequently be transported to the brain via their axons, or they can be transported via transcytosis across sustentacular cells into the underlying connective tissue (*lamina propria*) [16–18]. The extracellular pathway comprises paracellular diffusion into the *lamina propria* through intercellular clefts [18]. The drug's pathway through the mucosa mainly depends on its lipophilicity and the molecular weight [19]. For small and low molecular weight hydrophilic molecules, such as fluorescein-labelled isothiocyanate-dextran (FITC-dextran; 4.4 kD) or sodium fluorescein (0.37 kD), mainly the paracellular pathway is reported [20–25]. These nontoxic and easily detectable fluorescence labelled chemicals are widely used model substances for drug permeation studies [22,23,26]. In contrast, the transcellular transport is described for large molecules such as proteins [27].

To sum up, the nasal mucosa has become a focus of interest for drug application, to overcome the BBB issue. However, for research the olfactory mucosa as well as the dorsal part of the respiratory mucosa is rather difficult to access [28]. A feasible and simple solution may be the use of *post-mortem* porcine mucosa. The pig's nasal mucosa resembles well the human nasal histology and physiology [29]. The approach to use pigs as model organism for ex vivo mucosa-related experiments is well established [29–33]. The major problem using ex vivo mucosa explants is the limited lifespan of the tissue even under nutritional support. A promising alternative are in vitro models, using epithelial cells under controlled external experimental conditions, allowing a simplification to display only the first barrier without considering blood flow, systemic distribution and other *lamina propria*-related effects [26]. There are two different approaches available for the establishment of in vitro epithelial barrier models. First, there are standard tumour cell lines like the human nasal epithelial cell line RPMI 2650, and primary epithelial cell models can be cultivated as second option [34,35]. These two cell types vary greatly in their lifespan and their differentiation ability. In theory, tumour cell lines are immortal, whereas primary cells are limited to a certain number of cell divisions [36]. However, most tumour cells lack the ability to fully differentiate into, e.g., ciliated or mucus producing cells, while primary cell cultures are morphologically and physiologically very close to their native state [26,37]. As factors such as cilia and mucus can possibly influence the drug uptake time and the bioavailability due to, e.g., enzymes even in stagnant cell culture, these are important parameters to consider for drug permeation evaluations. This is especially important to evaluate drug potential in vitro for simulations of the whole nasal mucosa as it was recently built up by Na et al. (2017) with commercially available human nasal primary cells [38].

Epithelial cell features such as the presence of cilia, mucus secretion and tight junction formation are important factors that influence the bioavailability of nasally applied drugs [7]. Generally, a heterogenic mixture of different cell types is preferred for a valid model, since this better represents the mucosal epithelial assembly in respect to the permeation profile of drugs through the cellular barrier [37,39]. Furthermore, cultured tumour cells tend to build multilayers, as their growth is mostly not limited by contact inhibition after reaching confluency, whereas primary cells remain in monolayers. For epithelial barrier models, monolayers should be preferred as these resemble the natural epithelial structure more closely. Nevertheless, there are several studies proving the tumour cells line originating from nasal squamous epithelium (septum) RPMI 2650 is a suitable model for the nasal epithelium [20,21,26,37,40]. This cell line is closely related to normal human nasal epithelium in terms of its karyotype and the cytokeratin polypeptide pattern as well as the presence of mucus on the cell surface [34,35,41]. Also the metabolic features of the cellular barrier were shown to be vaguely similar to the one of excised human nasal tissue [42,43]. A special characteristic of nasal epithelial cells is their ability to adapt to growth under air–liquid interface (ALI) conditions. The vast majority of recently developed cellular airway models use this method developed by Tchao et al. (1989) [44]. Hereby, cells are grown on porous plastic membrane inserts in multiwell plates with basolateral medium supply whereas the apical side of the cellular layer is exposed to air. These ALI conditions naturally drive cells to form strong tight junctions and to start differentiating towards, e.g., ciliated or mucus producing cells [45–47]. Hereby, the lack of differentiation of tumour cells can be a major problem to mimic epithelial barriers. In an attempt to validate these different approaches, Mercier et al. (2018) postulated generally accepted key criteria that appropriate in vitro barrier cell models should meet [37]. These key criteria comprise a sufficient high transepithelial electrical resistance (TEER) coupled with the presence of specific tight junction proteins such as *zonula occludens* 1 (ZO-1) and adherents junction proteins such as E-cadherin when grown under ALI conditions [48,49]. Furthermore, cellular models must allow a measurement of paracellular permeability and ideally express drug transporters for transcellular permeation. The ability to measure paracellular permeation of small to medium sized model substances like sodium fluorescein or FITC-dextran and the screen for cilia and marker proteins such as ZO-1 and mucins allows a comparison of different models.

In our research group, we focus on region-specific model development and application methods for the olfactory mucosa and for nose-to-brain drug delivery. Flamm et al. recently published a method to deliver drugs directly to the olfactory region in mice [50]. Furthermore, Röhm et al. established a screening platform for aerosolizable protein formulations for intranasal drug delivery [51].

In accordance to former investigations, in this study an epithelial barrier model for intranasal delivery was established and characterised that used olfactory and respiratory primary cells. We considered criteria that strongly affect intranasal delivery such as mucin production and cilia formation important for mucociliary clearance. First, the differences between respiratory and olfactory primary cells were investigated. Second, the primary cell barrier models were evaluated and compared against the well-established tumour cell line RPMI 2650. The suitability and feasibility of primary cells as nasal epithelial barrier models for intranasal delivery studies was determined by immunofluorescence, molecular and biochemical investigations of marker proteins, TEER value determination and FITC-dextran permeation.

2. Materials and Methods

2.1. Cell Culture

2.1.1. Primary Cells

Primary cells were harvested from mucosal explants from the respiratory and olfactory region of 4–6-month-old slaughterhouse pigs. Respiratory tissue was dissected with a short *post-mortem* delay of below 1.5 h from the *concha nasalis ventralis* (*c.n. ventralis*) and olfactory tissue from the dorsal part of the *concha nasalis dorsalis* (*c.n. dorsalis*) and the *concha nasalis media* (*c.n. media*) (see Figure 1). The dissected mucosa explants were disinfected using Octenisept® (Schülke & Mayr GmbH, Norderstedt, Germany) and washed twice with PBS (cell culture grade, Gibco® Invitrogen, Darmstadt, Germany). The epithelial cells were isolated from the connective tissue by incubation for 1 h at 37 °C with 1.4 mg/mL pronase (Sigma-Aldrich, Taufkirchen, Germany) in EBSS (Gibco® Invitrogen, Darmstadt, Germany), 20 U Penicillin-20 µg Streptomycin (PenStrep (10,000 U), AppliChem, Darmstadt, Germany), and 300 I.U./mg gentamycinsulfate (≥590 I.U./mg, Carl Roth, Karlsruhe, Germany; pronase medium, Table 1). To obtain isolated cells the pronase–mucosa suspension was gently agitated, the liquid was carefully removed, and subsequently centrifuged at 700 rpm for 3 min. The supernatant was discarded and the remaining cells resuspended in appropriate volumes of primary culture adhesion medium (DMEM:F12 (1:1), 20% FBS, 2 mM Gln, 1% NEAA, 20 U Penicillin-20 µg Streptomycin, 300 I.U./mg gentamycinsulfate, Table 1). It should be noted that the cultivation in T175 cell culture flasks was highly inefficient compared to smaller formats.

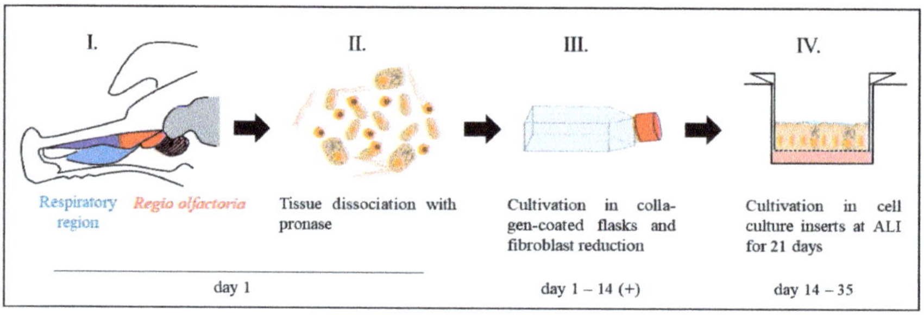

Figure 1. Workflow for primary cell isolation and cultivation under air–liquid interface (ALI) conditions. The mucosa was dissected either from the olfactory region (red) or the respiratory region (blue) (I.). Single cells were obtained by pronase digestion of the mucosa explants (II.). Single cells were cultivated in collagen-coated T75 cell culture flasks (III.). To reduce fibroblast overgrowth, the culture was shortly trypsinated (2–4 min) to get rid of the less adherent fibroblasts and select for epithelial cells. The cells were cultivated up to 80–90% confluence in the T75 flasks and transferred into cell culture inserts to grow under ALI conditions for 21 days (IV.).

The cells were seeded in collagen-coated cell culture T flasks (T flasks were incubated in advance with 0.05 mg/mL rat tail collagen solution (Primacyte, Schwerin, Germany) for 24 h at 37 °C) with a cell density of ~10^6 cells/mL and cultivated at 37 °C, 5% CO_2 and 95% rH. The medium was changed to primary culture medium (DMEM: F12 (1:1), 10% FBS, 2 mM Gln, 1% NEAA, 20 U Penicillin-20 µg Streptomycin, 300 I.U./mg gentamycinsulfate, Table 1) after 24 h. The exchange was necessary to contain fibroblast growth due to reduced serum concentration. Cells were regularly split by trypsination (Trypsin/EDTA, Biochrom, Berlin, Germany) for 10 min at 80% confluence.

Table 1. Composition of the cultivation media used in this work. EBSS = Earle's balanced salt solution; FBS = Foetal bovine serum; Gln = Glutamine; DMEM = Dulbecco's modified Eagle's medium; MEM = minimal essential medium.

Name	Composition
Primary culture adhesion medium	DMEM:F12 (1:1), 20% FBS, 2 mM Gln, 1% NEAA, 0.4 U/mL Penicillin-0.4 µg/mL Streptomycin, 0.6 I.U Gentamycinsulfate
Primary culture medium	DMEM:F12 (1:1), 10% FBS, 2 mM Gln, 1% NEAA, 0.4 U/mL Penicillin-0.4 µg/mL Streptomycin, 0.6 I.U Gentamycinsulfate
Pronase medium	EBSS + 1.4 mg/mL Pronase + 0.4 U/mL Penicillin-0.4 µg/mL Streptomycin, 0.6 I.U Gentamycinsulfate
RPMI 2650 medium	MEM, 10% FBS, 2 mM Gln, 0.4 U/mL Penicillin-0.4 µg/mL Streptomycin

For fibroblast depletion, the culture was incubated with trypsin/EDTA solution for 4 min at 37 °C twice a week. The fact that fibroblasts adherent less strongly to the cell culture flask surface was used to reduce fibroblasts by regular short-time trypsination steps The supernatant containing fibroblasts was discarded whereas the remaining adhered epithelial cells were washed with PBS. Fresh cultivation medium was added.

For seeding in cell culture inserts (ThinCert^{TM}, Greiner Bio-one, Frickenhausen, Germany), primary cells were isolated from T-flasks by a two-step trypsination to deplete remaining fibroblasts (Figure 1, IV). The cell culture was first treated with trypsin/EDTA solution for 4 min and the supernatant was removed. Subsequently, an additional trypsination step for 6 min was carried out to remove the epithelial cells. Membrane inserts were collagen coated as described before for the cell culture flasks and seeded with 1 × 10^5 cells per insert. The cells were cultured submerged for one day in the cell culture inserts. After 24 h, the apical medium was removed to cultivate cells under ALI conditions (260 µL medium per well) for 21 days. Cells were apically washed with 200 µL PBS and medium (260 µL/well) was changed every two days.

2.1.2. RPMI 2650 Cultivation

RPMI 2650 cells were cultivated in RPMI 2650 medium (MEM, 10% FBS, 2 mM Gln, 10 U Penicillin-10 µg Streptomycin, Table 1) at 37 °C, 5% CO_2 and 95% rH. Cells were regularly split at 80–90% confluency by a 5 min trypsination (trypsin/EDTA) treatment.

For permeation experiments, RPMI 2650 cells with passage numbers below 16 were seeded in cell culture inserts (ThinCert^{TM}, Greiner Bio-one, Frickenhausen, Germany) with a density of 1 × 10^5 cells per insert. After 24 h under submerged conditions, the apical volume was removed and the cells were cultivated under ALI conditions for 21 days. The ALI culture was washed every two days apically with 200 µL prewarmed PBS to remove diffused medium. Furthermore, 260 µL fresh medium was applied basolateral.

2.1.3. TEER Measurement

To evaluate the integrity of the cell layer the transepithelial electrical resistance measurement was used. Therefore, the cell culture inserts were filled apically with 350 µL MEM without phenol red (Gibco®, Invitrogen, Darmstadt, Germany) and basolateral filled with 500 µL MEM. For equilibration, cells were incubated for 20 min at 37 °C and 15 min cooled to room temperature (RT). The TEER value of each cell-covered membrane was determined in triplicates using an EVOM epithelial voltohmmeter and chopstick electrodes (World Precision Instruments, Sarasota, FL, USA). The raw data were processed by a blank subtraction (inserts without cells) and the multiplication by the growth area of the membrane (0.336 cm^2).

2.2. *Permeation*

To perform the permeation experiments, the medium was changed to 260 µL MEM without phenol red. For primary cells 10% FBS was added to the medium. To analyse fluorescein isothiocyanate-dextran solution (FITC-dextran) permeation, 100 µL of a solution with 500 µg/mL FITC-dextran (Sigma Aldrich, Taufkirchen, Germany) in PBS was applied on the apical surface of the cell layer. The experiments were carried out under normal cell culture conditions at 37 °C, 5% CO_2 and 95% rH. Permeation was studied for 24 h. A volume of 20 µL was taken as sample from the basolateral compartment at 0.5 h, 1 h, 2 h, 3 h, 4 h, 6 h, 8 h and 12 h. To keep the basolateral volume constant, 20 µL fresh medium was added. The samples were diluted 1:5 with PBS and analysed via fluorescence spectrometry (Spectra Max M5e, Molecular Devices, San Jose, CA, USA) at 490/520 nm.

2.3. *Immunofluorescence Staining*

Immunofluorescence (IF) staining was performed as described previously [31], with the difference that the cell-covered membranes were stained directly in the cell culture insert and transferred to an object slide as last step. Briefly, slides were washed with PBS pH 7.4 three times for 5 min, followed by blocking with block solution (4% BSA, 0.5% Saponin and 10% normal goat serum in PBS pH 7.4) for at least 2 h or overnight. The primary antibodies (Table 2) were diluted 1:100 in blocking buffer without normal goat serum and incubated in the cell culture insert for 24 to 48 h at 4 °C. Subsequently, the slices were washed (5 min, 10 min, and 15 min) and incubated with the adequate secondary antibodies (Table 2, 1:500 diluted in PBS pH 7.4) for 2 h. After washing again three times slides were mounted with Fluoroshield™ mounting medium containing DAPI (4′,6-diamidin-2-phenylindol; Sigma-Aldrich, Taufkirchen, Germany).

Table 2. List of antibodies used in this study. FITC: fluorescein-isocyanate; HRP: horseradish peroxidase; ZO-1: zonula occludens-1.

Antibody	Antigen	Immunogen	Host	Source, Cat. #
Anti-MUC5AC Antibody (45M1)	Peptide core of gastric mucin M1 (Mucin 5AC)	M1 mucin	mouse	Novus biologicals, Centennial, CO, USA, Cat. #NBP2-15196
Anti-ZO-1 (ZMD. 437)	ZO-1	synthetic peptide derived from the *N*-terminal region of human, dog, mouse, and rat ZO-1	rabbit	Thermo Fisher Scientific, Dreieich, Germany, Cat. #40-2300
Anti-acetylated tubulin (6-11B-1)	Acetylated tubulin	acetylated tubulin from the outer arm of *Strongylocentrotus purpuratus*	mouse	Sigma Aldrich, Taufkirchen, Germany, Cat. #T7451
Anti-β Actin (AC-15)	β Actin	not specified	mouse	Sigma Aldrich, Taufkirchen, Germany, Cat. #A5441
Anti-murine IgG-Alexa Fluor® 488	whole molecule mouse IgG	not specified	Goat	Jackson Immuno Research Europe Ltd., Cambridgeshire, UK, Cat. #115-545-003
Anti-rabbit IgG-Rhodamine Red™-X	whole molecule rabbit IgG	not specified	donkey	Jackson Immuno Research Europe Ltd., Cambridgeshire, UK, Cat. #711-295-152
Anti-murine IgG-HRP	whole molecule mouse IgG	not specified	goat	Sigma Aldrich, Taufkirchen, Germany, Cat. #AP5278
Anti-rabbit IgG-HRP	Whole molecule rabbit IgG	not specified	Goat	Jackson Immuno Research Europe Ltd., Cambridgeshire, UK, Cat. #111-035-003

2.4. *Cryosectioning and Colorimetric Staining of Cell Culture Insert Membranes*

Cell culture inserts were fixed in 4% paraformaldehyde for 10 min, cryo-conserved in 30% sucrose overnight and stored at 4 °C until sectioning. Before sectioning, the membranes were embedded in gelatine (7.5% gelatine and 30% sucrose in PBS [52]) to allow precise adjustment for sectioning. The membranes were cut in 14 µm slices using a cryostat at −25 °C (HM525 NX, Thermo Fisher

Scientific, Dreieich, Germany) and mounted on Superfrost® Plus Micro slides (VWR, Darmstadt, Germany). For hematoxylin-eosin staining, slides were washed with distilled water for 1 min, followed by a 5 min hematoxylin staining step. The slides were destained under running tap water for 3 min, and counterstained with Eosin Y-0.5% acidic acid (Sigma-Aldrich, Taufkirchen, Germany) for 5 min. After an additional destaining step (0.5 min), the slides were dehydrated (75% ethanol, 96% ethanol, 100% ethanol, Xylene; 2 min each) and mounted in Eukitt®Quick hardening mounting medium (Sigma-Aldrich, Taufkirchen, Germany).

2.5. *Reverse Transcription PCR (RT PCR)*

The transcript analysis was carried out as described before [31]. Briefly, the total RNA was isolated from cell lysates using TRIzol (Invitrogen™, Dreieich, Germany) according to the manufacturer's instructions. For reverse transcription, 1 µg of total RNA was used. Via oligo dT15 primers (Bioron, Ludwigshafen, Germany) and 400 U Reverase® (Bioron, Ludwigshafen, Germany) mRNA was transcribed to cDNA.

For PCR analysis of the cDNA library 1 µg cDNA, 1 µM primers (see Table 3), 25 mM $MgCl_2$, 2.5 mM dNTP Mix and 0.5 U/µL Taq polymerase were diluted in 10× Taq-PCR buffer (all Bioron, Ludwigshafen, Germany).

Table 3. Primer sequences of the targets *MUC5AC* and β-actin for RT-PCR.

mRNA Targets	Forward Primer (5′-3′)	Reverse Primer (5′-3′)
MUC5AC	CCGGGCCTGTGCAACTA	GTTCCCAAACTCGATAGGGC
β-actin	GACACCAGGGCGTGATGG	GCAGCTCGTAGCTCTTCTCC

The PCR reaction was performed with the following parameters: initial denaturation for 30 s at 95 °C, 40 cycles with denaturation at 95 °C for 30 s, annealing at 60 °C for 30 s, elongation at 72 °C for 60 s and a final elongation at 72 °C for 10 min. The PCR products were analysed using agarose gel electrophoresis.

2.6. *Dot Blot*

MUC5AC expression was analysed via dot blot as the high molecular weight protein is difficult to analyse via western blot [53]. Samples were collected from apical washing of ALI cultures with PBS as well as from tissue and cell lysates. The lysates were obtained as described before [31]. For dot blot analysis, 2 µL of these samples were applied onto a nitrocellulose membrane. The membrane was blocked by incubation with 5% skimmed milk powder (in PBS pH 7.4, 0.1% Tween20 (PBST)) for 2 h at RT. The anti-MUC5AC primary antibody (Table 2) was diluted 1:5000 in PBST and incubated for 24 h at 4 °C. The membrane was washed 4 times with PBST, and subsequently incubated with the HRP-coupled secondary antibody (dilution 1:4000 in PBST, Table 2) for 1h at RT. After another 4 washing steps, the membrane was developed by using the chemiluminescence substrate Immobilon® (Merck Millipore, Darmstadt, Germany). The analysis was carried out by Fusion FX Imaging systems (VILBER Lourmat, Collégien, France) and Image J (java.version: 1.8.0_171, National Institute of Health, Bethesda, MD, USA).

2.7. *Western Blot*

The western blot was performed as described before [31]. Briefly, the cell lysates were made by using chilled RIPA cell lysis buffer (10 mM Tris-Cl, pH 8.0; 1 mM EDTA, 0.5 mM EGTA, 1% Triton X-100; 0.1% sodium deoxycholate, 0.1% SDS and 140 mM NaCl; and protease inhibitor mix (Thermo Fisher Scientific, Dreieich, Germany) and gentle agitation. Equal volumes of homogenized tissue were loaded, separated in a 12.5% SDS PAGE and blotted onto a nitrocellulose membrane (Carl Roth, Karlsruhe, Germany). The membrane was blocked (5% skimmed milk powder in PBS/0.1% Tween20, pH 7.4).

Primary antibodies (Table 2, 1:5000) were incubated overnight at 4 °C. Secondary antibodies were diluted 1:100,000 (Anti-rabbit IgG-HRP) and 1:4000 (Anti-murine IgG-HRP). Signal detection was carried out by developing the membrane with the chemiluminescence substrate Immobilon® (Merck Millipore, Darmstadt, Germany) according to the manufacturer's instructions.

2.8. Statistics

Data were assessed for significance using unpaired *t*-test (GraphPad Prism 8) comparing the primary nasal cells to the RPMI 2650 cell line. The exact repeat numbers are divided into technical replicates (*n*) and biological replicates (*N*) meaning different pigs and are addressed in the subtitles of the graphs.

3. Results

The increasing interest in intranasal delivery requires the development and the evaluation of specialized in vitro models to reduce the use of laboratory animals for drug transport and permeation studies. These models should display the main parameters of the mucosa that influence the drug permeation. Here, we defined the most important parameters as the growth in defined monolayers, the formation of cilia (drug clearance), tightly connected cell layers and the physico-chemical environment (mucus). The actual state of the art model for nasal epithelial cells is the tumour cell line RPMI 2650 [21,39]. As an alternative to tumour cells lines, primary cell models are moving more and more into focus in many fields. Here, the aim of this study was to establish a robust protocol for nasal primary epithelial cell cultivation and to develop an appropriate primary epithelial cell model. This model should especially display the first epithelial barrier during intranasal drug delivery, and hence will be compared to the standard in vitro cellular model.

3.1. Evaluation of Nasal Primary Cells and RPMI 2650 Concerning Olfactory Mucosa Model Characteristica-Monolayer, Tight Growth, Cilia and Mucus Production

A general observation of primary nasal cell cultivation was the slow growth of epithelial cells along with faster growth of undesired fibroblasts and the rapid growth of bacterial contaminations (Figure 2A,B). For fibroblast growth we have determined the content of serum in the cell culture medium to be the most influencing factor. Primary cells displayed poor adherence to the collagen coated cell culture flasks at serum levels of 10%. Increasing the serum concentration to 20% resulted in primary cells that were adherent on collagen-treated cell culture flasks after 3–4 h. However, a serum content of 30% increased, above all, the growth of fibroblasts and decreased the growth of epithelial cells. Therefore, despite the positive effects on adhesion to cell flask surface the negative effects on fibroblasts' growth counterbalanced these benefits. Hence, seeding the cells in 20% serum content followed by a change to 10% serum after adhesion (5 h) turned out to be method of choice to culture the primary porcine nasal epithelial cells.

Morphologically, a majority of respiratory epithelial primary cells (REPC) and olfactory epithelial primary cells (OEPC) appeared in the typical flat and squared cobblestone shape with sizes varying from 10 μm to 100 μm (Figure 2C,D). Particularly in REPC cultures, some cells were detected that contained larger vesicles (Figure 2E). These cells had morphological similarities to goblet or gland cells. Ciliated cells were observed in OEPC and REPC, however REPC showed considerably more cells with motile cilia, whereas in OEPC primary, non-motile cilia were also present (Figure 2F). The motile cilia showed a median length of 10 μm with a beating frequency of >5 Hz (Supplementary Materials Video S1).

To investigate the general cell growth of primary nasal epithelial cells and the tumour cell line RPMI 2650 after 21 days under ALI conditions, colorimetric staining (Hematoxylin/Eosin, Figure 3A–C) of cryosections of cell-covered membranes was performed. OEPC and REPC grew in monolayers under ALI conditions whereas the RPMI 2650 cells formed up to 10 layers. In contrast to OEPC, REPC cultures consisted of larger and smaller cells resulting in an uneven monocellular layer surface (Figure 3B). Additionally, the distribution of cilia and the formation of tight junctions were determined using immunofluorescence (IF) staining against acetylated α-tubulin (cilia marker [54]) and ZO-1 (tight junction marker [55]) (Figure 3D–F). OEPC demonstrated the highest signal for the cilia marker acetylated α-tubulin. IF against ZO-1 showed that REPC formed the highest number of tight junctions. Only a weak signal for both marker proteins was observed in RPMI 2650.

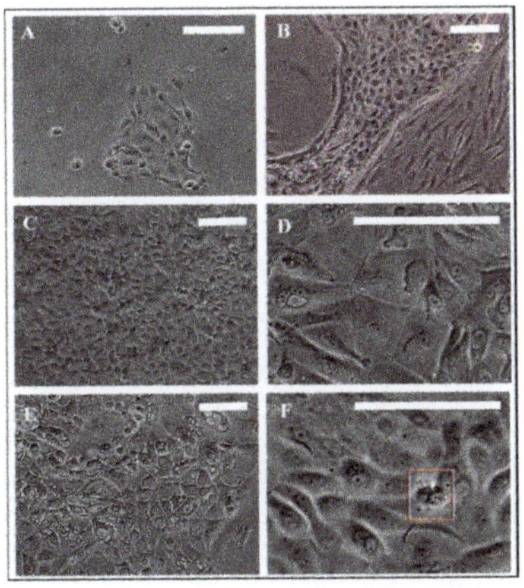

Figure 2. Morphology of porcine primary nasal epithelial cells. (**A**) Olfactory primary cells, 24 h in culture. (**B**) Fibroblast contamination in primary epithelial cells. (**C**,**D**) Morphological appearance of OEPC and REPC: small round or cobblestone shaped and big flat epithelial cells; cell membranes are clearly visible. (**E**) REPC: vesicular cells. (**F**) Red box: Ciliated cells can be motile or non-motile (beating frequency of >5 Hz). Cilia length was ~10 μm. OEPC: olfactory epithelial primary cells; REPC: respiratory epithelial primary cells; Scale bars: 100 μm.

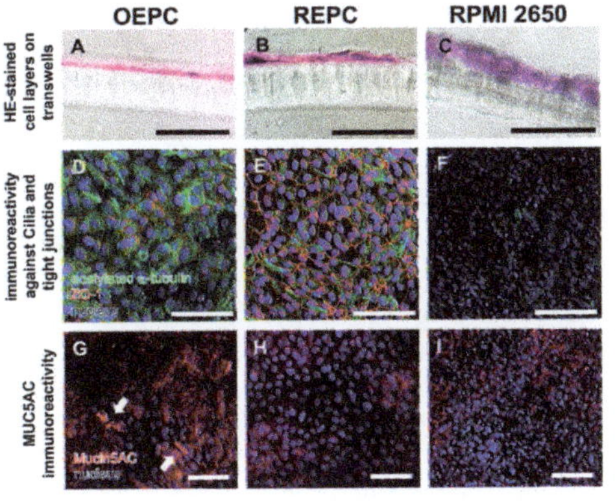

Figure 3. Characterisation of olfactory and respiratory primary cells in comparison to the standard cell line RPMI 2650. Monolayers are necessary to evaluate transport over the epithelial layer in the nasal mucosa: 14 μm sections were made of olfactory epithelium primary cells (OEPC, **A**), respiratory epithelial primary cells (REPC, **B**) and RPMI 2650 (**C**) grown on a cell insert membrane for 21 days. Morphological features such as tight junctions and the formation of cilia are important influencing factors in investigations of drug permeation and clearance studies. Acetylated α-tubulin is a common marker for cilia [54]. IF double-staining of acetylated α-tubulin and the tight junction marker zonula occludens-1 (ZO-1) of ALI cultures of OEPC (**D**), REPC (**E**) and RPMI 2650 (**F**) after 21 days of incubation were made. An additional feature of mucosal cells is the ability to produce mucus. The marker protein mucin 5AC was used in this work, because the olfactory mucosa is to be simulated above all to investigate nose-to-brain transport. Again, IF was performed in OEPC (**G**), REPC (**H**) and RPMI 2650 (**I**). Scale bars: 100 μm.

Beyond cilia and tight junctions, the physico-chemical environment of the epithelium also has an influence on intranasally administered substances. Therefore, the production of the mucin MUC5AC was investigated using IF (Figure 3G–I), RT-PCR (Figure 4A) and dot blot (Figure 4B). Both the dot blot cell lysate analysis, as well as the IF analysis, showed that OEPC have the highest immunoreactivity against MUC5AC. In the RT PCR no significant difference between *c.n. media* and the nasal primary epithelial cells could be observed. In contrast, the RPMI 2650 cells showed a significantly lower *MUC5AC* transcript level and no signal in the dot blot. Immunolabelling against MUC5AC resulted in weak signals in RPMI 2650 and REPC compared to OEPC.

Furthermore, we analysed the presence of secreted MUC5AC on the apical surface of the OEPC ALI cultures by washing the luminal compartment each 3 to 4 days. The supernatant of the washing buffer was then analysed by dot blot for immunoreactivity against MUC5AC (Figure 4C). At day 6 under ALI conditions the highest signal was observed. After 10 days the protein level appeared to stay at a constant level up to 20 days.

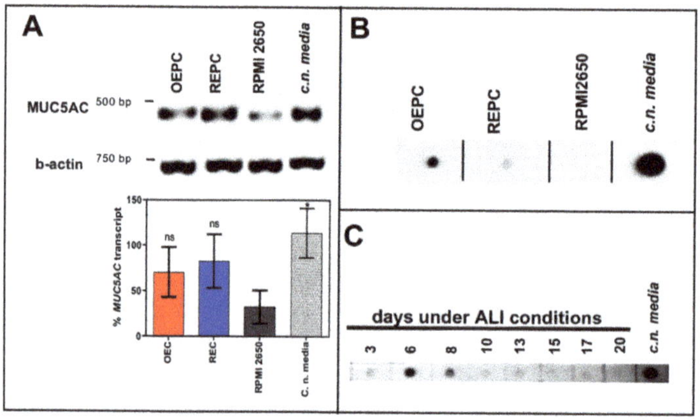

Figure 4. Mucin MUC5AC expression and immunoreactivity in primary cells of the nasal cavity. (**A**) Transcription analysis (RT-PCR) of *MUC5AC* gene in olfactory epithelial primary cells (OEPC), respiratory epithelial primary cells (REPC), tumour cell line RPMI 2650 and the *concha nasalis media* (*c.n. media*). MUC5AC transcript signal was referenced to beta-actin transcript signal. The significance was calculated by comparison of the OEPC, REPC and RPMI 2650 data with the *c.n. media* transcription data using an unpaired t-test. * $p < 0.05$; $n = 4$; error bars represent mean ± SD. (**B**) Dot blot analysis of MUC5AC protein in lysates of OEPC, REPC, RPMI 2650 and *c.n. media*. All OEPC and REPC cultures shown in (**A**,**B**) were cultivated for 14 days in vitro in T flasks. (**C**) Immunoreactivity against MUC5AC in OEPC that were first cultured for 7 days in T flask with a minimum confluency of 70% then under ALI conditions additional 20 days. Apically secreted mucus was collected at the days indicated corresponding to a mucin production of 2 to 3 days. Statistical analysis: unpaired *t*-test, * $p < 0.05$ compared to the standard model RPMI 2650.

3.2. FITC-Dextran Permeation Thought RPMI 2650 and Nasal Primary Cell Barriers

The analysis of the TEER values of 21 days ALI cultures of RPMI 2650, OEPC and REPC ALI revealed that REPC with 846 ± 550 Ω cm^2 displayed the highest TEER values followed by OEPC (648 ± 371 Ω cm^2) and last RPMI 2650 (66 ± 5 Ω cm^2; Figure 5A). Analogous to the permeation data, the nasal primary cells showed considerably higher variability displayed as higher standard deviations compared to RPMI 2650. The highest TEER value measured for OEPC was 1000 Ω cm^2, whereas REPC reached a TEER value of 1600 Ω cm^2.

To evaluate cell barrier function, TEER values of the cell layers after 21 days under ALI conditions were determined and permeation of FITC-dextran was analysed from the luminal/apical to the abluminal/basolateral compartment. Determination of permeability was performed over 24 h with different sampling times. The results were compared to a diffusion control without cells (cell culture insert membrane only). Here, RPMI 2650 showed the highest permeability at all times investigated in comparison to OEPC and REPC. After 24 h, 11.7 ± 0.9% of FITC-dextran permeated through RPMI 2650 cells, whereas OEPC showed a permeation of 3.1 ± 2.0% and REPC resulted in 3.7 ± 2.0% FITC-dextran (Figure 5B). An equilibrium is obtained for FITC-dextran in the luminal and abluminal compartment of the insert. It should be noted that this equilibrium establishes earlier in the control insert without cells and limits the maximal flux of molecules in this static system (see Supplementary Materials Figure S1A,B). As the published literature usually does not acknowledge this fact, we normalized these results to the maximal concentration of FITC-dextran that is able to diffuse to the abluminal compartment within the same time (Supplementary Materials Figure S1B). The molecular flux was calculated for each cell type investigated and displayed as the amount of FITC-dextran that was able to cross the epithelial barrier per time and distance (Figure 5C). Here, it is obvious that RPMI 2650 cells are rather leaky and display a significantly higher flux compared to OEPC and REPC.

In these experiments, we observed a correlation of permeability and the TEER value for OEPC and REPC (Figure 5D). For RPMI 2650 cells no correlation occurred as the measured TEER values showed a very narrow range from 55 Ω cm^2 up to 75 Ω cm^2. Hereby, a higher TEER value did not result in a lower flux. In contrast to REPC and OEPC, a higher TEER value resulted in a lower flux. We determined a minimum TEER value of about 300 Ω cm^2 that appears to be necessary for primary cells to achieve reproducible permeability results. Hence, all cell-layered inserts used for the flux experiments with FITC-dextran had to meet this criterion.

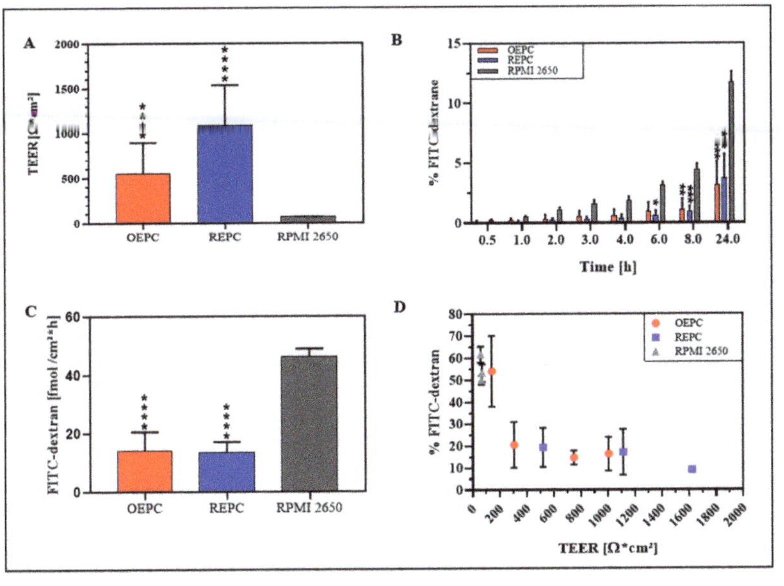

Figure 5. Comparison of FITC-dextran permeation and TEER value of nasal primary cells vs. RPMI 2650. (**A**) TEER values of OEPC, REPC and RPMI 2650 after 21 days ALI. (**B**) FITC-dextran permeation data: Values of FITC-dextran permeated through a cell layer were normalized to total of FITC-dextran. (**C**) Flux of FITC-dextran through an OEPC, REPC and RPMI 2650 layer after 24 h. (**D**) Correlation of percentage FITC-dextran permeation after 24 h and TEER value. Statistical analysis: unpaired *t*-test, * $p < 0.05$, ** $p < 0.01$, *** $p < 0.001$, **** $p < 0.0001$; all compared to the standard model RPMI 2650. OEPC: olfactory epithelial primary cells; REPC: respiratory epithelial primary cells; TEER: Transepithelial electrical resistance. FITC-dextran: Fluorescein isothiocyanate-dextran; $N = 4$, $n = 21$; error bars represent mean ± SD of all repetitions.

4. Discussion

The increasing importance of intranasal drug delivery leads to a growing need for adequate models to test permeation of drugs through the different nasal epithelia. In the olfactory region, sensory neurons located in the epithelium were implicated to be involved in drug transport: either with a direct uptake by dendritic knobs or by epithelial or gland cells, depending highly on the molecule [5,56,57]. To display the pathway and transport of drugs to the brain in vivo studies are necessary where the intracellular transport in neurons can be determined using mucosa explants ex vivo [31,57]. However, studies on nasal olfactory epithelium are harder to perform in vivo and ex vivo as this area is rather inaccessible and difficult to reach [26,58]. In vitro models are robust tools to simulate active or passive transport through the first epithelial barrier without modelling neuronal transport. Nevertheless, in vitro cell systems can be seen as simplified models of the first epithelial barrier. They can be used to study specific transporters for high molecular weight molecules and unspecific diffusion of small molecules [59–61]. Nevertheless, considering transport mechanism and cell type specific characteristic it is necessary to set stringent criteria for these cellular models. In the nasal mucosa the epithelial cells form the first barrier for intranasal drug delivery. They are responsible for either uptake or clearance of the administered drugs [6,62]. Parameters that influence the uptake of substances with higher molecular weight are mainly the cell type (e.g., gland cells, sustentacular cells, immune-related microfold cells, etc.) and the expression of specific transport proteins such as aminopeptidases for peptide permeation studies [26,31,42,63]. Furthermore, the drug clearance alongside of nasal mucosal barriers is highly associated with cilia formation and beating [62]. In the olfactory epithelium the cells are described to form non-motile cilia. However, it was demonstrated that so called respiratory mucosa patches exist in the *regio olfactoria* that consist of epithelial cells with beating cilia. Consequently, it is harder to define the clearance rate of the olfactory mucosa only [10]. In addition, for drug stability the physico-chemical environment has to be considered which includes mucus production and the growth under ALI conditions [64]. According to these parameters that influence drug permeation on nasal mucosa, we set the following criteria for the in vitro model: tight epithelial barrier and tight junction formation, cellular monolayer, cilia formation and mucus production.

Standard nasal permeation models are often based on the tumour cell line RPMI 2650. This permanent cell line is derived from an anaplastic squamous cell carcinoma of the nasal septum, which is part of the respiratory region of the nose. Its human origin makes this cell line advantageous in terms of species comparability. Moreover, this cell line was described to mimic the human nasal mucosa and is well investigated in terms of expression of mucus-like material, aminopeptidases, ABC and SLS transporters, cytokeratin patterns, tight junctions and the growth under ALI conditions [26,41,65,66]. The good characterisation makes the RPMI 2650 cell line an established model for intranasal delivery. However, differences in the handling of this cell line have been reported to highly impact growth and protein expression, which may question the reliability of this model [37,67]. Also, we noticed that the permeation rate, as well as the TEER values, depends on the passage count of the RPMI 2650 cells; the higher the passage the lower the TEER value and the higher the permeation rate (unpublished observation). Thus, the nasal mucosa has a heterogeneous composition of several cell types such as ciliated, goblet, columnar and basal cells whereas the RPMI 2650 cell line is only composed of squamous cells which has a limiting impact on the physiology of this model. Furthermore, like most tumour cell lines also the RPMI 2650 cell line suffers from drawbacks such as genetic instability, lack of differentiation and unstable protein expression profiles [37]. The approach that probably mimics best the in vivo situation are ex vivo tissue explants from the olfactory or the respiratory mucosa. Such explants contain all features and the correct cellular composition of the nasal mucosa. However, to take specimen from human olfactory mucosa is associated with a reasonable risk. Yet, porcine nasal tissue is highly similar to the human nasal mucosa and therefore explants are available from slaughterhouse pigs [29]. However, so far there is no cultivation protocol available for long time studies with those tissue samples. We found out 8 h to be the maximum cultivation time before the tissue integrity is highly compromised in histological sections [31]. Consequently, for long-term studies, e.g., to analyse the influence of drug formulations or drug uptake from sustained release, this model is limited.

Hence, we compared the long-term culture of primary nasal cells derived from porcine respiratory and olfactory mucosa with the standard model RPMI 2650 to gain more information about the relevance of potential artefacts due to the use of a cancer cell lines.

4.1. Comparison of OEPC and REPC: Differences in Barrier Formation and Marker Protein Expression

Our results show that there are actually differences between OEPC and REPC in terms of tight junction formation and, consequently, in TEER values, as well as in the expression of the secretory mucin MUC5AC. The permeation experiment, especially the sample time points below 8 h, resulted in a lower permeation through REPC in comparison to OEPC. This can also be related to the higher TEER values of REPC as shown in the correlation of TEER values to permeated FITC-dextran (Figure 5A). The TEER values are amongst others influenced by the thickness of the cell layer and the formation of tight junctions [68]. Both primary cell types form only monolayers that are comparable in size to mucosal epithelial cells. Thus, we conclude that REPC form a tighter barrier in comparison to the OEPC, also shown in the immunostaining of the tight junctions (ZO-1). This is in accordance with the common knowledge of the in vivo olfactory and respiratory epithelium as the epithelial cells in the respiratory epithelium are strongly connected via tight junctions, whereas epithelial cells in the olfactory epithelium are sealed to the olfactory sensory neurons neither by tight nor by gap junctions [69]. Thus, our results concerning the ZO-1 signal of the REPC are comparable to the human nasal primary cell model MucilAir^{TM} [40]. In contrast to the respiratory epithelium, olfactory sensory neurons are located in the olfactory epithelium surrounded by sustentacular cells. The olfactory sensory neurons are sealed to those cells and span from the apical surface to the olfactory bulb [9]. Neurogenesis of olfactory neurons is known to be a continuous process to replace lost neurons in the epithelial layer, which leaves a gap in the epithelial layer until the new neuron takes its place [70]. Therefore, it might be possible that the lower number of tight junctions and the lower TEER value seen in the OEPC is correlated to the neurogenesis and the connection between the sustentacular cells and the olfactory sensory neurons. Pezzulo et al. (2010) described TEER values ranging from 700 to 1200 Ω cm^2 for primary cells originated from the tracheal and bronchial tract, which are similar to the TEER values for REPC in the present study [47]. However, for human nasal epithelial cells TEER values up to 3155 Ω cm^2 are reported [71,72]. In contrast, ex vivo TEER measurements of excised human nasal mucosa range from 60 to 180 Ω cm^2 [37]. The discrepancy between the excised mucosa and the primary cells could possibly be explained by damage to the axons of the olfactory sensory neurons during the dissection. These tight junctions are highly dependent on the presence of certain ions. Ca^{2+} ions were reported to be necessary to form tight barriers and a high electrical resistance. It might be possible that the excised mucosa suffers from ion depletion when it is no longer connected to blood flow and therefore shows lower TEER [73]. TEER measurements in rabbit airway epithelium in vivo resulted in a higher TEER value of 260–320 Ω cm^2 [74]. Besides the preparation and the tight junction formation, the cellular composition of the tissue, e.g., the presence of olfactory sensory neurons, plays an important role in the TEER measurement. Furthermore, gland ducts such as from Bowman's glands will also lower the TEER value. In the primary culture we did not observe whole glands and neurons. Hence, the higher TEER values could be a result of lacking gland ducts and void neurons.

The ability to form cilia is shared by REPC and OEPC as seen in the immunoreactivity against acetylated α-tubulin. Different groups postulated that acetylated α-tubulin might be a selective marker for primary (non-motile) and motile cilia of polarized cells [75–77]. In culture, the cilia had a mean length of 10 μm and beating frequencies of >5 Hz. According to the literature this indicates healthy cilia formation [78–80]. In the culture flasks beating cilia were seen in patch-like clusters whereas under ALI conditions the cell layer showed a confluent signal for acetylated α-tubulin. Nevertheless, to further set up valid experiments for drug clearance studies the differentiation of the epithelial cells towards cells with motile cilia, for example by using retinoic acid, should be optimized [81]. Bateman et al.

(2013) for example published a porcine tracheal epithelial cell model to investigate virus infection and replication. They used a complex medium containing epidermal growth factor (EGF) for proliferation and retinoic acid to promote ciliogenesis [81]; similar to the present study they obtained ciliated cells. In contrast, those cells grew in multilayers with TEER values increasing up to 12 days to 800 Ω cm^2 and a subsequent decrease to 250–300 Ω cm^2 after 18 days. We do not observe this effect as we found TEER values around 850 Ω cm^2 for REPC and 650 Ω cm^2 for OEPC after 21 days of ALI cultivation. As we cultured the cells in serum-containing medium without the additional supplementation of EGF or retinoic acid there must be other factors influencing the barrier function that are present in the serum but not in serum-free bronchial epithelial growth medium. As the use of serum-free medium is rather expensive, it is of high interest to further investigate the barrier formation influencing factors present in vivo and in the serum. In contrast to our findings and the findings of Bateman et al. (2013), Delgado-Ortega et al. (2014) reported a leaky barrier model for cells from newborn pig trachea [81,82]. Probably here again the influence of the in vivo cell location and maybe also the age of the donor on the barrier model plays a role. In terms of intranasal drug delivery, using both cell types—REPC and OEPC—for barrier investigations with donor pigs with a similar age can give more detailed information of the permeation in vivo and the most efficient application site in terms of bioavailability.

In terms of mucus expression, the REPC showed lower levels of the secretory mucin MUC5AC expression but transcript levels similar to the OEPC. It has already been described in the literature that transcript expression and the actual protein amount do not always correlate [66]. The reduction of MUC5AC production in OEPC results most probably from lower amounts of glands cells over time as they do not grow under ALI conditions in vivo but in gland formations. A lack of necessary factors such as calcium can be excluded as the medium is regularly exchanged and its composition is constant. The plateau in MUC5AC production after 10 days is possibly due to the fact that also the epithelial cells (sustentacular cells) themselves are capable of producing mucus in vivo [83]. The mucin subtype MUC5AC is known to be mainly expressed in Bowman's glands of the olfactory mucosa and is therefore a rather poor marker for respiratory epithelial cells [64,84,85]. We screened for this marker protein as we are particularly interested in modelling the *regio olfactoria* for intranasal drug delivery, and therefore the application of drugs at the olfactory epithelium. The absence of MUC5AC does not mean the cells lack mucus production in general, but only this specific mucin type that is known to be highly expressed in the olfactory mucosa [84]. Aust et al. (1997) found MUC5AC to be expressed only in subpopulations of epithelial cells in the human inferior turbinate [86]. Furthermore, recent studies postulated MUC5AC to be expressed in higher amounts in the olfactory mucosa in comparison to the respiratory mucosa which is in accordance with the results of our study. Those data show clear differences in mucin type expression (MUC1, MUC2, MUC5AC and MUC5B) patterns between respiratory and olfactory epithelial cells [87]. Thus, this study supports our hypothesis that the choice of the cellular model should be based on the site of drug application. MUC5AC expression was previously found in human primary nasal epithelial cells after 21 days of cultivation under ALI conditions which is in accordance to our results with 21 day-ALI culture of OEPC [46]. Also the cultivation conditions were comparable despite of the use of serum-containing media, which however had no obvious disadvantageous effect on the porcine primary epithelial culture.

Nevertheless, taken together there are several differences concerning REPC and OEPC that can influence the drug permeation. It thus appears to be advantageous to use the cell type of the area of interest for the drug application.

4.2. *Primary Cell Model Evaluation: RPMI 2650 vs. Primary Nasal Cell Barrier*

The most prominent disadvantages of primary cells are the high variability between different batches and the slow and limited growth in comparison to tumour cell lines. Because of those, and other reasons, the current standard model for nasal mucosal permeation is the tumour cell line RPMI 2650 [26]. The TEER values and permeation rates measured here fit the results of other groups, showing again that the RPMI 2650 is highly reproducible [20,88,89]. Yet, several differences were observed between RPMI 2650 and primary nasal cells that influence the permeation of molecules. Probably the most influencing factor is the lower TEER value of the RPMI 2650 cells. This goes along with the low expression of tight junction marker ZO-1 in comparison with the primary nasal cells. The paracellular permeation of molecules is highly restricted by these tight junctions in the epithelial barrier [49]. There is a high variance in the TEER values yielded for RPMI 2650 under ALI conditions in the literature from the large range observed (41 Ω cm^2 to 270 Ω cm^2), which most probably depends on the size of the cellular multilayer and passage number. Compared to primary cells, RPMI 2650 forms leaky multilayered barriers as described before [22,37,39,88,90]. Mercier et al. (2018) described this discrepancy by the strong dependency on culture conditions and the experience of the operators. Therefore TEER values of 70 to 100 Ω cm^2 are regarded as valid for cultures grown 21 days under ALI conditions [37]. Returning to the above-mentioned criteria for a suitable in vitro model for intranasal drug delivery, the primary nasal epithelial cells demonstrated a 10 to 13 fold higher TEER value and considerable higher signals for the ZO-1 marker protein. However, the TEER values yielded for the RPMI 2650 cells are similar to the TEER values observed in excised human nasal mucosa (TEER value: 60 to 180 Ω cm^2) as described above. In addition to the possible explanations of the low TEER value of the ex vivo explants discussed above, it is in general rather questionable to compare TEER value of a single-component system such as the homogenous RPMI 2650 model or even the heterogeneous primary cell models with TEER values from multi-component systems like excised nasal mucosa, as the latter contains components such as e.g., glands and neurons that influence TEER values.

In terms of the production of MUC5AC the RPMI 2650 showed the lowest transcript level and very low signal in the immunofluorescence staining. This is in accordance with previous studies from Berger et al. (1999) who reported discrepancies between the transcript expression and the protein expression in tumour cell lines [66]. In their work they state that synthesis, maturation and secretion are most probably regulated post-transcriptionally. In general, RPMI 2650 are mostly undifferentiated, homogenous epithelial cells that originate from squamous epithelium which is part of the respiratory mucosa. Together with the lack of differentiation to form goblet cells, the lower expression of MUC5AC in comparison to OEPC is therefore explainable [20]. However, in current literature the mucus secretion pattern of RPMI 2650 cells shows discrepancies concerning the expression level and the mucin type as previously described by Mercier et al. (2018) [37]. This might be due to different culture conditions or the use of cells with different ages. To fully evaluate the mucin protein expression of the RPMI 2650 model further investigations are required.

In respect to the last criterion—the formation of cilia—our results showed very low and diffuse signals for acetylated tubulin in RPMI 2650 in comparison to well defined and fibril-shaped structures in OEPC and REPC. Several groups described the RPMI 2650's inability to differentiate and form cilia which we can support in our present study [37,90]. Furthermore, the tendency of RPMI 2650 to form homogenous multilayers is highly disadvantageous when heterogeneous epithelial layers should be simulated. Also, these cells are not polarized as described for epithelial cells but rather undifferentiated round-shaped stapled cells without a growth direction [91]. It is also questionable whether the cells located in the intermediate layers show the same phenotype as the cells grown at the ALI interface. Especially for transport mechanisms when polarized cells are required the tumour cell model is unfavoured whereas the primary cell model sufficiently fits those criteria.

5. Conclusions

In the present study, we demonstrate, for the first time, differences between primary epithelial cells from the olfactory and respiratory nasal region. Consequently, we highly recommend using primary cells originated from the equivalent tissue type for model intranasal application of drugs on the olfactory mucosa. In terms of comparability of porcine and human primary epithelial cells we obtained similar results for ciliogenesis, and mucus production as well as similar TEER values [40,71,92]. As the TEER value is the state-of-the-art parameter to evaluate barrier models, the pig is corroborated as an adequate surrogate for human tissue. Furthermore, for the investigation of complex drug transport pathways primary cell culture should be favoured instead of RPMI 2650 tumour cell lines as those cells have different physiological features due to their growth in multilayers and their lack of differentiation ability. In general, we assume the heterogeneity of the primary cells to be advantageous compared to the RPMI 2650 as we see differences in uptake rates of molecules in different cells in our preliminary data. In conclusion, we strongly recommend the observation of several factors such as the analyte molecule, receptor or transporter expression, the importance of adequate physico-chemical environment and the impact of cilia to choose the correct model for the permeation investigation. The primary cell model is a cheap and robust alternative to the tumour cell line RPMI2650 with the advantage of having all features of epithelial cells in vivo and lacking the throwbacks of tumour cells such as genetic instability. It thus has the potential to replace the tumour cell model even in early assessment of drug development as drug efficacy and toxicity can be determined with a higher comparability to the in vivo situation. However, to fully evaluate the full potential of the primary cell model ex vivo experiments should be performed to compare the permeation of a paracellular transported molecule such as FITC-dextran. Furthermore, our study focused on the barrier function and cell type specific features of the cellular models. We only studied the paracellular transport. For further evaluation, the permeation of transcellular transported molecules such as mannitol should also be assessed.

Supplementary Materials: The following are available online at http://www.mdpi.com/1999-4923/11/8/367/s1, Figure S1: Permeation data referenced to control without cells vs. referencing to total FITC-dextran, Video S1: Beating cilia in submerged OEPC culture.

Author Contributions: Conceptualization, S.L., P.S. and K.S.; Methodology, S.L., P.S., J.F., H.L. and K.S.; Validation, J.F., S.L. and P.S.; Investigation, S.L. and P.S.; Resources, K.S.; Visualization, S.L. and K.S.; Writing—Original Draft Preparation, S.L., P.S. and K.S.; Writing—Review & Editing, J.F., H.L. and B.M.; Supervision, K.S. and B.M.; Project Administration, K.S.; Funding Acquisition, K.S.

Funding: This study was supported by an EU grant "N2B-patch" under the European Framework Programme for Research and Innovation Horizon 2020 (Grant No. 721,098; www.n2b-patch.eu), from a grant "FcRn in Drug Delivery" funded by the Baden-Württemberg State Ministry of Science, Research and Arts, from a grant "ALIVE" funded by the Federal Ministry for Economic Affairs and Energy (BMWi) and from fellowships from the Stiftung der Deutschen Wirtschaft (to S.L. and J.F.).

Acknowledgments: A special thanks to Rene Handrick from Biberach University for excellent advisory and technical support whenever needed. Also, special thanks to Angelika Rück, Dominik Biesinger and Sviatlana Kalinina from the core facility for confocal and multiphoton microscopy of the Ulm University. Moreover, for excellent support in microscopical questions the authors want to thank Stefan Weigel from the Technical University Munich.

Conflicts of Interest: The authors declare no conflicts of interest.

References

1. Barnabas, W. Drug targeting strategies into the brain for treating neurological diseases. *J. Neurosci. Methods* **2019**, *311*, 133–146. [CrossRef] [PubMed]
2. Illum, L. Nasal drug delivery: New developments and strategies. *Drug Discov. Today* **2002**, *7*, 1184–1189. [CrossRef]
3. Neuwelt, E.; Abbott, N.J.; Abrey, L.; Banks, W.A.; Blakley, B.; Engelhardt, B.; Grammas, P.; Nedergaard, M.; Nutt, J.; Pardridge, W.; et al. Strategies to advance translational research into brain barriers. *Lancet Neurol.* **2008**, *7*, 84–96. [CrossRef]
4. Gao, H. Progress and perspectives on targeting nanoparticles for brain drug delivery. *Acta Pharm. Sin. B* **2016**, *6*, 268–286. [CrossRef] [PubMed]
5. Illum, L. Nasal drug delivery—Possibilities, problems and solutions. *J. Control. Release* **2003**, *87*, 187–198. [CrossRef]
6. Costantino, H.R.; Illum, L.; Brandt, G.; Johnson, P.H.; Quay, S.C. Intranasal delivery: Physicochemical and therapeutic aspects. *Int. J. Pharm.* **2007**, *337*, 1–24. [CrossRef]
7. Lochhead, J.J.; Thorne, R.G. Intranasal delivery of biologics to the central nervous system. *Adv. Drug Deliv. Rev.* **2012**, *64*, 614–628. [CrossRef]

8. Harkema, J.R.; Carey, S.A.; Wagner, J.G. The Nose Revisited: A Brief Review of the Comparative Structure, Function, and Toxicologic Pathology of the Nasal Epithelium. *Toxicol. Pathol.* **2006**, *34*, 252–269. [CrossRef]
9. Gizurarson, S. Anatomical and histological factors affecting intranasal drug and vaccine delivery. *Curr. Drug Deliv.* **2012**, *9*, 566–582. [CrossRef]
10. Paik, S.; Lehman, M.; Seiden, A.M.; Duncan, H.J. Olfactory Biopsy. *Arch. Otolaryngol. Head Neck Surg.* **1992**, *118*, 731–738. [CrossRef]
11. Marttin, E.; Schipper, N.G.M.; Coos Verhoef, J.; Merkus, F.W.H.M. Nasal mucociliary clearance as a factor in nasal drug delivery. *Adv. Drug Deliv. Rev.* **1998**, *29*, 13–38. [CrossRef]
12. Graziadei, P.P.; Graziadei, G.M. Neurogenesis and neuron regeneration in the olfactory system of mammals. I. Morphological aspects of differentiation and structural organization of the olfactory sensory neurons. *J. Neurocytol* **1979**, *8*, 1–18. [CrossRef]
13. Morrison, E.E.; Costanzo, R.M. Morphology of olfactory epithelium in humans and other vertebrates. *Microsc. Res. Tech.* **1992**, *23*, 49–61. [CrossRef]
14. Mittal, D.; Ali, A.; Md, S.; Baboota, S.; Sahni, J.K.; Ali, J. Insights into direct nose to brain delivery: Current status and future perspective. *Drug Deliv.* **2014**, *21*, 75–86. [CrossRef]
15. Jafek, B.W. Ultrastructure of human nasal mucosa. *Laryngoscope* **1983**, *93*, 1576–1599. [CrossRef]
16. Anton, F.; Peppel, P. Central projections of trigeminal primary afferents innervating the nasal mucosa: A horseradish peroxidase study in the rat. *Neuroscience* **1991**, *41*, 617–628. [CrossRef]
17. Deatly, A.M.; Haase, A.T.; Fewster, P.H.; Lewis, E.; Ball, M.J. Human herpes virus infections and Alzheimer's disease. *Neuropathol. Appl. Neurobiol.* **1990**, *16*, 213–223. [CrossRef]
18. Balin, B.J.; Broadwell, R.D.; Salcmant, M.; El-Kalliny, M. Entry of peripherally administered protein to the CNS in mouse, rat and squirrel monkey. *J. Comp. Neurol.* **1986**, *251*, 260–280. [CrossRef]
19. Bahadur, S.; Pathak, K. Physicochemical and physiological considerations for efficient nose-to-brain targeting. *Expert Opin. Drug Deliv.* **2011**, *9*, 19–31. [CrossRef]
20. Wengst, A.; Reichl, S. RPMI 2650 epithelial model and three-dimensional reconstructed human nasal mucosa as in vitro models for nasal permeation studies. *Eur. J. Pharm. Biopharm.* **2010**, *74*, 290–297. [CrossRef]
21. Gonçalves, V.S.S.; Matias, A.A.; Poejo, J.; Serra, A.T.; Duarte, C.M.M. Application of RPMI 2650 as a cell model to evaluate solid formulations for intranasal delivery of drugs. *Int. J. Pharm.* **2016**, *515*, 1–10. [CrossRef]
22. Kürti, L.; Veszelka, S.; Bocsik, A.; Dung, N.T.K.; Ózsvári, B.; Puskás, L.G.; Kittel, Á.; Szabó-Révész, P.; Deli, M.A. The effect of sucrose esters on a culture model of the nasal barrier. *Toxicol. Vitr.* **2012**, *26*, 445–454. [CrossRef]
23. Nakamura, K.; Maitani, Y.; Takayama, K. The enhancing effect of nasal absorption of FITC-dextran 4400 by β-sitosterol β-D-glucoside in rabbits. *J. Control. Release* **2002**, *79*, 147–155. [CrossRef]
24. Grainger, C.I.; Greenwell, L.L.; Martin, G.P.; Forbes, B. The permeability of large molecular weight solutes following particle delivery to air-interfaced cells that model the respiratory mucosa. *Eur. J. Pharm. Biopharm.* **2009**, *71*, 318–324. [CrossRef]
25. Nicolazzo, J.A.; Reed, B.L.; Finnin, B.C. The Effect of Various in Vitro Conditions on the Permeability Characteristics of the Buccal Mucosa. *J. Pharm. Sci.* **2003**, *92*, 2399–2410. [CrossRef]
26. Schmidt, M.C.; Peter, H.; Lang, S.R.; Ditzinger, G.; Merkle, H.P. In vitro cell models to study nasal mucosal permeability and metabolism. *Adv. Drug Deliv. Rev.* **1998**, *29*, 51–79. [CrossRef]
27. Kim, K.-J.; Malik, A.B. Protein transport across the lung epithelial barrier. *Am. J. Physiol. Cell. Mol. Physiol.* **2015**, *284*, L247–L259. [CrossRef]
28. Moran, D.T.; Rowley, J.C., 3rd; Jafek, B.W. Electron microscopy of human olfactory epithelium reveals a new cell type: The microvillar cell. *Brain Res.* **1982**, *253*, 39–46. [CrossRef]
29. Wadell, C.; Björk, E.; Camber, O. Nasal drug delivery—Evaluation of an in vitro model using porcine nasal mucosa. *Eur. J. Pharm. Sci.* **1999**, *7*, 197–206. [CrossRef]
30. Samson, G.; De La Calera, A.G.; Dupuis-Girod, S.; Faure, F.; Decullier, E.; Paintaud, G.; Vignault, C.; Scoazec, J.Y.; Pivot, C.; Plauchu, H.; et al. Ex vivo study of bevacizumab transport through porcine nasal mucosa. *Eur. J. Pharm. Biopharm.* **2012**, *80*, 465–469. [CrossRef]
31. Ladel, S.; Flamm, J.; Zadeh, A.S.; Filzwieser, D.; Walter, J.C.; Schlossbauer, P.; Kinscherf, R.; Lischka, K.; Luksch, H.; Schindowski, K. Allogenic fc domain-facilitated uptake of IgG in nasal Lamina propria: Friend or foe for intranasal CNS delivery? *Pharmaceutics* **2018**, *10*, 107. [CrossRef]

32. Glorieux, S.; Van den Broeck, W.; van der Meulen, K.M.; Van Reeth, K.; Favoreel, H.W.; Nauwynck, H.J. In vitro culture of porcine respiratory nasal mucosa explants for studying the interaction of porcine viruses with the respiratory tract. *J. Virol. Methods* **2007**, *142*, 105–112. [CrossRef]
33. Tulinski, P.; Fluit, A.C.; van Putten, J.P.M.; de Bruin, A.; Glorieux, S.; Wagenaar, J.A.; Duim, B. An Ex Vivo Porcine Nasal Mucosa Explants Model to Study MRSA Colonization. *PLoS ONE* **2013**, *8*, e53783. [CrossRef]
34. Moore, G.E.; Sandberg, A.A. Studies of a human tumor cell line with a diploid karyotype. *Cancer* **1964**, *17*, 170–175. [CrossRef]
35. Moorhead, P.S. Human tumor cell line with a quasi-diploid karyotype (RPMI 2650). *Exp. Cell Res.* **1965**, *39*, 190–196. [CrossRef]
36. Hayflick, L.; Moorhead, P.S. The serial cultivation of human cell strains. *Exp. Cell Res.* **1961**, *621*, 585–621. [CrossRef]
37. Mercier, C.; Perek, N.; Delavenne, X. Is RPMI 2650 a Suitable In Vitro Nasal Model for Drug Transport Studies? *Eur. J. Drug Metab. Pharmacokinet.* **2018**, *43*, 13–24. [CrossRef]
38. Na, K.; Lee, M.; Shin, H.; Chung, S. In vitro nasal mucosa gland-like structure formation on a chip. *Lab Chip* **2017**. [CrossRef]
39. Mercier, C.; Hodin, S.; He, Z.; Perek, N.; Delavenne, X. Pharmacological Characterization of the RPMI 2650 Model as a Relevant Tool for Assessing the Permeability of Intranasal Drugs. *Mol. Pharm.* **2018**, *15*, 2246–2256. [CrossRef]
40. Mercier, C.; Jacqueroux, E.; He, Z.; Hodin, S.; Perek, N.; Boudard, D.; Delavenne, X.; Constant, S. Pharmacological characterization of the 3D MucilAir™ nasal model. *Eur. J. Pharm. Biopharm.* **2019**, *139*, 186–196. [CrossRef]
41. Moll, R.; Krepler, R.; Franke, W.W. Complex Cytokeratin Polypeptide Patterns Observed in Certain Human Carcinomas. *Differentiation* **1982**, *23*, 256–269. [CrossRef]
42. Peter, H.G. *Cell Culture Sheets to Study Nasal Peptide Metabolism the Human Nasal RPMI 2650 Cell Line Model*; ETH Zürich: Zürich, Switzerland, 1996.
43. Stratford, R.E.; Lee, V.H.L. Aminopeptidase activity in homogenates of various absorptive mucosae m the albino rabbit: Implications in peptide delivery. *Int. J. Pharm.* **1986**, *30*, 73–82. [CrossRef]
44. Tchao, R. Epithelial cell interaction in air-liquid interface culture. *Vitro Cell. Dev. Biol.* **1989**, *25*, 460–465. [CrossRef]
45. de Jong, P.M.; van Sterkenburg, M.A.; Hesseling, S.C.; Kempenaar, J.A.; Mulder, A.A.; Mommaas, A.M.; Dijkman, J.H.; Ponec, M. Ciliogenesis in human bronchial epithelial cells cultured at the air-liquid interface. *Am. J. Respir. Cell Mol. Biol.* **1994**, *10*, 271–277. [CrossRef]
46. Lee, M.K.; Yoo, J.W.; Lin, H.; Kim, Y.S.; Kim, D.D.; Choi, Y.M.; Park, S.K.; Lee, C.H.; Roh, H.J. Air-liquid interface culture of serially passaged human nasal epithelial cell monolayer for in vitro drug transport studies. *Drug Deliv. J. Deliv. Target. Ther. Agents* **2005**, *12*, 305–311. [CrossRef]
47. Pezzulo, A.A.; Starner, T.D.; Scheetz, T.E.; Traver, G.L.; Tilley, A.E.; Harvey, B.-G.; Crystal, R.G.; McCray, P.B.; Zabner, J. The air-liquid interface and use of primary cell cultures are important to recapitulate the transcriptional profile of in vivo airway epithelia. *Am. J. Physiol. Cell. Mol. Physiol.* **2010**, *300*, L25–L31. [CrossRef]
48. Chen, S.; Einspanier, R.; Schoen, J. Transepithelial electrical resistance (TEER): A functional parameter to monitor the quality of oviduct epithelial cells cultured on filter supports. *Histochem. Cell Biol.* **2015**, *144*, 509–515. [CrossRef]
49. Van Itallie, C.M.; Anderson, J.M. Claudins and Epithelial Paracellular Transport. *Annu. Rev. Physiol.* **2005**, *68*, 403–429. [CrossRef]
50. Flamm, J.; Boscher, M.; Maigler, F.; Akana, C.; Lindemann, J.; Kleiner, S.; Sommer, F.; Schindowski, K. Standardized refined intranasal admini- stration for region-specific intranasal drug deposition in mice established with 3D rapid prototypes under 3R criteria. *Berl. Munch. Tierarztl. Wochenschr.* **2018**, *131*, 408–416. [CrossRef]
51. Röhm, M.; Carle, S.; Maigler, F.; Flamm, J.; Kramer, V.; Mavoungou, C.; Schmid, O.; Schindowski, K.; Schmid, O.; Mavoungou, C.; et al. A comprehensive screening platform for aerosolizable protein formulations for intranasal and pulmonary drug delivery. *Int. J. Pharm.* **2017**, *532*, 537–546. [CrossRef]
52. Nieder, B.; Wagner, H.; Luksch, H. Development of output connections from the inferior colliculus to the optic tectum in barn owls. *J. Comp. Neurol.* **2003**, *464*, 511–524. [CrossRef]

53. Troxler, R.F.; Offner, G.D.; Nunes, D.P.; Oppenheim, F.G.; Iontcheva, I. Molecular characterization of a major high molecular weight mucin from human sublingual gland. *Glycobiology* **2007**, *7*, 965–973. [CrossRef]
54. Shah, A.S.; Ben-Shahar, Y.; Moninger, T.O.; Kline, J.N.; Welsh, M.J. Motile Cilia of Human Airway Epithelia Are Chemosensory. *Science* **2009**, *325*, 1131–1134. [CrossRef]
55. Heijink, I.H.; Brandenburg, S.M.; Noordhoek, J.A.; Postma, D.S.; Slebos, D.J.; Van Oosterhout, A.J.M. Characterisation of cell adhesion in airway epithelial cell types using electric cell-substrate impedance sensing. *Eur. Respir. J.* **2010**, *35*, 894–903. [CrossRef]
56. Crowe, T.P.; Greenlee, M.H.W.; Kanthasamy, A.G.; Hsu, W.H. Mechanism of intranasal drug delivery directly to the brain. *Life Sci.* **2018**, *195*, 44–52. [CrossRef]
57. Baker, H.; Spencer, R.F. Transneuronal transport of peroxidase-conjugated wheat germ agglutinin (WGA-HRP) from the olfactory epithelium to the brain of the adult rat. *Exp. Brain Res.* **1986**, *63*, 461–473. [CrossRef]
58. Jorissen, M.; Van Der Schueren, B.; Van Den Berghe, H.; Cassiman, J.J. Contribution of in vitro culture methods for respiratory epithelial cells to the study of the physiology of the respiratory tract. *Eur. Respir. J.* **1991**, *4*, 210–217.
59. Bhowmick, R.; Gappa-Fahlenkamp, H. Cells and Culture Systems Used to Model the Small Airway Epithelium. *Lung* **2016**, *194*, 419–428. [CrossRef]
60. Even-Tzur, N.; Jaffa, A.; Gordon, Z.; Gottlieb, R.; Kloog, Y.; Einav, S.; Wolf, M.; Elad, D. Air-liquid interface culture of nasal epithelial cells on denuded amniotic membranes. *Cell. Mol. Bioeng.* **2010**, *3*, 307–318. [CrossRef]
61. Jeliazkova-Mecheva, V.V.; Bobilya, D.J. A porcine astrocyte/endothelial cell co-culture model of the blood-brain barrier. *Brain Res. Protoc.* **2003**, *12*, 91–98. [CrossRef]
62. Pardeshi, C.V.; Belgamwar, V.S. Direct nose to brain drug delivery via integrated nerve pathways bypassing the blood–brain barrier: An excellent platform for brain targeting. *Expert Opin. Drug Deliv.* **2013**, *10*, 957–972. [CrossRef]
63. Rath, T.; Kuo, T.T.; Baker, K.; Qiao, S.W.; Kobayashi, K.; Yoshida, M.; Roopenian, D.; Fiebiger, E.; Lencer, W.I.; Blumberg, R.S. The immunologic functions of the neonatal FC receptor for IGG. *J. Clin. Immunol.* **2013**, *33*, 9–17. [CrossRef]
64. Gänger, S.; Schindowski, K. Tailoring formulations for intranasal Nose-to-Brain delivery via the olfactory area: A review on physico-chemical characteristics and mucociliary clearance of the nasal olfactory mucosa. *Pharmaceutics* **2018**, *10*, 116. [CrossRef]
65. Dolberg, A.M.; Reichl, S. Expression of P-glycoprotein in excised human nasal mucosa and optimized models of RPMI 2650 cells. *Int. J. Pharm.* **2016**, *508*, 22–33. [CrossRef]
66. Berger, J.T.; Voynow, J.A.; Peters, K.W.; Rose, M.C. Respiratory carcinoma cell lines MUC genes and glycoconjugates. *Am. J. Respir. Cell Mol. Biol.* **1999**, *20*, 500–510. [CrossRef]
67. Kreft, M.E.; Lasi, E.; Kristan, K. The Characterization of the Human Nasal Epithelial Cell Line RPMI 2650 Under Different Culture Conditions and Their Optimization for an Appropriate in vitro Nasal Model. *Pharm. Res.* **2015**, *32*, 665–679. [CrossRef]
68. Srinivasan, B.; Kolli, A.R.; Esch, M.B.; Abaci, H.E.; Shuler, L.; Hickman, J.J.; Srinivasan, B.; Kolli, A.R.; Esch, M.B.; Abaci, H.E.; et al. TEER Measurement Techniques for in vitro barrier model systems. *J. Lab. Autom.* **2016**, *20*, 107–126. [CrossRef]
69. Engström, B.; Ekblom, A.; Hansson, P. The Olfactory and Respiratory Epithelium in Rhesus and Squirrel Monkeys Studied with Freeze-fracture Technique. *Acta Otolaryngol.* **2009**, *108*, 259–267. [CrossRef]
70. Holcomb, J.D.; Graham, S.; Calof, A.L. Neuronal homeostasis in mammalian olfactory epithelium: A review. *Am. J. Rhinol.* **1996**, *10*, 125–134. [CrossRef]
71. Yoo, J.-W.; Kim, Y.-S.; Lee, S.-H.; Lee, M.-K.; Roh, H.-J.; Jhun, B.-H.; Lee, C.-H.; Kim, D.-D. Serially passaged human nasal epithelial cell monolayer for in vitro drug transport studies. *Pharm. Res.* **2003**, *20*, 1690–1696. [CrossRef]
72. Kim, N.; Hee, D.; Suh, M.; Ho, J.; Oh, S.; Kyun, M. Effect of lipopolysaccharide on diesel exhaust particle-induced junctional dysfunction in primary human nasal epithelial cells. *Environ. Pollut.* **2019**, *248*, 736–742. [CrossRef]
73. Farshori, P.; Kachar, B. Redistribution and Phosphorylation of Occludin During Opening and Resealing of Tight Junctions in Cultured Epithelial Cells. *J. Membr. Biol.* **1999**, *170*, 147–156. [CrossRef]

74. Bhat, M.; Toledo-Velasquez, D.; Wang, L.; Malanga, C.J.; Ma, J.K.; Rojanasakul, Y. Regulation of tight junction permeability by calcium mediators and cell cytoskeleton in rabbit tracheal epithelium. *Pharm. Res.* **1993**, *10*, 991–997. [CrossRef]
75. Piperno, G.; Fuller, M.T. Monoclonal antibodies specific for an acetylated form of alpha-tubulin recognize the antigen in cilia and flagella from a variety of organisms. *J. Cell Biol.* **1985**, *101*, 2085–2094. [CrossRef]
76. Quinones, G.B.; Danowski, B.A.; Devaraj, A.; Singh, V.; Ligon, L.A. The posttranslational modification of tubulin undergoes a switch from detyrosination to acetylation as epithelial cells become polarized. *Mol. Biol. Cell* **2011**, *22*, 1045–1057. [CrossRef]
77. Zholos, A.V.; Atherton-Watson, H.; Elborn, J.S.; Ennis, M.; de Courcey, F.; Danahay, H.L.; Canning, P.; Williams, M.T.S. Development of primary human nasal epithelial cell cultures for the study of cystic fibrosis pathophysiology. *Am. J. Physiol. Physiol.* **2012**, *303*, C1173–C1179. [CrossRef]
78. Jorissen, M.; Bessems, A. Normal ciliary beat frequency after ciliogenesis in nasal epithelial cells cultured sequentially as monolayer and in suspension. *Acta Otolaryngol.* **1995**, *115*, 66–70. [CrossRef]
79. Salathe, M.; Bookman, R.J. Coupling of [Ca2+] i and ciliary beating in cultured tracheal epithelial cells. *J. Cell Sci.* **1995**, *108*, 431–440.
80. Yager, J.; Chen, T.M.; Dulfano, M.J. Measurement of frequency of ciliary beats of human respiratory epithelium. *Chest* **1978**, *73*, 627–633. [CrossRef]
81. Bateman, A.C.; Karasin, A.I.; Olsen, C.W. Differentiated swine airway epithelial cell cultures for the investigation of influenza A virus infection and replication. *Influ. Other Respir. Viruses* **2013**, *7*, 139–150. [CrossRef]
82. Delgado-Ortega, M.; Olivier, M.; Sizaret, P.-Y.; Simon, G.; Meurens, F. Newborn pig trachea cell line cultured in air-liquid interface conditions allows a partial in vitro representation of the porcine upper airway tissue. *BMC Cell Biol.* **2014**, *15*, 14. [CrossRef]
83. Getchell, M.L.; Getchell, T.V. Fine structural aspects of secretion and extrinsic innervation in the olfactory mucosa. *Microsc. Res. Tech.* **1992**, *23*, 111–127. [CrossRef]
84. Solbu, T.T.; Holen, T. Aquaporin pathways and mucin secretion of bowman's glands might protect the olfactory mucosa. *Chem. Senses* **2012**, *37*, 35–46. [CrossRef]
85. Kim, C.-H.; Song, K.S.; Kim, S.-S.; Kim, H.-U.; Seong, J.-K.; Yoon, J.-H. Expression of MUC5AC mRNA in the Goblet Cells of Human Nasal Mucosa. *Laryngoscope* **2000**, *110*, 2110–2113. [CrossRef]
86. Aust, M.R.; Madsen, C.S.; Jennings, A.; Kasperbauer, J.L.; Gendler, S.J. Mucin mRNA Expression in Normal and Vasomotor Inferior Turbinates. *Am. J. Rhinol.* **1997**, *11*, 293–302. [CrossRef]
87. Kennel, C.; Gould, E.A.; Larson, E.D.; Salcedo, E.; Vickery, T.W.; Restrepo, D.; Ramakrishnan, V.R. Differential Expression of Mucins in Murine Olfactory Versus Respiratory Epithelium. *bioRxiv* **2019**, 1–32. [CrossRef]
88. Reichl, S.; Becker, K. Cultivation of RPMI 2650 cells as an in-vitro model for human transmucosal nasal drug absorption studies: Optimization of selected culture conditions. *J. Pharm. Pharmacol.* **2012**, *64*, 1621–1630. [CrossRef]
89. Pozzoli, M.; Sonvico, F.; Ong, H.X.; Traini, D.; Bebawy, M.; Young, P.M. Optimization of RPMI 2650 Cells as a Model for Nasal Mucosa. *Respir. Drug* **2014**, *2*, 739–742.
90. De Fraissinette, A.; Brun, R.; Felix, H.; Vonderscher, J.; Rummelt, A. Evaluation of the human cell line RPMI 2650 as an in vitro nasal model. *Rhinology* **1995**, *33*, 194–198.
91. Werner, U.; Kissel, T. In-vitro cell culture models of the nasal epithelium: A comparative histochemical investigation of their suitability for drug transport studies. *Pharm. Res.* **1996**, *13*, 978–988. [CrossRef]
92. Yoon, J.-H.; Moon, H.-J.; Seong, J.-K.; Kim, C.-H.; Lee, J.-J.; Choi, J.Y.; Song, M.S.; Kim, S.-H. Mucociliary differentiation according to time in human nasal epithelial cell culture. *Differentiation* **2002**, *70*, 77–83. [CrossRef]

pharmaceutics

MDPI

Article

Validation of an Ex Vivo Permeation Method for the Intestinal Permeability of Different BCS Drugs and Its Correlation with Caco-2 In Vitro Experiments

Aroha B. Sánchez [1], Ana C. Calpena [1], Mireia Mallandrich [1] and Beatriz Clares [2,*]

1 Department of Pharmacy and Pharmaceutical Technology and Physical Chemistry, Faculty of Pharmacy and Food Sciences, University of Barcelona, 08028 Barcelona, Spain; aroha_89_1@hotmail.com (A.B.S.); anacalpena@ub.edu (A.C.C.); mireia.mallandrich@ub.edu (M.M.)

2 Department of Pharmacy and Pharmaceutical Technology, Faculty of Pharmacy, University of Granada, 18071 Granada, Spain

* Correspondence: beatrizclares@ugr.es; Tel.: +34-958246664

Received: 21 October 2019; Accepted: 27 November 2019; Published: 29 November 2019

Abstract: The absorption study of drugs through different biological membranes constitutes an essential step in the development of new pharmaceutical dosage forms. Concerning orally administered forms, methods based on monolayer cell culture of Caco-2 (Caucasian colon adenocarcinoma) have been developed to emulate intestinal mucosa in permeability studies. Although it is widely accepted, it has disadvantages, such as high costs or high technical complexity, and limitations related to the simplified structure of the monolayer or the class of molecules that can be permeated according to the transport mechanisms. The aim of this work was to develop a new ex vivo methodology which allows the evaluation of the intestinal apparent permeability coefficient (P_{app}) while using fewer resources and to assess the correlation with Caco-2. To this end, pig (*Sus scrofa*) duodenum segments were mounted in Franz diffusion cells and used to permeate four different drugs: ketorolac tromethamine (Kt), melatonin (Mel), hydrochlorothiazide (Htz), and furosemide (Fur). No statistically significant differences ($p > 0.05$) were observed corelating P_{app} values from Franz diffusion cells and Caco-2 cell experiments for Kt, Htz, and Fur. However, there were statistically significant differences ($p < 0.05$) correlating P_{app} values and Mel. The difference is explained by the role of Mel in the duodenal epithelial paracellular permeability reduction. Ex vivo permeation may be an equivalent method to Caco-2 for drugs that do not produce intestinal membrane phenomena that could affect absorption.

Keywords: Franz cells; Caco-2 cells; intestinal permeability; apparent permeability coefficient

1. Introduction

At the earliest stages of drug product or new pharmaceutical dosage form development, in vitro permeation through Caco-2 (human epithelial colorectal adenocarcinoma cell monolayer line) is widely accepted to estimate the intestinal apparent permeability coefficient (P_{app}). As described in Figure 1, Caco-2 is a donor–receptor compartment apparatus separated by a cell monolayer grown on a porous polycarbonate filter. P_{app}, defined as the flux of a substance permeating a membrane from the donor to receptor compartment normalized by the membrane surface and initial concentration in the donor chamber [1], is usually obtained based on a two-compartmental model approach; however, some authors have developed an alternative definition of a P_{app} index for three-compartment models describing the membrane as well as donor and receiver compartments [2]. This index may be predictive of oral bioavailability, showing an acceptable correlation with a human one, especially for drugs absorbed by passive diffusion. Carrier-mediated absorption drugs are not so easily extrapolated and require a scaling factor because of the low expression of carriers in this cell line [3]. Both undifferentiated and differentiated Caco-2 models have been developed, with the undifferentiated being more susceptible than the differentiated and, therefore, effective in cytotoxicity and cytoprotective studies [4–8]. More complex differentiated models are suitable for studying transport mechanisms and efficacy of substances [9–12].

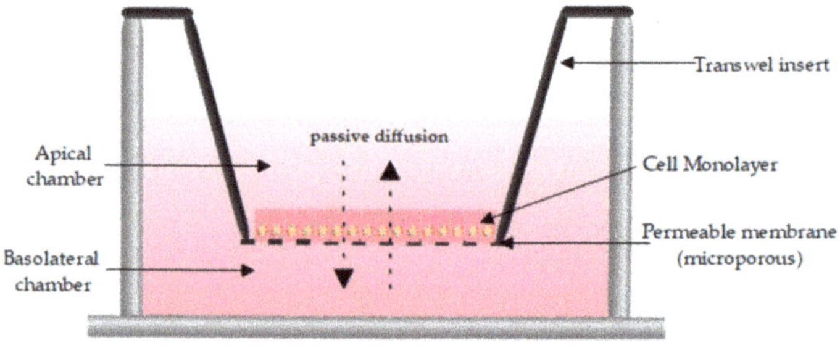

Figure 1. Caco-2 diagram formed by a transwell insert preloaded with polycarbonate membrane inserts with a known pore size. Original figure drawn in Edraw Max 9.4.

Focusing on the relevance of Caco-2 in the design, optimization, and selection of potential candidates in the development of oral drugs, cell monolayer lines have been used for the study of improving the oral absorption of highly lipophilic and poorly water-soluble drugs [13]; lipid-based self-emulsifying drug delivery systems [14]; and for the evaluation of new oral formulations based on nanotechnology, such as solid lipid nanoparticles (SLNps) [15], or bioadhesive drug delivery systems, such as chitosan-modified nanoparticles [10]. Caco-2 has been optimized, although it exhibits variability attributable to biological methods [16], and the significance of emulating physiological conditions to improve in vitro experiments, for example, using bile acids, surfactants, or plasma proteins, is well known for providing a better in vitro–in vivo correlation [17]. Because of their usefulness and relevance, international institutions such the FDA accept in vitro permeability studies across Caco-2 to classify the permeability of drug substances according to the Biopharmaceutics Classification System (BCS) proposed in the International Council for Harmonization of Technical Requirements for Pharmaceuticals for Human Use (ICH) guidelines [18].

Regardless of it being an established method, the Caco-2 cell model has some disadvantages, such as its high cost or the need for highly specialized staff, which can be restrictive in both academic and private sector environments, where the optimization of resources is a paramount task. Other limitations are related to the structural and functional differences between a monolayer of cells and a biological membrane, such as the intestinal one. In contrast to Caco-2, which is a monolayer of cells, the gastrointestinal (GI) tract is composed of four main layers: tunica mucosa (mucous layer), tunica submucosa (submucous layer), tunica muscularis (muscle layer), and tunica serosa (serous layer) (Figure 2), which are additionally composed of different layers. Tunica mucosa is the most proximal layer to the lumen, which contains a thick layer of mucous covering the epithelial cells that are linked together by intercellular linkages, and an underlying layer called the lamina propria. The small intestine epithelium is a simple columnar epithelium, typical of regions of high secretion and absorption functions. The cells of simple columnar epithelium form finger-like projections called villi. In addition to intestinal villi, the surface of the cells contains microvilli, which collectively increase the surface area of the lumen by 400–600-fold [19].

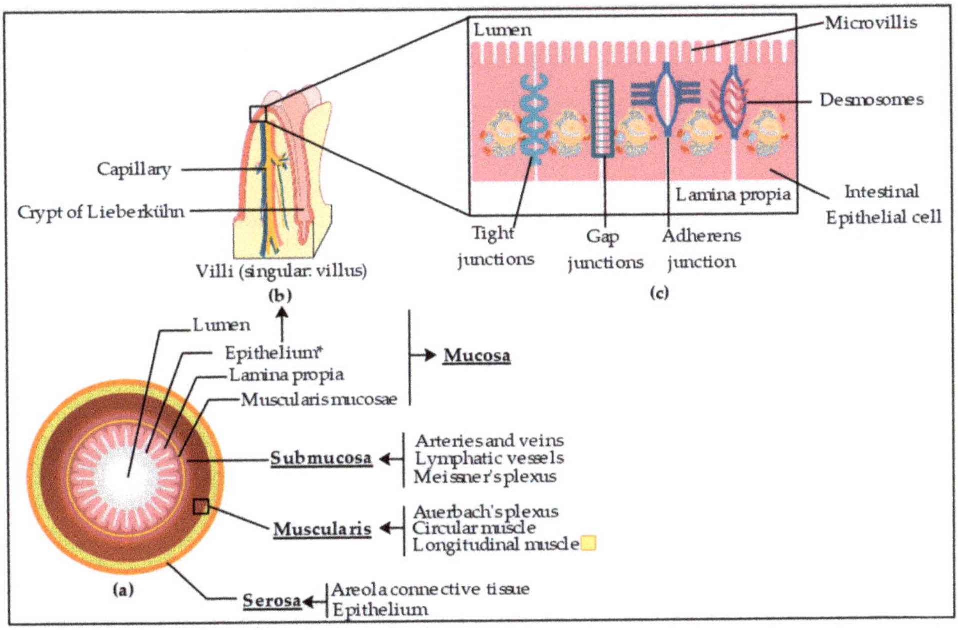

Figure 2. (**a**) Basic structure of small intestine layers, (**b**) a simplified schematic representation of villi, and (**c**) epithelium composed of intestinal epithelial cells including types of intercellular junctions. Original figure drawn from Edraw Max 9.4.

Other important features of GI epithelium are the structural linkages between cells that connect the plasma membrane of neighboring cells. Four classes of intercellular junctions have been described: gap junctions (selective for small molecules such as ions, second messengers, and metabolites), tight junctions (paracellular barrier regulating the movement of water and solutes between epithelial layers), adherens junctions (help to seal the space between cells), and desmosomes (help to maintain shear forces and mechanical stress) (Figure 2c) [19].

GI membrane complexity demonstrates that drug absorption across this barrier is a multipathway process which could be classified as transcellular and paracellular. The most important is the transcellular route, whereby compounds go across the cells by traversing the cell membrane following passive diffusion or carrier-mediated transport (active transport, facilitated diffusion, absorption limited by P-glycoprotein or other efflux transports, intestinal first-pass metabolism followed by absorption of parent and metabolite- and receptor-mediated transport). There is also a paracellular passive diffusion via the junction route [20].

As explained above, the use of Caco-2, although predictive, is a very simplified approach to the GI membrane, so the use of ex vivo methods may lead to more accurate predictions of P_{app}. Nejdfors et al. studied the permeability of C-mannitol, fluorescein isothiocyanate (FITC)–dextran 4400, a-lactalbumin, ovalbumin, and FITC–dextran through different intestinal regions of humans, rats, and pigs using small diffusion chambers of 5 mL and 1.74 cm^2 of exposed tissue area (Navicyte, San Diego, CA, USA) [21]. Differences between intestinal regions and species were detected, and a good correlation between humans and pigs was also observed, mainly for the polyol mannitol in the jejunum and ileum. P_{app} was not compared with Caco-2 values, but the experiment showed that the use of intestinal membranes may be predictive of oral bioavailability.

Considering all the above, the aim of this work was to develop an alternative technique to Caco-2 and small diffusion cells, which allows the evaluation of the intestinal absorption rate or P_{app} for different drugs while using fewer resources and to assess the correlation with Caco-2. The purpose was to provide a new suitable and economical ex vivo method to test and compare new oral formulations or modified release systems, including therapeutic and higher drug concentrations than Caco-2.

2. Materials and Methods

2.1. Chemical and Reagents

Melatonin (Mel), ketorolac tromethamine (Kt), hydrochlorothiazide (Htz), ammonium dihydrogen phosphate ($NH_4H_2PO_4$), phosphoric acid (H_3PO_4), disodium hydrogen phosphate (Na_2HPO_4), and tetrahydrofuran (THF) were purchased from Sigma-Aldrich (Madrid, Spain). Furosemide (Fur) was supplied by Acofarma (Barcelona, Spain). Potassium dihydrogen phosphate (KH_2PO_4), potassium hydroxide (KOH), acetonitrile (ACN), and methanol (MeOH) were purchased from Panreac Quimica (Barcelona, Spain). Hanks' balanced salt solution (HBSS) was purchased from Merck S.L. (Barcelona, Spain). Double-distilled water was obtained from a Milli-Q® Gradient A10 system apparatus (Millipore Iberica S.A.U., Madrid, Spain).

2.2. Instrumentation

HPLC-UV Procedure

The HPLC equipment consisted of a Waters® Alliance 2695 Separation Module (Waters Co., Milford, MA, USA) with a 2996 Photodiode Array Detector (DAD) at a wavelength range of 190–800 nm and sensitivity settings from 0.0001 to 2.0000 absorbance units. HPLC parameters are summarized in Table 1 and listed below.

Table 1. Summary of HPLC parameters for ketorolac tromethamine (Kt), melatonin (Mel), furosemide (Fur), and hydrochlorothiazide (Htz).

Molecule	Column	UV-ʎ (nm)	Mobil Phase	Flow Rate (mL/min)	IV² (μL)
Kt	C8, 150 × 2.1 mm, 5 μm	313	NH_4H_2PO:THF (70:30)	0.25	100
Mel	C18, 100 × 4.6 mm, 3.5 μm	223	H_2O:MeOH (55:45)	0.60	100
Fur	C18, 150 × 3.9 mm, 4 μm	230	KH_2PO_4:CAN (80:20)	0.50	100
Htz	100 ODS2, 100 × 4.6, 3 μm	224	NaH_2PO_4:MeOH:THF¹	0.70	100

¹ Hydrochlorothiazide HPLC mobile phase consisted of a gradient elution of two solutions: A (93:6:1) and B (47.6:47.6:4.8). ² IV: injection volume.

Kt analysis was conducted with a reverse-phase column C8 (150 × 2.1 mm) packed up with 5 μm (Kromasil®, Teknokroma Anlítica, SA; Barcelona, Spain), with a UV detector set up at 313 nm. The mobile phase, previously filtered by a 0.45 μm nylon membrane filter (Technokroma, Barcelona, Spain) and degassed by sonication, consisted of a 70:30 ratio of NH_4H_2PO (5.75 g/L; pH 3) to THF under isocratic elution at a flow rate of 0.25 mL/min. The injection volume was 10 μL.

Mel analysis was performed with a reverse-phase column C18 (150 × 4.6 mm) packed up with 3.5 μm (SunFire®, Waters Co., Milford, MA, USA), with a UV detector set up at 223 nm. The mobile phase, previously filtered by a 0.45 μm nylon membrane filter (Technokroma and degassed by sonication, consisted of a 55:45 ratio of double-distilled water to MeOH under isocratic elution at a flow rate of 0.6 mL/min. The injection volume was 100 μL.

Fur analysis was carried out with a reverse-phase column C18 (150 × 3.9 mm) packed up with 4 μm (Nova-Pack®, Waters Co., Milford, MA, USA), with a UV detector set up at 230 nm. The mobile phase, previously filtered by a 0.45 μm nylon membrane filter (Technokroma, Barcelona, Spain) and degassed by sonication, consisted of an 80:20 ratio of KH_2PO_4 (0.01M; pH of 6.3, adjusted with KOH 10%) to ACN under isocratic elution at a flow rate of 0.5 mL/min. The injection volume was 100 μL.

Htz analysis was conducted with a reverse-phase column ultrabase 100 ODS2 analytical column (100 × 4.6 nm; diameter of 3 μm (Akady, Spain)) with a UV detector set up at 224 nm. The mobile phase, previously filtered by a 0.45 µm nylon membrane filter (Technokroma, Barcelona, Spain) and degassed by sonication, consisted of a gradient elution of two solutions (A and B) at a flow rate of 0.7 mL/min. Solution A consisted of 940 mL of NaH_2PO_4 (35.8 g/L; pH of 3.2, adjusted with H_3PO_4) with 60 mL of MeOH and 10 mL of THF. Solution B consisted of 500 mL of NaH_2PO_4 (35.8 g/L; pH of 3.2, adjusted with H_3PO_4) with 500 mL of MeOH and 50 mL of THF. The percentage of B was 20% at time 0, 20% at 4 min, 80% at 10 min, 80% at 12 min, 20% at 13 min, and 20% at 20 min. The injection volume was 100 µL.

2.3. *Validation and Verification of Analytical Methods*

Previously validated HPLC-UV methods were selected for the analysis of the four assayed analytes. Considering that the samples were obtained from biological sources, the specificity was studied.

Specificity, expressed by the ICH guidelines as the ability to assess an analyte in the presence of components which may be expected to be present, was evaluated by the absence of interference of the phosphate-buffered saline (PBS; pH 7.4) and other components from biological membranes used as a blank at the retention times shown by the different standard solutions.

2.4. *Ex Vivo Permeation Studies through Pig Intestine*

2.4.1. Franz Cell System and Intestinal Membrane

Ex vivo permeation was performed in the duodenum, the most proximal portion of the small intestine, of young female pigs (*Sus scrofa*). Animals were sacrificed for other purposes in the Animal Facility at Bellvitge Campus (University of Barcelona, Spain) (no. 7428) (Date of approval: 10 January 2019), and intestinal samples were obtained according to the 3R (reduction, refinement and replacement) principle.

The duodenum was excised, cleaned, and preserved in HBBS at 5 ± 3 °C for 12 h. Then, 6 × 6 cm pieces were cut, mounted on the metal ring of the Franz cells as shown in Figure 3, and the remaining corners were trimmed.

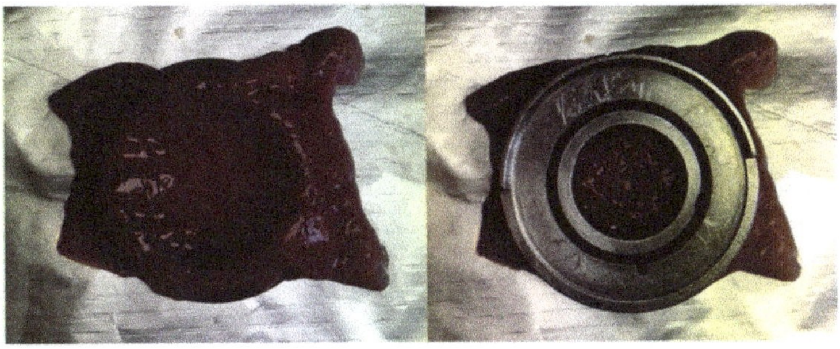

Figure 3. Piece of proximal small intestine of young female pigs (*Sus scrofa*) mounted on the metal ring of the Franz cells.

To avoid damage to the biological intestinal membrane, 0.02 M PBS (pH 7.4) was prepared as a receiving medium. The composition was 0.6 g of KH_2PO_4 and 3.17 g of Na_2HPO_4 per liter of double-distilled water. The pH value was adjusted with H_3PO_4 or NaOH.

The ex vivo permeation study was performed in Franz diffusion cells (Vidra Foc Barcelona, Spain), where duodenum portions were placed between the receptor and donor compartments with the basal side in contact with the receiving medium and the apical side in contact with the donor chamber, avoiding bubble formation. The diffusion area was 2.54 cm^2. A representative chart of the Franz cell system is shown in Figure 4.

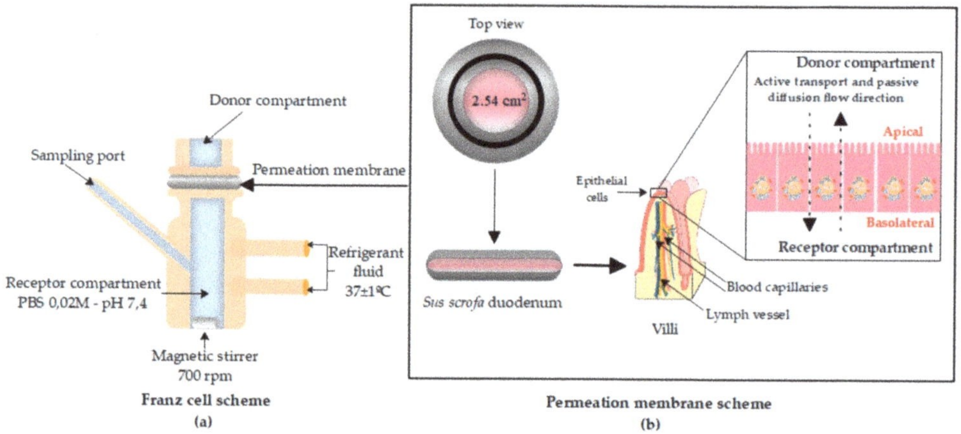

Figure 4. Franz cell scheme (**a**), including permeation membrane model (**b**) formed by the proximal small intestine of young female pigs (*S. scrofa*) that were opened with an incision and positioned with the area corresponding to the microvilli of the enterocytes in contact with the donor compartment and the basolateral part in contact with the receptor compartment. Original figure drawn in Edraw Max 9.4.

Homogeneity during experiments was ensured by a small Teflon®-coated magnetic stir bar at 700 rpm. The diffusion cells were previously incubated in a water bath to equalize the temperature in all cells (37 ± 1 °C).

2.4.2. Donor Solution Preparation and Sampling Method

Four drugs according to BCS classification were randomly selected. Table 2 includes the chemical structure; pKa (negative base 10 logarithm of the acid dissociation constant (Ka) of a solution) values; and tested formulations, including solvents, drug concentration, and pH.

Table 2. Name, structure, Biopharmaceutics Classification System (BCS) type, and pKa (negative base 10 logarithm of the acid dissociation constant (Ka) of a solution) of selected drugs for permeation experiments.

Molecule	Structure	BCS	pKa	Dissolution Media	Concentration
Ketorolac tromethamine	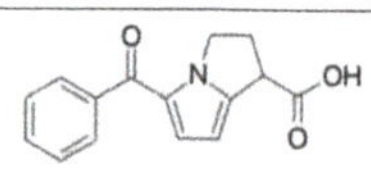	Class 1	3.5 [22]	PBS pH 7.4	1 mg/mL
Melatonin	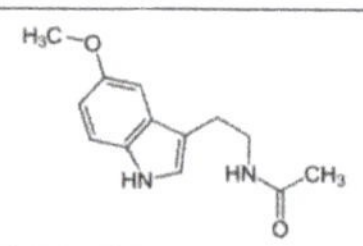	Class 2	16.5 [23]	PBS pH 7.4	0.8 mg/mL
Hydrochlorothiazide	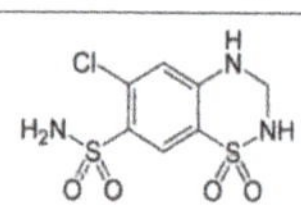	Class 3	7.9 [24]	PBS pH 7.4	0.05 mg/mL
Furosemide	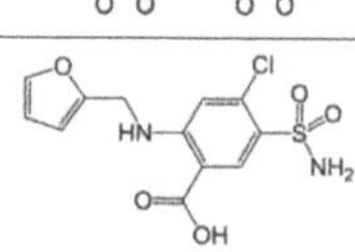	Class 4	3.9 [25]	PBS pH 7.4	0.6 mg/mL

All the drugs were dissolved by stirring at 30 ± 0.1 °C in PBS (pH 7.4) to guarantee the biocompatibility to the permeation membrane. Infinite dose conditions were ensured in all experiments.

The donor compartment was then sealed by parafilm to prevent water evaporation. All the experiments were carried out under sink conditions, ensuring that the drug concentration in the receptor compartment was negligible compared to the donor one.

Samples of 300 µL were collected via a sampling port from the middle of the receptor compartment at preselected time intervals (30, 60, 90, 120, 180, 240, 300, 360, and 420 min) for 7 h. The removed sample volume was immediately replaced with the same volume of tempered fresh receiving medium of each molecule with great care to avoid trapping air beneath the membrane.

2.4.3. Sample Analysis

The cumulative amount of the different BCS drugs through the small intestine membrane from the acceptor compartment was monitored by a validated HPLC-UV methodology. Results are reported as mean ± SD of five experiments for each drug.

2.4.4. Data Treatment and Statistical Analysis

Our permeability model has the same structure as the two-compartment classic model, composed of donor (apical) and receptor (basal) chambers, both separated by the permeation membrane. So, apical-to-basal P_{app} was calculated based on classic parameters according to Equation (1):

$$P_{app} = (dQ/dt) / (C_0 \times A) \quad (1)$$

where (dQ/dt) is the transport rate or flux (J) (µg/min) across the biological membrane, C_0 (µg/mL) is the initial concentration of the drug in the donor chamber, and A is the surface area (cm^2) of the permeation membrane.

The cumulative amount (Q) (µg) permeated through porcine duodenum was obtained by multiplying the acquired concentration (µg/mL) of each drug at the receptor chamber and the volume (mL) of the receptor chamber. J (µg/min) was calculated as the slope at the steady state obtained by linear regression analysis (GraphPad Prism® software, v. 5.01, GraphPad Software Inc., San Diego, CA, USA) of Q as a function of time (min). Then, P_{app} (cm/min) was calculated according to Equation (1) by dividing the J (µg/min), the permeation area (A) (2.54 cm^2), and the initial drug concentration (C_0) (µg/mL = µg/cm^3) in the donor chamber. Finally, the units were expressed in centimeters per second for comparison with the obtained results in the Caco-2 experiments. It was assumed that under sink conditions, the drug concentration in the receptor compartment is negligible compared to the donor compartment.

Obtained experimental data were analyzed by unpaired Student's *t*-test to compare P_{app} values for both bibliographic Caco-2 results and experimental data obtained with Franz cells. A p-value < 0.05 was established as an indicator of statistically significant differences.

3. Results and Discussion

3.1. Obtained Kinetics Parameters and P_{app} Calculation

Table 3 shows the kinetics parameters of Kt, Mel, Fur, and Htz. Cumulative permeated drug was measured. Then, the flux and flux normalized by the permeation area (2.54 cm^2) were calculated.

Table 3. Permeation parameters for Kt, Mel, Fur, and Htz solutions in vertical Franz cells ($n = 5$).

Permeation Parameters	Kt	Mel	Htz	Fur
Flux (µg/min)	0.855 ± 0.069	0.683 ± 0.016	0.004 ± 0.0002	0.0180 ± 0.0018
Flux/sup (µg/(cm/min))	0.336 ± 0.027	0.268 ± 0.006	0.0015 ± 0.0001	0.0071 ± 0.0001
Co (µg/mL)	1000	800	50	600
P_{app} (x10^{-6}) (cm/s)	5.609 ± 0.452	5.598 ± 0.130	0.487 ± 0.026	0.196 ± 0.020

Abbreviation: P_{app}—apparent permeability coefficient.

Figure 5 shows Kt, Mel, Htz, and Fur cumulative permeated amounts in micrograms as a function of time (min).

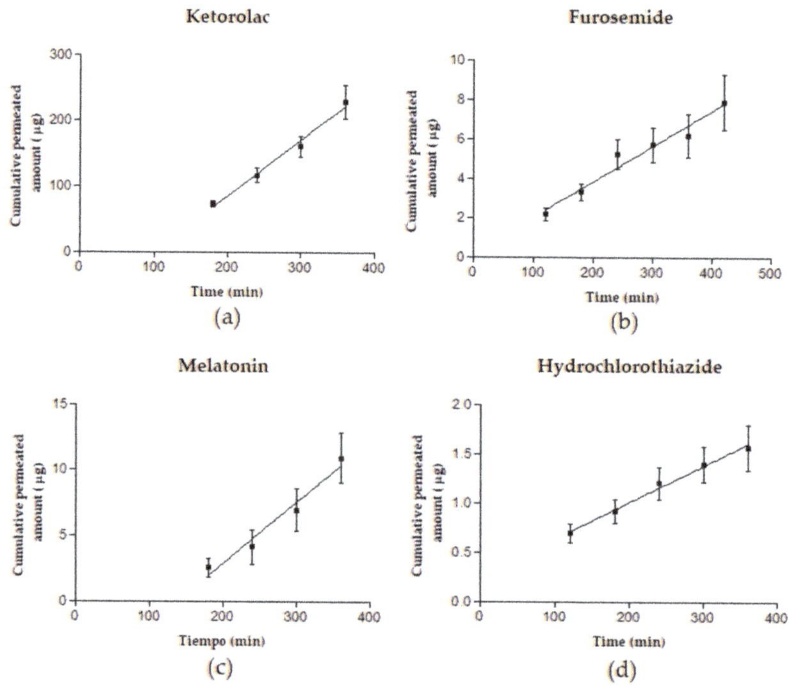

Figure 5. Cumulative permeated amounts (µg) as a function of time (min) of ketorolac tromethamine (**a**), melatonin (**b**), furosemide (**c**), and hydrochlorothiazide (**d**). Results are reported as mean ± SD ($n = 5$).

3.2. *Specificity*

Under the assay conditions described in the methodology section for each analyte, the mean retention times of Kt, Fur, Mel, and Htz were 9.45, 3.03, 5.05, and 9.5 min, respectively. The selectivity of the selected analytical method was confirmed by the studied chromatograms (Table S1), where Kt, Fur, Mel, and Htz peaks did not overlap with any other of the endogenous components of the medium. Blanks were obtained at time T_0, from the receptor compartment, after incubation of diffusion cells and before adding the drugs. Therefore, the method is considered specific for the detection and quantification of the four molecules.

3.3. Correlation between P_{app} in Caco-2 versus Franz Cells

After a literature search, different Caco-2 permeation studies and P_{app} values were found for the tested drugs, which are summarized in Table 4.

Table 4. P_{app} values expressed as mean ± SD (cm/s) from different literature datasets about Kt, Mel, Fur, and Htz in Caco-2 experiments.

Drug	P_{app} ($\times 10^{-6}$) (cm/s) (AP → BL)) Caco-2 Cells
Kt	8.30 ± 5.20 (n = 6) (HBBS[1] pH 7.4) [26]
Mel	12.50 ± 0.01 (n = 3) (HEPES[2] pH 7.4) [27]
Htz	0.51 ± 0.02 (n = 3) (HBBS[1] pH 7.4) [28]
Fur	0.19 ± 0.01 (n = 3) (KBR[3] pH 7.4) [29]

[1] Hank's balanced salts solution; [2] 4-(2-hydroxyethyl)-1-piperazineethanesulfonic acid; [3] Krebs–Ringer modified buffer (KBR).

Table 5 shows that no statistically significant differences ($p > 0.05$) were observed correlating Franz diffusion cells and Caco-2 cell experiments P_{app} values for Kt, Htz, and Fur. However, there were statistically significant differences ($p < 0.05$) correlating P_{app} values for Mel. Figure 6 shows a plot of statistical correlation.

Table 5. Statistical correlation between both Franz diffusion cells and Caco-2 experiments. P_{app} values are expressed as mean ± SD (cm/s) for Kt, Mel, Fur, and Htz.

Drug	P_{app} ($\times 10^{-6}$) (cm/s) (AP → BL)		
	Franz Diffusion Cells (PBS pH 7.4)	Caco-2 (pH 7.4) [1]	Unpaired t-Test (p)
Kt	5.61 ± 0.45 (n = 5)	8.3 ± 5.2 (n = 6) [26]	0.28 ($p > 0.05$)
Mel	5.60 ± 0.13 (n = 5)	12.50 ± 0.01 (n = 3) [27]	0.0001 ($p < 0.05$) *
Htz	0.49 ± 0.03 (n = 5)	0.42 ± 0.33 (n = 3) [28]	0.30 ($p > 0.05$)
Fur	0.20 ± 0.020 (n = 5)	0.19 ± 0.01 (n = 3) [29]	0.87 ($p > 0.05$)

[1] Caco-2 experiments were carried out in Hank's balanced salts solution (pH 7.4) for Kt and Htz, HEPES (pH 7.4) for Mel, and KBR pH 7.4 for Fur.

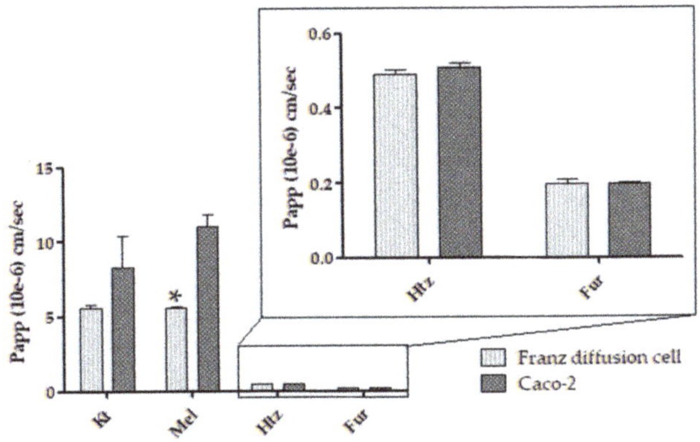

Figure 6. Comparative P_{app} between both Franz diffusion cells and Caco-2 cell culture for Kt, Mel, Htz, and Fur. Data are expressed as mean ± SD × 10^{-6} (cm/s). * $p < 0.05$.

P_{app} statistical correlation for both Franz diffusion cells and Caco-2 cell culture indicates that the ex vivo permeation is an equivalent method to Caco-2 for Kt, Htz, and Fur, which are BCS classes 1, 3, and 4, respectively. Regarding Mel, the obtained Franz diffusion cell P_{app} values showed statistically significant differences, with the ex vivo permeation data being 1.79 times lower than Caco-2. This could be related to the intestinal accumulation of Mel described in mammals [30], where Mel intestinal receptors MT1 and MT2 are involved in multiple roles, such as regulation of gastrointestinal motility and epithelial permeability [31]. It could also be explained by the fact that Mel, although it exhibits protein-facilitated transport [32], reduces the duodenal epithelial paracellular permeability [33]. This may justify the P_{app} value of $2.31 \pm 0.12 \times 10^{-6}$ cm/s ($n = 4$) for Mel obtained by other authors in an ex vivo permeation through rat jejunum in small diffusion chambers [34], which is also a lower and statistically different value than Caco-2. The difference may be also associated with the cytotoxicity in Caco-2 promoted by Mel concentrations of 1.56 and 0.78 μg/mL [35]. The ultrastructural damage in a simple structure such as a monolayer of cells would increase permeability through tight junctions, leading to an increased P_{app} value. Both circumstances would explain the differences. Drug solubility is not a limiting factor when applying Franz diffusion cells through an intestinal ex vivo membrane since Fur, which is BCS class 4, shows a good correlation. In contrast, intestinal membrane phenomena that modify intestinal permeation, such as accumulation or metabolism (among others), may hinder this method in the case of Mel.

4. Conclusions

A new ex vivo technique based on permeation through pig small intestinal membrane was developed. It allows the prediction of absorption rate or P_{app} and apical-to-basal permeation for different BCS drugs. This ex vivo method requires fewer economic resources than other in vitro techniques for P_{app} determination, providing a new suitable process to test and compare new oral formulations or modified release systems, including therapeutic and higher drug concentrations than Caco-2. Application of this method requires determining if the drug produces intestinal membrane phenomena that could affect the absorption process.

Supplementary Materials: The following are available online at http://www.mdpi.com/1999-4923/11/12/638/s1, Table S1: HPLC-UV chromatograms of blank vs. standard solutions of ketorolac, furosemide, melatonin and Htz.

Author Contributions: Conceptualization, B.C., A.C.C., and A.B.S.; methodology, B.C., A.C.C., and A.B.S.; software, A.C.C. and A.B.S.; formal analysis, A.B.S.; investigation, A.B.S.; data curation, A.B.S. and A.C.C.; writing—original draft preparation, A.B.S.; writing—review and editing, B.C. and A.B.S.; supervision, B.C., M.M., and A.C.C.; project administration, B.C. and A.C.C.

Funding: This research received no external funding.

Acknowledgments: We would like to thank Álvaro Gimeno Sandig, Director of Animal House Bellvitge of University of Barcelona and Lidia Gómez Segura from the Ethical Committee of Animal Experimentation at the University of Barcelona for their contribution in the intestinal samples obtaining process.

Conflicts of Interest: The authors declare no conflict of interest.

References

1. Sugano, K. *Biopharmaceutics Modeling and Simulations: Theory, Practice, Methods, and Applications*, 1st ed.; John Wiley & Sons, Inc.: Hoboken, NJ, USA, 2012; pp. 66–102.
2. Palumbo, P.; Picchini, U.; Beck, B.; Gelder, J.; Van Delbar, N. A general approach to the apparent permeability index. *J. Pharm. Pharm.* **2008**, *35*, 235–248. [CrossRef] [PubMed]
3. Lennernäs, H.; Palm, K.; Fagerholm, U.; Artursson, P. Comparison between active and passive drug transport in human intestinal epithelial (Caco-2) cells in vitro and human jejunum in vivo. *Int. J. Pharm.* **1996**, *127*, 103–107. [CrossRef]
4. Jiang, K.; Huang, C.; Jiao, R.; Bai, W.; Zheng, J.; Ou, S. Adducts formed during protein digestion decreased the toxicity of five carbonyl compounds against Caco-2 cells. *J. Hazard. Mater.* **2019**, *363*, 26–33. [CrossRef] [PubMed]

5. Jiang, L.; Li, Z.; Xie, Y.; Liu, L.; Cao, Y. Cyanidin chloride modestly protects Caco-2 cells from ZnO nanoparticle exposure probably through the induction of autophagy. *Food Chem. Toxicol.* **2019**, *127*, 251–259. [CrossRef] [PubMed]
6. Juan-García, A.; Montesano, D.; Mañes, J.; Juan, C. Cytoprotective effects of carotenoids-rich extract from Lycium barbarum L. on the beauvericin-induced cytotoxicity on Caco-2 cells. *Food Chem. Toxicol.* **2019**, *133*, 110798. [CrossRef] [PubMed]
7. Gómez, L.J.; Gómez, N.A.; Zapata, J.E.; López-García, G.; Cilla, A. In-vitro antioxidant capacity and cytoprotective/cytotoxic effects upon Caco-2 cells of red tilapia (Oreochromis spp.) viscera hydrolysates. *Food Res. Int.* **2019**, *120*, 52–61. [CrossRef] [PubMed]
8. Li, Q.; Liang, X.; Guo, N.; Hu, L.; E, M.P.; Wu, Y.; Wang, K. Protective effects of Bee pollen extract on the Caco-2 intestinal barrier dysfunctions induced by dextran sulfate sodium. *Biomed. Pharmacother.* **2019**, *117*, 109200. [CrossRef]
9. Ciappellano, S.G.; Tedesco, E.; Marco Venturini, M.; Benetti, F. In vitro toxicity assessment of oral nanocarriers. *Adv. Drug Deliv. Rev.* **2016**, *106*, 381–401. [CrossRef]
10. Dou, T.; Wang, J.; Han, C.; Shao, X.; Zhang, J.; Lu, W. Cellular uptake and transport characteristics of chitosan modified nanoparticles in Caco-2 cell monolayers. *Int. J. Biol. Macromol.* **2019**, *138*, 791–799. [CrossRef]
11. Yan, X.; Cao, S.; Li, Y.; Xiao, P.; Huang, Z.; Li, H.; Ma, Y. Internalization and subcellular transport mechanisms of different curcumin loaded nanocarriers across Caco-2 cell model. *J. Drug Deliv. Sci. Technol.* **2019**, *52*, 660–669. [CrossRef]
12. Lu, C.; Fu, K.; Cao, K.; Wei, J.; Zhou, J.; Zhao, D.; Zhang, Y. Permeability and transport mechanism of trihexyphenidyl hydrochloride in Caco-2 cell monolayer model with a validated UPLC-MS/MS method. *J. Pharm. Biomed. Anal.* **2019**, 112924. [CrossRef] [PubMed]
13. Xue-Qing, C.; Stefanski, K.; Shen, H.; Huang, C.; Caporuscio, C.; Yang, W.; Lam, P.; Su, C.; Gudmundsson, O.; Hageman, M. Oral delivery of highly lipophilic poorly water-soluble drugs: spray-dried dispersions to improve oral absorption and enable high-dose toxicology studies of a P2Y1 antagonist. *J. Pharm. Sci.* **2014**, *12*, 3924–3931.
14. Bu, P.; Ji, Y.; Narayanan, S.; Dalrymple, D.; Cheng, X.; Serajuddin, A.T. Assessment of cell viability and permeation enhancement in presence of lipid-based self-emulsifying drug delivery systems using Caco-2 cell model: Polysorbate 80 as the surfactant. *Eur. J. Pharm. Sci.* **2017**, *99*, 350–360. [CrossRef] [PubMed]
15. Chai, G.; Xu, Y.; Chen, S.; Cheng, B.; Hu, F.; You, J.; Du, Y.Z.; Yuan, H. Transport mechanisms of solid lipid nanoparticles across Caco-2 cell monolayers and their related cytotoxicology. *ACS Appl. Mater. Interfaces* **2016**, *8*, 5929–5940. [CrossRef] [PubMed]
16. Markowska, M.; Oberle, R.; Juzwin, S.; Hsu, C.P.; Gryszkiewicz, M.; Streeter, A.J. Optimizing Caco-2 cell monolayers to increase throughput in drug intestinal absorption analysis. *J. Pharm. Toxicol. Methods* **2001**, *46*, 51–55. [CrossRef]
17. Yamashita, S.; Furubayashi, T.; Kataoka, M.; Sakane, T.; Sezaki, H.; Tokuda, H. Optimized conditions for prediction of intestinal drug permeability using Caco-2 cells. *Eur. J. Pharm. Sci.* **2000**, *10*, 195–204. [CrossRef]
18. Committee for Human Medicinal Products. *ICH Guideline M9 on Biopharmaceutics Classification System Based Biowaivers*; European Medicines Agency: Amsterdam, The Netherlands, 2018.
19. Welcome, M.O. *Gastrointestinal Physiology*, 1st ed.; Springer: Basel, Switzerland, 2018; pp. 53–219.
20. Balimane, P.V.; Chong, S. Evaluation of permeability and P-glycoprotein interactions: Industry outlook. In *Biopharmaceutics Applications in Drug Development*, 1st ed.; Krishna, R., Yu, L., Eds.; Springer Science + Business Media, LLC: New York, NY, USA, 2008; pp. 101–105.
21. Nejdfors, P.; Ekelund, M.; Jeppsson, B. Mucosal in vitro permeability in the intestinal tract of the pig, the rat, and man: Species- and region-related differences. *Scand. J. Gastroenterol.* **2009**, *35*, 501–507.
22. United States Pharmacopeial Convention. *USP DI*, 21th ed.; United States Pharmacopeial Convention: Rockville, MD, USA, 2006; Volume 1, Parts 1–2; Volumes 2–3; pp. 39–40.
23. Römsing, S. Development and Validation of Bioanalytical Methods, Application to Melatonin and Selected Anti-Infective Drugs. Ph.D. Thesis, Uppsala University, Uppsala, Sweden, 16 November 2010.
24. O'Neil, M.J.; Smith, A.; Henckelman, P.E.; Budavari, S. *The Merck Index: An Encyclopedia of Chemicals, Drugs, and Biologicals*, 13th ed.; Merck & Co., Inc.: Kenilworth, NJ, USA, 2001; p. 854.
25. Khan, S.J.; Ongerth, J.E. Modelling of pharmaceutical residues in Australian sewage by quantities of use and fugacity calculations. *Chemosphere* **2004**, *54*, 355–367. [CrossRef]

26. Jung, S.J.; Choi, S.O.; Um, S.Y.; Kim, J.I.; Choo, H.Y.; Choi, S.Y.; Chung, S.Y. Prediction of the permeability of drugs through study on quantitative structure-permeability relationship. *J. Pharm. Biomed. Anal.* **2016**, *41*, 469–475. [CrossRef]
27. Yeleswaram, K.; Lg, M.; Jo, K.; Schabdach, D. Pharmacokinetics and oral bioavailability of exogenous melatonin in preclinical animal models and clinical implications. *J. Pineal Res.* **1997**, *22*, 45–51. [CrossRef] [PubMed]
28. Yazdanian, M.; Glynn, S.L.; Wright, J.L.; Hawi, A. Correlating partitioning and Caco-2 cell permeability of structurally diverse small molecular weight compounds. *Pharm. Res.* **1998**, *15*, 1490. [CrossRef] [PubMed]
29. Bock, U.; Flototto, T.; Haltner, E. Validation of cell culture models for the intestine and the blood-brain barrier and comparison of drug permeation. *ALTEX* **2004**, *21* (Suppl. 3), 57–64.
30. Lee, P.P.N.; Shiu, S.Y.W.; Chow, P.H.; Pang, S.F. regional and diurnal studies of melatonin and melatonin binding sites in the duck gastro-intestinal tract. *Neurosignals* **1995**, *4*, 212–224. [CrossRef] [PubMed]
31. Söderquist, F.; Hellström, P.M.; Cunningham, J.L. Human gastroenteropancreatic expression of melatonin and its receptors MT1 and MT2. *PLoS ONE* **2015**, *10*, e0120195. [CrossRef] [PubMed]
32. Mayo, J.C.; Aguado, A.; Cernuda-Cernuda, R.; Álvarez, A.; Cepas, V.; González, I.Q.; Rosa, M.S. Melatonin uptake by cells: An answer to its relationship with glucose? *Molecules* **2018**, *23*, 1999. [CrossRef] [PubMed]
33. Res, P. Melatonin decreases duodenal epithelial paracellular permeability via a nicotinic receptor – dependent pathway in rats *in vivo*. *J. Pineal Res.* **2013**, *54*, 282–291.
34. Tran, H.T.; Tran, P.H.; Lee, B. New findings on melatonin absorption and alterations by pharmaceutical excipients using the Ussing chamber technique with mounted rat gastrointestinal segments. *Int. J. Pharm.* **2009**, *378*, 9–16. [CrossRef]
35. Batista, A.P.; Da Silva, T.G.; Teixeira, A.A.; De Medeiros, P.L.; Teixeira, V.W.; Alves, L.C.; Dos Santos, F.A.; Silva, E.C. Ultrastructural aspects of melatonin cytotoxicity on Caco-2 cells in vitro. *Micron* **2014**, *59*, 17–23. [CrossRef]

pharmaceutics

Article

Encapsulation in Polymeric Nanoparticles Enhances the Enzymatic Stability and the Permeability of the GLP-1 Analog, Liraglutide, Across a Culture Model of Intestinal Permeability

Ruba Ismail [1], Alexandra Bocsik [2], Gábor Katona [1], Ilona Gróf [2,3], Mária A. Deli [2,4] and Ildikó Csóka [1,*]

1 Institute of Pharmaceutical Technology and Regulatory Affairs, Faculty of Pharmacy, University of Szeged, H-6720 Szeged, Hungary; ismail.ruba@pharm.u-szeged.hu (R.I.); katona@pharm.u-szeged.hu (G.K.)
2 Institute of Biophysics, Biological Research Centre H-6726 Szeged, Hungary; bocsik.alexandra@brc.mta.hu (A.B.); grof.ilona@brc.mta.hu (I.G.); deli.maria@brc.mta.hu (M.A.D.)
3 Doctoral School of Biology, University of Szeged, H-6726 Szeged, Hungary
4 Department of Cell Biology and Molecular Medicine, University of Szeged, H-6720 Szeged, Hungary
* Correspondence: csoka@pharm.u-szeged.hu; Tel.: +36-62-546116

Received: 16 October 2019; Accepted: 11 November 2019; Published: 12 November 2019

Abstract: The potential of poly (lactic-*co*-glycolic acid) nanoparticles (PLGA NPs) to overcome the intestinal barrier that limits oral liraglutide delivery was evaluated. Liraglutide-loaded PLGA NPs were prepared by the double emulsion solvent evaporation method. In vitro release kinetics and enzymatic degradation studies were conducted, mimicking the gastrointestinal environment. The permeability of liraglutide solution, liraglutide-loaded PLGA NPs, and liraglutide in the presence of the absorption enhancer PN159 peptide was tested on the Caco-2 cell model. Liraglutide release from PLGA NPs showed a biphasic release pattern with a burst effect of less than 15%. The PLGA nanosystem protected the encapsulated liraglutide from the conditions simulating the gastric environment. The permeability of liraglutide encapsulated in PLGA NPs was 1.5-fold higher (24×10^{-6} cm/s) across Caco-2 cells as compared to liraglutide solution. PLGA NPs were as effective at elevating liraglutide penetration as the tight junction-opening PN159 peptide. No morphological changes were seen in the intercellular junctions of Caco-2 cells after treatment with liraglutide-PLGA NPs, confirming the lack of a paracellular component in the transport mechanism. PLGA NPs, by protecting liraglutide from enzyme degradation and enhancing its permeability across intestinal epithelium, hold great potential as carriers for oral GLP-1 analog delivery.

Keywords: liraglutide; GLP-1 analog; oral peptide delivery; enzymatic barrier; intestinal permeability; PLGA nanoparticles; Caco-2 cells

1. Introduction

The worldwide prevalence of type 2 diabetes mellitus (T2DM) has been increasing dramatically and has become a serious issue at an alarming rate. Because the incretin effect has been proven to be severely reduced or lost in relatively lean type 2 diabetic patients, incretin-based therapy, especially glucagon-like peptide 1 (GLP-1) receptor agonists, is now widely investigated for T2DM [1]. Long acting GLP-1 analogs have been developed to overcome the clinical limitations of the native GLP-1 due to its short circulating half-life [2,3].

Liraglutide (Lira; MW: 3751.2 Da) is an acylated derivative of GLP-1 that shares 97% homology to the native GLP-1, with two modifications: an Arg34Lys substitution, and a fatty acid side chain

(16-carbon palmitate) attached to Lys26 via a glutamic acid linker [4,5]. Lira retains the physiological activities of GLP-1 with a considerably longer half-life (approximately 13 h) that supports once-daily dosing. Subcutaneous Lira has been EMA and FDA approved for T2DM treatment, and soon after was approved for chronic weight management [6]. As Lira is limited to administration parenterally, the development of a patient-friendly delivery should be aimed for. Herein, oral administration is the most attractive choice as this route likely mimics physiological GLP-1 secretion in addition to ensuring good patient adherence to the treatment [7,8]. Moreover, oral delivery appears to be feasible for Lira due to the relatively large safety window of GLP1 analogs compared to insulin [9]. However, the oral delivery of Lira is still challenging due to low stability along the gastrointestinal (GI) tract and poor intestinal permeability that result in low oral bioavailability [10].

The encapsulation of peptides into nanocarrier systems, especially polymeric NPs, has arisen as a very promising alternative carrier system that has greater stability in biological fluids and during storage when compared to lipid-based nanosystems [11]. Polymeric NPs can not only protect the encapsulated peptide from the harsh environment in the GI tract but also control drug release and enhance its intracellular uptake [12,13]. Poly (lactic-*co*-glycolic acid) (PLGA) nanoparticles showed potential results as nanocarriers designed for the oral delivery of insulin and exenatide [14–17].

To enhance the delivery of protein or peptide molecules across biological barriers, several strategies can be used. One of them is the opening of intercellular tight junctions [18]. Another possibility is the use of cell penetrating peptides (CPPs). CPPs comprise a family of functional carrier peptides consisting of 5–30 amino acid residues that have been reported to have great potential in enhancing the peptide drugs permeability across the intestinal epithelium. Amphipathic CPPs, such as penetratin, are among the most widely promising ones. Many recent studies proved that the non-covalent intermolecular interaction between penetratin and insulin could be clinically promising as an absorption enhancer for successful oral insulin delivery [19–22]. Our group has recently demonstrated, that a permeability enhancing 18-mer amphiphilic peptide, PN159, also known as KLAL or MAP, has a dual action: by acting on claudin transmembrane tight junction proteins it opens the paracellular route in both epithelial and blood–brain barrier models [23] and at the same time it has cell permeabilizing and penetrating properties [24].

As realizing the dream of administering antidiabetic peptides such as liraglutide orally is still an elusive goal, we reported in our previous paper the potential of implementing the quality by design methodology from the early stage of product formulation, especially when dealing with complex nanosystems such as polymeric NPs designed for peptide delivery. We have successfully optimized the formulation of Lira encapsulated in hydrophobic PLGA NPs. For the present work, the purpose was to evaluate the effectiveness of our previously developed PLGA NPs in protecting the encapsulated peptide from the harsh environment in the GI tract. As there is no available literature regarding the intestinal permeability of Lira solution, we also evaluated the permeability of Lira through the Caco-2 cell model. Moreover, we investigated the potential of the optimized PLGA NPs in enhancing the permeability of encapsulated Lira through the Caco-2 cell model and compared it with the permeability enhancer PN159 peptide.

2. Materials and Methods

2.1. Materials

Liraglutide was purchased from Xi'an Health Biochem Technology Co., Ltd. (Shaanxi, China), Poly(lactide-*co*-glycolide) (PLGA 50:50, Mw = 30,000–60,000 Da), and PVA (MOWIOL 4-98, Mw ~ 27,000 Da) were purchased from Sigma Aldrich (Munich, Germany). D-mannitol was purchased from Molar Chemicals Ltd. (Budapest, Hungary). Sodium acetate anhydrous was purchased from Scharlau Chemie S.A. (Barcelona, Spain). Ethyl acetate was from REANAL Labor (Budapest, Hungary). Pepsin from porcine gastric mucosa, powder (≥400 units/mg protein) and pancreatin from porcine pancreas (≥3× USP specifications) were purchased from Sigma Aldrich (Budapest, Hungary). All other chemicals in the study were of analytical reagent grade.

2.2. *Human Caco-2 Intestinal Epithelial Cell Line*

The Caco-2 intestinal epithelial cell line was purchased from ATCC (cat.no. HTB-37) at passage 60. The cells were grown, as previously reported [24], in Dulbecco's modified Eagle's medium (Gibco, Life Technologies, Carlsbad, CA, USA) supplemented with 10% fetal bovine serum (Pan-Biotech GmbH, Aidenbach, Germany) and 50 μg/mL gentamycin in a humidified incubator with 5% CO_2 at 37 °C. All plastic surfaces were coated with 0.05% rat tail collagen in sterile distilled water before cell seeding in culture dishes and the medium was changed every 2 days.

2.3. *Preparation of PLGA NPs*

The Lira loaded PLGA NPs were prepared by the double emulsion solvent evaporation method and then lyophilized as described previously [25]. Following the initial risk assessment-based study [26], the Plackett–Burman screening design of the experiment was applied by our research group to optimize the lyophilized Lira loaded PLGA NP formulation regarding four critical quality attributes namely: particle size, polydispersity index, encapsulation efficiency and zeta potential [25]. The optimized formula is shown in Table 1.

Table 1. Optimized Lira loaded PLGA NPs; where PVA is the polyvinyl alcohol-stabilizer, and W2/O is the outer aqueous phase to organic phase ratio.

Formulation Parameters	
PLGA amount	60 mg
Lira amount	5 mg
2nd sonication time	0.5 min
PVA amount	1.48%
Lyoprotectant type	Mannitol
Lyoprotectant amount	5%
W2/O ratio	5

2.4. *Preparation of Lira and PN159 Solutions*

PN159 peptide (NH2-KLALKLALKALKAALKLA-amide) was previously synthesized and purified as reported by our research group [23,24]. To prepare Lira and PN159 solutions, liraglutide was dissolved in cell culture medium or Ringer-buffer (136 mM NaCl, 0.9 mM $CaCl_2$, 0.5 mM $MgCl_2$, 2.7 mM KCl, 1.5 mM KH_2PO_4, 10 mM NaH_2PO_4, pH 7.4) in a plastic vial. PN159 was weighed in a plastic vial and dissolved in Ringer buffer as well. Equal volumes of Lira and PN159 solutions were gently mixed at room temperature. The final concentrations of liraglutide and PN159 were 100 μM and 3 μM, respectively, in all experiments.

2.5. *In Vitro Drug Release Study*

The in vitro release behavior of Lira from the prepared PLGA nanoparticles was assessed by dispersing Lira-loaded PLGA NPs (corresponding to 500 μg of Lira) in 10 mL of simulated gastric fluid without enzymes (SGFsp: 0.1 N HCl at pH 1.2). After 2 h, the NPs were centrifuged and transferred to simulated intestinal fluid without enzymes (SIFsp: phosphate buffer saline at pH 7.4).

The beaker was placed over a magnetic stirrer (100 rpm) and the temperature was kept at 37 ± 1 °C throughout the experiment. At specified time points (0, 0.5, 1, 2, 3, 4, 6 h), an aliquot of 500 μL was withdrawn from the release medium and replenished with the same volume of fresh preheated medium. Samples were centrifuged at 16,500× *g* and 4 °C for 10 min, and the Lira concentrations in the supernatant were determined by HPLC. The cumulative percentage of Lira released was calculated and then plotted versus time. All experiments were conducted in triplicate.

2.6. Release Kinetics Studies

To understand the in vitro release from PLGA NPs, the data were fitted into various kinetic models, which were the zero order (where the drug release rate is independent of its concentration), and first order model (where the drug release rate is concentration-dependent), and Higuchi (which describes the drug release from an insoluble matrix as a square root of the time-dependent process based on Fickian diffusion). The best-fit model was selected based on the coefficient of correlation R^2. Then the release mechanism was further confirmed by fitting the release data in the Korsmeyer–Peppas equation, where the exponent (η) value was used to describe the release mechanism of the drug through the PLGA matrix; where $0.45 \leq \eta$ corresponds to the Fickian diffusion mechanism, $0.45 < \eta < 0.89$ corresponds to anomalous (non-Fickian) transport, $\eta = 0.89$ corresponds to case II (relaxational) transport, and $\eta > 0.89$ corresponds to super case II transport.

2.7. Enzymatic Degradation Study

Stability analysis in the presence of pepsin and pancreatin was conducted and compared between native Lira and Lira loaded in PLGA NPs. Five hundred micrograms of native Lira, or the amount of NPs containing an equivalent amount of Lira, were added to 2 mL of pepsin-containing simulated gastric fluid SGF (3.2 g pepsin, 2 g of sodium chloride, 7 mL HCl, mixed and diluted with water to 1 L, pH = 1.2) or pancreatin-containing simulated intestinal fluid SIF (10 g pancreatin, 6.8 g KH_2PO_4, mixed and adjusted with 0.2 N NaOH then diluted with water to 1 L, pH = 6.8) and incubated at 37 °C under stirring at a speed of 100 rpm. The SGF and SIF were prepared as per USP specifications (Test Solutions, United States Pharmacopeia 35, NF 30, 2012). The same methodology was followed to assess the native Lira stability in SGFsp and SIFsp. The samples (750 µL) were withdrawn at specified time intervals for 2 h, and an equal volume of ice-cold reagent was added: 0.1 M NaOH for SGF and 0.1 M HCl for SIF, to stop the enzymatic reaction. The samples were centrifuged at 16,500× *g* and 4 °C for 10 min and the supernatant was analyzed by HPLC to calculate the residual Lira. All incubations were done in triplicates. Lira recovery in the withdrawn samples was calculated using the following equation:

$$\text{Lira recovery\%} = (\text{Remaining Lira amount}/\text{theoretical Lira amount}) \times 100$$

2.8. Treatment of Caco-2 Cells

The concentration of stock solutions for cell culture experiments were 1 mM for both the therapeutic peptide liraglutide and the PN159 peptide, which was used as a reference absorption enhancer [23,24]. The working solutions were diluted in cell culture medium or Ringer-buffer depending on the experiments. The final concentration of liraglutide encapsulated in the PLGA NPs was 100 µM and was diluted directly before using. Liraglutide was examined at 100 µM, while PN159 was examined at 3 µM concentration both for cell viability and permeability.

2.9. Cell Viability Measurement by Impedance

Impedance was measured at 10 kHz by an RTCA SP instrument (RTCA-SP instrument, ACEA Biosciences, San Diego, CA, USA). We have successfully tested the cellular effects of peptides and pharmaceutical excipients by impedance kinetics [23,27,28]. For background measurements, 50 µL cell culture medium was added to the wells. This was followed by seeding the cells at a density of 6×10^3 cells/well to 96-well plate with gold electrodes (E-plate 96, ACEA Biosciences) coated with collagen. Cells were cultured for 5 days in a CO_2 incubator at 37 °C and monitored every 10 min until the end of experiments. Cells were treated at the beginning of the plateau phase of growth. Lira, Lira-loaded PLGA NPs, blank PLGA NPs (without cargo), Lira and PN159 solution, and PN159 peptide were diluted in cell culture medium and the effects were followed for 24 h. Triton X-100 detergent (1 mg/mL) was used as a reference compound to obtain total cell toxicity. Cell index was defined as R_n-R_b at each time point of measurement, where R_n is the cell–electrode impedance of the well when it contains cells and R_b is the background impedance of the well with the medium alone.

2.10. Permeability Study on the Caco-2 Model

Transepithelial electrical resistance (TEER) reflects the overall tightness of cell layers of biological barriers. TEER monitoring was performed by an EVOM volt-ohmmeter (World Precision Instruments, Sarasota, FL, USA) combined with STX-2 electrodes. The final TEER was expressed relative to the surface area of the monolayers as $\Omega \times cm^2$ after subtraction of TEER values of cell free inserts.

Caco-2 cells were seeded onto Transwell inserts (polycarbonate membrane, 3 μm pore size, 1.12 cm^2 surface area; Corning Life Sciences, Tewksbury, MA, USA) and cultured for three weeks [29,30]. For transport experiments, the inserts were transferred to 12-well plates containing 1.5 mL Ringer-buffer in the acceptor (lower/basal) compartments. In the donor (upper/apical) compartments, the culture medium was replaced by 0.5 mL Ringer-buffer containing treatment solutions of Lira, Lira loaded in PLGA NPs, and Lira and PN159 solution at the concentration of 100 μM for liraglutide for 1 h. Treatment solutions from both compartments were collected and the Lira level was detected by HPLC.

The apparent permeability coefficients (P_{app}) were calculated as described previously [23]. Briefly, the cleared volume was calculated from the concentration difference of the tracer in the acceptor compartment ($\Delta[C]_A$) after 1 h and the donor compartments at 0 h ($[C]_D$), the volume of the acceptor compartment (V_A; 1.5 mL) and the surface area available for permeability (A; 1.1 cm^2) using this equation:

$$P_{app}(cm/s) = \frac{\Delta[C]_A \times V_A}{A \times [C]_D \times \Delta t}$$

Recovery (mass balance) was calculated according to the equation:

$$\%Recovery = \frac{C_f^D V^D + C_f^A V^A}{C_0^D V^D} \times 100\%$$

where C_0^D and C_f^D are the initial and final concentrations of the compound in the donor compartment, respectively; C_0^A is the final concentration in the acceptor compartment; and V^D and V^A are the volumes of the solutions in the donor and acceptor compartments [29].

2.11. Immunohistochemistry

Aiming to investigate the morphological changes in interepithelial junctions, immunostaining for the junctional proteins, zonula occludens protein-1 (ZO-1) and β-catenin, was carried out. Cells were grown on glass coverslips (Menzel-Glaser, Braunschweig, Germany) at a density of 4×10^4 cells/coverslips for 4 days and treated with Lira (100 μM), Lira loaded in PLGA NPs, Lira and PN159 solution, and PN159 peptide (3 μM) solutions for 1 h. After the treatment, the coverslips were washed with phosphate buffer (PBS) and the cells were fixed with 3% paraformaldehyde solution for 15 min at room temperature and incubated in 0.2% TX-100 solution for permeabilization. The cells were blocked with 3% bovine serum albumin in PBS and incubated with the rabbit primary antibodies, anti-ZO-1 and anti-β-catenin, overnight. Incubation with secondary Cy3-labeled anti-rabbit antibody lasted for 1 h. Hoechst dye 33342 was used to stain the cell nuclei. After mounting the samples (Fluoromount-G; Southern Biotech, Birmingham, AL, USA), the staining was visualized by a Visitron spinning disk confocal system (Visitron Systems GmbH, Puchheim, Germany).

2.12. Chromatographic Equipment and Conditions

Lira was analyzed by a reversed phase HPLC (Agilent 1200, San Diego, CA, USA) method that was previously developed and validated in our lab [25]. A Kinetex® C18 column with dimensions of (5 μm, 150 × 4.6 mm, (Phenomenex, Torrance, CA, USA) was used as the stationary phase. The mobile phase comprised 0.02 M aqueous KH_2PO_4 solution (pH = 7.0, solvent A) and acetonitrile (solvent B) was pumped in a gradient mode from 80:20 (A:B, *v*/*v*) to 30:70 (A:B, *v*/*v*) in 12 min then back to 80:20 (*v*/*v*) between 12.1–15 min, at a flow rate of 1.5 mL/min. Fifty microliters of the sample was injected. The wavelength of UV detection was 214 nm. The retention time of Lira was 8.65 min.

The regression of the linearity (R^2) of the Lira calibration curve was 0.996.

2.13. *Statistical Analysis*

All data presented are means ± SD. The values were compared using the analysis of variance (ANOVA) followed by the Dunnett test or the Bonferroni test, using GraphPad Prism 5.0 software (GraphPad Software Inc., San Diego, CA, USA). Changes were considered statistically significant at $p < 0.05$.

3. Results

3.1. *In Vitro Release of Lira from PLGA NPs*

The release behavior data presented in (Figure 1) showed a biphasic release pattern starting by a moderate initial burst release during the first 2 h in SGFsp, where 14.2 ± 0.86% of Lira was released from the NPs. This was followed by a slow release profile until 6 h in the SIFsp, where only 18.5 ± 2.39% of cumulative Lira release was reached.

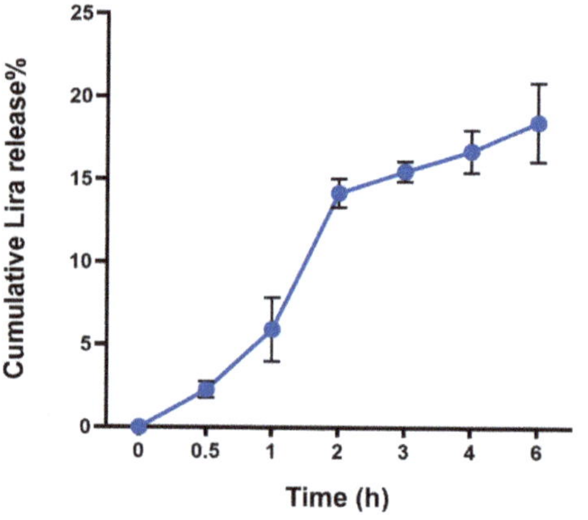

Figure 1. Cumulative in vitro release profile of liraglutide (Lira) from PLGA NPs.

3.2. *Release Kinetics Studies*

Based on the best fit with the highest correlation (R^2) value, it is concluded that Lira release from PLGA NPs follows the zero-order model (R^2 = 0.999) in SGFsp (pH = 1.2). When the release data is fitted into the Korsemeyer–Peppas equation (R^2 = 0.999), the exponent (n) value is 1.316, which is consistent with the zero-order release mechanism (Table 2).

Regarding the following 4 h in SIFsp, the release mechanism follows the Higuchi model (R^2 = 0.998), indicating diffusion-controlled release, and the exponent (n) value of the Korsmeyer–Peppas equation is 0.254, indicating that the release mechanism from PLGA NPs follows the Fickian diffusion mechanism (Table 2).

Table 2. Release kinetics data of Lira from PLGA NPs, in SGF and SIF (without enzymes).

Kinetic Model	SGF			SIF		
	N	*k*	R^2	*n*	*k*	R^2
Zero order	7.9857	1.84	0.9991	0.9836	12.658	0.9925
First order	0.038	2.01	0.998	0.0052	1.9418	0.9934
Higuchi	17.049	10.262	0.9854	4.1563	8.3559	0.9983
Korsmeyer-Peppas	1.3162	0.762	0.9993	0.2456	1.0696	0.9996

3.3. *Enzymatic Degradation Study*

It is obvious that only 1.9 ± 0.46% and 9.2 ± 0.7% of the free Lira was recovered after 30 min incubation in SGF and SIF, respectively. Lira was completely degraded after incubation for 1 h in SIF, while only 5.7 ± 0.53% Lira recovery occurred after 2 h incubation with SGF. On the other hand, the encapsulation of Lira into PLGA NPs was able to successfully protect 71.2 ± 1.49% and 87.6 ± 1.3% of Lira from degradation in the SGF and SIF at the end of the 2-h incubation, respectively (Figure 2). PLGA nanoparticles were claimed in previous research papers to be able to provide a protective and stable environment to encapsulate peptide drugs [11,31]. Encapsulation of GLP-1 into PLGA nanosystems could successfully shelter the peptide from the harsh environment of the simulated conditions of the stomach with sustained GLP-1 release [12].

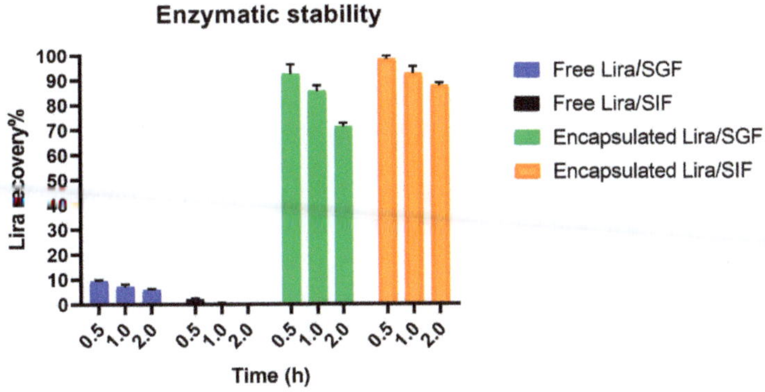

Figure 2. Enzymatic stability of liraglutide (Lira) encapsulated in PLGA NPs in both SGF and SIF mediums, with free Lira as a control.

3.4. *Cell Viability Assay*

Treatments with Lira, Lira loaded in PLGA NPs, Lira with PN159 peptide, unloaded PLGA NPs or PN159 peptide did not change the cell index values measured by impedance, a sensitive method to detect cellular effects, indicating good cell viability (Figure 3). Figure 3A shows the kinetics of the cellular effects of the treatment solutions, while the columns in Figure 3B show the effect of treatments at the 1-h time point. When cells were lysed with the detergent Triton X-100 the impedance dropped to zero (Figure 3B).

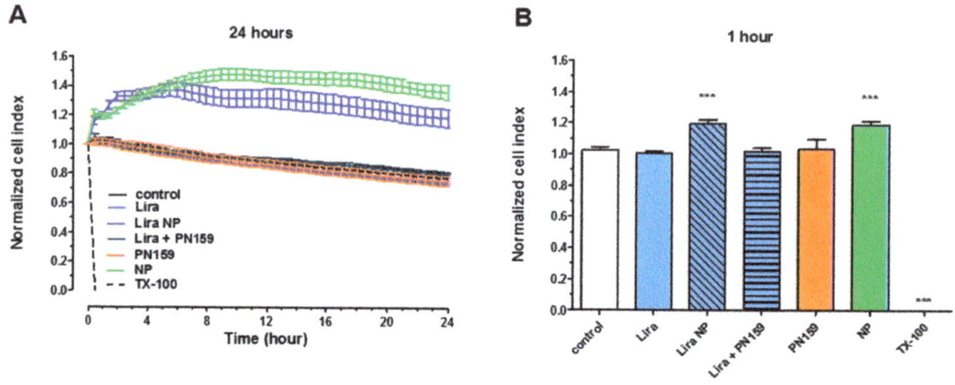

Figure 3. Cell viability kinetics for 24 h (**A**) and results of a 1-h treatment (**B**) of Caco-2 intestinal epithelial cells with liraglutide (Lira), NPs, liraglutide in NPs, liraglutide with PN159 peptide, and PN159 peptide measured by impedance. Values are presented as means ± SD, n = 6–12. Statistical analysis: Analysis of Variance (ANOVA) followed by Dunnett's test. NPs, nanoparticles; TX-100, Triton X-100. *** $p < 0.001$ compared to control.

3.5. Permeability Study on the Caco-2 Intestinal Barrier Model

Caco-2 monolayers showed high TEER values (893 ± 135 $\Omega \times cm^2$, n = 20) before permeability experiments indicating tight barrier properties. Because of the tight barrier the permeability was low for the marker molecule fluorescein (P_{app}: below 0.5×10^{-6} cm/s) as in our previous study [29]. The free Lira at a donor concentration of 100 μM showed good penetration as the P_{app} was 16×10^{-6} cm/s (Figure 4). The permeability of Lira encapsulated in NPs, 24×10^{-6} cm/s, was significantly higher than that for Lira solution. An increased Lira permeability (28×10^{-6} cm/s) was measured in the presence of PN159 peptide, our reference absorption enhancer (Figure 4A). There was no statistical difference between the liraglutide permeability values of the Lira-NP and Lira + PN159 groups. In contrast, the only group where the TEER values dropped after the 1-h treatment was the one containing PN159 peptide (Figure 4B) indicating opening of the paracellular pathway in agreement with observations from our previous studies [23,24]. Liraglutide alone or encapsulated in PLGA NPs did not change the ionic permeability (Figure 4B), suggesting no toxic effect on differentiated Caco-2 cells in concordance with the viability data (Figure 3) and no effect on the paracellular pathway.

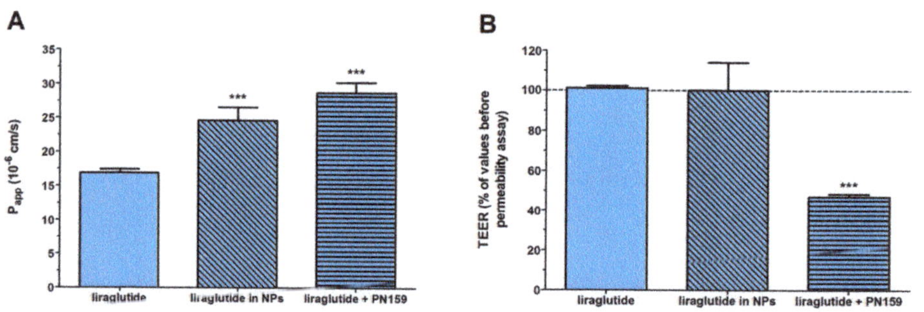

Figure 4. Evaluation of permeability of liraglutide (100 μM) across Caco-2 epithelial cell layers treated with different liraglutide formulations for 1 h (**A**). Changes in transepithelial electrical resistance (TEER) values of Caco-2 cell layers after 1-h treatment with different liraglutide formulations as compared to TEER values before treatment (**B**). Values are presented as means ± SD, n = 4. Statistical analysis: ANOVA followed by Bonferroni test, *** $p < 0.001$ compared to liraglutide group.

There was a good recovery for Lira after the permeability experiments and we found no significant differences between the recovery values of the different investigated Lira groups (Table 3).

Table 3. Recovery (mass balance) calculation after liraglutide permeability on Caco-2 cells.

Liraglutide	Recovery (%) Mean ± SD
Liraglutide	80.9 ± 1.6
Liraglutide in NPs	75.3 ± 2.3
Liraglutide + PN159	81.3 ± 6.9

3.6. *Immunohistochemistry*

The Caco-2 intestinal epithelial cells formed confluent layers visualized by the localization of the junctional proteins ZO-1 and β-catenin. The cells were attached to each other without gaps and had similar immunostaining patterns. An intact, belt-shaped continuous localization around the cell borders for the junctional proteins was observed both in the control and the treated groups. Only the PN159 peptide treated group showed a visible change in the staining pattern of β-catenin adherens junctional protein (Figure 5).

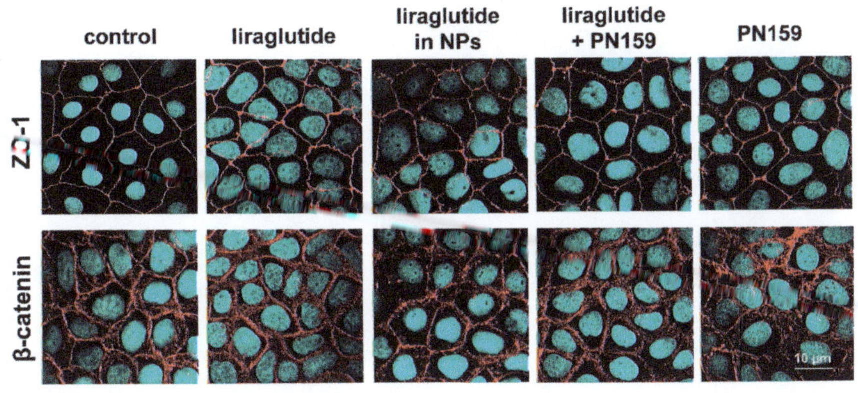

Figure 5. Effects of liraglutide, liraglutide in NPs, liraglutide and PN159 peptide together, and PN159 peptide on the junctional morphology of Caco-2 epithelial cells. Immunostaining for zonula occludens-1 (ZO-1), and β-catenin junction proteins after a 1-h treatment. Red color: immunostaining for junctional proteins. Blue color: staining of cell nuclei. Bar: 10 μm.

4. Discussion

GLP-1 analogs represent a unique class of antidiabetic peptide drugs with potential clinical benefits over existing therapies for T2DM treatment [32]. Lira, a lipophilic long-acting GLP-1 analog, is still subcutaneously administered. Since the oral delivery of Lira can bypass the inconvenience zone of patients, a smart carrier system that can tackle the challenges hindering the oral peptide delivery has been aimed for. In a previous work, we formulated and statistically optimized the formulation and process parameters affecting the quality of Lira loaded PLGA NPs that are designed for oral delivery. Spherical shaped NPs with homogeneous distribution, 188.95 nm particle size and 51.81% encapsulation efficiency were obtained [25]. As a follow-up study, the aim of this work was to investigate the potential of the developed PLGA NPs in overcoming the main barriers limiting the oral peptide delivery, namely, the harsh environment through the GI tract and the absorption membrane barrier.

The behavior of drug release from polymeric NPs is a complex process attributed to diffusion followed by degradation and influenced by the drug physicochemical properties in addition to various formulation and process variables [33]. The Lira release from PLGA NPs showed a biphasic release pattern, which was frequently reported for polymeric NPs by previous papers [12,34]. At the burst release phase (where less than 15% of Lira was released), PLGA NPs are exposed to the gastric media and the surface of the NPs is hydrated. Then, non-capsulated Lira, or Lira which exists close to the surface having weak interactions with it is easily accessible by hydration and is released in the media. Potentially, less than 15% of Lira was released from NPs in the gastric media following a zero-order model. At the second phase of slow Lira release, the degradation of the polymer matrix took place, leading to diffusion of the encapsulated Lira, and the release mechanism followed the Higuchi model, which further confirms the diffusion-controlled release. These results prove that the PLGA nanosystem is able to hinder the release of encapsulated peptides in gastric simulating conditions without enzymes (SGFsp), and later sustain the peptide release in intestinal simulating conditions without enzymes (SIFsp).

The enzymatic stability results were compared between the free Lira and Lira encapsulated in NPs. There was no Lira detected after the 2-h incubation of free Lira with SGF or SIF, which is due to the presence of amino acids, especially the aromatic ones, in the Lira structure, which makes it vulnerable to pepsin–pancreatin digestion [35]. The results revealed that PLGA NPs were effective in protecting 71% and 87% of Lira from pepsin and pancreatin digestion after a 2-h incubation with SGF and SIF, respectively. These findings confirm that PLGA NPs can provide a physical barrier between the encapsulated Lira and the harsh environment in the GI tract, and thereby are promising for obtaining higher oral peptide bioavailability [12,36].

There was no decrease in cell impedance kinetics regarding the five treatment groups; Lira, Lira loaded in PLGA NPs, Lira and PN159 peptide in solution, unloaded PLGA NPs or PN159 peptide, during the 24-h long treatment. These findings proved the biocompatibility of PLGA NPs and showed that the composition of the nanosystem did not contribute to toxicity in Caco-2 cells. This is in accordance with previous reports where PLGA nanosystems larger than 100 nm did not trigger any toxic effects at different concentrations [37–39]. Regarding PN159 peptide, the absence of cytotoxic effects at the concentration of 3 μM is also in accordance with our previous report [24,40].

To further evaluate the potential of the prepared nanosystem, permeability studies on the Caco-2 cell model were conducted. Lira showed a good apparent permeability through the cell model compared to what was reported for the native GLP-1 or exenatide, which could be due to the 16-carbon fatty acid chain that is attached to lysine at position 26 via a glutamic acid spacer [9]. This acylation leads to higher Lira hydrophobicity [41], which can enhance the intracellular permeability of this GLP-1 analog when compared to the native GLP-1 or exenatide [42]. The potential of the optimized polymeric NPs was also investigated to further enhance Lira permeability. Lira encapsulated in NPs showed 1.5-fold higher apparent permeability as compared to Lira alone. Because PLGA NPs are more lipophilic compared to the free peptide drug, their transport across the lipid membrane of Caco-2 cells is better. Furthermore, these optimized polymeric NPs showed a smaller size than 200 nm, and it was previously reported that NPs within the 100–200 nm size range showed the best properties for cellular uptake, while smaller-sized (50 nm) or larger-sized (≥500 nm) particles resulted in reduced uptake [43]. Moreover, the prepared NPs were spherical with a smooth surface, as previously confirmed by scanning electron microscopy [25], which is better for cellular uptake when compared to needle-shaped ones. The ability of PLGA NPs to enhance the in vitro permeability of the hydrophilic drug alendronate [44], salmon calcitonin [45], bovine serum albumin [46], and insulin [15,17] through the Caco-2 cell model has been reported in the literature. In the presence of PN159 peptide, our reference absorption enhancer, Lira permeability also increased. This finding is consistent with our results, and results from the literature, on the reversible tight junction opening by the PN159 peptide in biological barrier models, which effectively improves the permeability of different drugs and hydrophilic marker molecules through the paracellular pathway [23,24,40]. In addition, it was reported that this amphipathic CPP can strongly bind to and interact with biological membranes due to electrostatic and hydrophobic interactions [47,48], and we also confirmed the cell permeabilizing and penetration effects of PN159 on Caco-2 cells [23].

The immunostaining for the junctional proteins, ZO-1 and β-catenin, showed that Lira-PLGA NPs did not change the morphology of interepithelial junctions. The PN159 peptide treated group showed a change in the staining pattern of the β-catenin adherens junctional protein, which is also in accordance with our previous results [23,24]. We suppose that an increase in both the paracellular transport and the membrane permeability in Caco-2 cells by PN159 can contribute to the enhanced Lira permeability.

5. Conclusions

Being a relatively new GLP-1 analog, there are no reported studies for Caco-2 permeability of liraglutide alone or encapsulated into a carrier system. In this study we found that the developed PLGA nanosystem could efficiently protect the encapsulated liraglutide from the conditions simulating the harsh environment in the GI tract. This polymeric system seems to be promising as it can also enhance the permeability 1.5-fold compared to free liraglutide solution. These findings reveal that encapsulation in a polymeric nanosystem holds promise for oral GLP-1 analog delivery.

Author Contributions: Conceptualization, R.I., M.A.D. and I.C.; methodology, R.I, A.B, G.K. and M.A.D.; software, R.A., A.B. and G.K.; validation, R.I. and A.B.; formal analysis, R.I., A.B., G.K. and I.G.; investigation, R.I., A.B., G.K. and I.G.; resources, R.I. and A.B.; data curation, R.I., A.B., G.K. and I.G.; writing—original draft preparation, R.I. and A.B.; writing—review and editing, R.I., I.G., M.A.D and I.C.; visualization, R.I. and I.G.; supervision, M.A.D. and I.C.; project administration, I.Cs.; funding acquisition, I.C.

Funding: This work was supported by the Ministry of Human Capacities, Hungary (grant 20391-3/2018/FEKUSTRAT), the National Research, Development and Innovation Office, Hungary (GINOP-2.2.1-15-2016-00007) and the EU-funded Hungarian grant EFOP-3.6.1-16-2016-00008.

Conflicts of Interest: The authors declare no conflict of interest.

References

1. Hinnen, D. Glucagon-like peptide 1 receptor agonists for type 2 diabetes. *Diabetes Spectr.* **2017**, *30*, 202–210. [CrossRef] [PubMed]
2. Gupta, V. Glucagon-like peptide-1 analogues: An overview. *Indian J. Endocrinol. Metab.* **2013**, *17*, 413–421. [CrossRef] [PubMed]
3. Zhang, X.; Liu, S.; Li, Y.; Wang, Y.; Tian, M.; Liu, G. Long-term effectiveness and cost-effectiveness of metformin combined with liraglutide or exenatide for type 2 diabetes mellitus based on the CORE diabetes model study. *PLoS ONE* **2016**, *11*, e0156393. [CrossRef] [PubMed]
4. Araújo, F.; Fonte, P.; Santos, H.A.; Sarmento, B. Oral delivery of glucagon-like peptide-1 and analogs: Alternatives for diabetes control? *J. Diabetes Sci. Technol.* **2012**, *6*, 1486–1497. [CrossRef] [PubMed]
5. Tella, S.H.; Rendell, M.S. Glucagon-like polypeptide agonists in type 2 diabetes mellitus: Efficacy and tolerability, a balance. *Ther. Adv. Endocrinol. Metab.* **2015**, *6*, 109–134. [CrossRef] [PubMed]
6. Crane, J.; McGowan, B. The GLP-1 agonist, liraglutide, as a pharmacotherapy for obesity. *Ther. Adv. Chronic Dis.* **2016**, *7*, 92–107. [CrossRef] [PubMed]
7. Lin, Y.; Krogh-Andersen, K.; Pelletier, J.; Marcotte, H.; Östenson, C.G.; Hammarström, L. Oral delivery of pentameric glucagon-like peptide-1 by recombinant lactobacillus in diabetic rats. *PLoS ONE* **2016**, *11*, e0162733. [CrossRef] [PubMed]
8. Shamekhi, F.; Tamjid, E.; Khajeh, K. Development of chitosan coated calcium-alginate nanocapsules for oral delivery of liraglutide to diabetic patients. *Int. J. Biol. Macromol.* **2018**, *120*, 460–467. [CrossRef] [PubMed]
9. Knudsen, L.B.; Lau, J. The discovery and development of liraglutide and semaglutide. *Front. Endocrinol.* **2019**, *10*, 155. [CrossRef] [PubMed]
10. Ma, S.; We, L.; Yang, H.; Deng, S.; Jevnikar, A.M. Emerging technologies to achieve oral delivery of GLP-1 and GLP-1 analogs for treatment of Type 2 diabetes mellitus (T2DM). *Can. J. Biotechnol.* **2017**, *1*, 1–10. [CrossRef]
11. Araujo, F.; Shrestha, N.; Gomes, M.J.; Herranz-Blanco, B.; Liu, D.; Hirvonen, J.; Granja, P.L.; Santos, H.A.; Sarmento, B. In vivo dual-delivery of glucagon like peptide-1 (GLP-1) and dipeptidyl peptidase-4 (DPP4) inhibitor through composites prepared by microfluidics for diabetes therapy. *Nanoscale* **2016**, *1*, 10706–10713. [CrossRef] [PubMed]

12. Araújo, F.; Shrestha, N.; Shahbazi, M.A.; Fonte, P.; Mäkilä, E.M.; Salonen, J.J.; Hirvonen, J.T.; Granja, P.L.; Santos, H.A.; Sarmento, B. The impact of nanoparticles on the mucosal translocation and transport of GLP-1 across the intestinal epithelium. *Biomaterials* **2014**, *35*, 9199–9207. [CrossRef] [PubMed]
13. Ismail, R.; Csóka, I. Novel strategies in the oral delivery of antidiabetic peptide drugs—Insulin, GLP 1 and its analogs. *Eur. J. Pharm. Biopharm.* **2017**, *115*, 257–267. [CrossRef] [PubMed]
14. Zhang, X.; Sun, M.; Zheng, A.; Cao, D.; Bi, Y.; Sun, J. Preparation and characterization of insulin-loaded bioadhesive PLGA nanoparticles for oral administration. *Eur. J. Pharm. Sci.* **2012**, *45*, 632–638. [CrossRef] [PubMed]
15. Reix, N.; Parat, A.; Seyfritz, E.; van der Werf, R.; Epure, V.; Ebel, N.; Danicher, L.; Marchioni, E.; Jeandidier, N.; Pinget, M.; et al. In vitro uptake evaluation in Caco-2 cells and in vivo results in diabetic rats of insulin-loaded PLGA nanoparticles. *Int. J. Pharm.* **2012**, *437*, 213–220. [CrossRef] [PubMed]
16. Wang, M.; Zhang, Y.; Feng, J.; Gu, T.; Dong, Q.; Yang, X.; Sun, Y.; Wu, Y.; Chen, Y.; Kong, W. Preparation, characterization, and in vitro and in vivo investigation of chitosan-coated poly (D,L-lactide-*co*-glycolide) nanoparticles for intestinal delivery of exendin-4. *Int. J. Nanomed.* **2013**, *8*, 1141–1154. [CrossRef]
17. Sheng, J.; Han, L.; Qin, J.; Ru, G.; Li, R.; Wu, L.; Cui, D.; Yang, P.; He, Y.; Wang, J. N-trimethyl chitosan chloride-coated PLGA nanoparticles overcoming multiple barriers to oral insulin absorption. *ACS Appl. Mater. Interfaces* **2015**, *7*, 15430–15441. [CrossRef] [PubMed]
18. Deli, M.A. Potential use of tight junction modulators to reversibly open membranous barriers and improve drug delivery. *Biochim. Biophys. Acta Biomembr.* **2009**, *1788*, 892–910. [CrossRef] [PubMed]
19. Kamei, N.; Morishita, M.; Eda, Y.; Ida, N.; Nishio, R.; Takayama, K. Usefulness of cell-penetrating peptides to improve intestinal insulin absorption. *J. Control Release* **2008**, *132*, 21–25. [CrossRef] [PubMed]
20. Khafagy, E.S.; Morishita, M.; Kamei, N.; Eda, Y.; Ikeno, Y.; Takayama, K. Efficiency of cell-penetrating peptides on the nasal and intestinal absorption of therapeutic peptides and proteins. *Int. J. Pharm.* **2009**, *381*, 49–55. [CrossRef] [PubMed]
21. Kamei, N.; Nielsen, E.J.B.; Khafagy, E.S.; Takeda-Morishita, M. Noninvasive insulin delivery: The great potential of cell-penetrating peptides. *Ther. Deliv.* **2013**, *4*, 315–326. [CrossRef] [PubMed]
22. Nielsen, E.J.B.; Yoshida, S.; Kamei, N.; Iwamae, R.; Khafagy, E.S.; Olsen, J.; Rahbek, U.L.; Pedersen, B.L.; Takayama, K.; Takeda-Morishita, M. In vivo proof of concept of oral insulin delivery based on a co-administration strategy with the cell-penetrating peptide penetratin. *J. Control Release* **2014**, *189*, 19–24. [CrossRef] [PubMed]
23. Bocsik, A.; Walter, F.R.; Gyebrovszki, A.; Fülöp, L.; Blasig, I.; Dabrowski, S.; Ötvös, F.; Tóth, A.; Rákhely, G.; Veszelka, S.; et al. Reversible opening of intercellular junctions of intestinal epithelial and brain endothelial cells with tight junction modulator peptides. *J. Pharm. Sci.* **2016**, *105*, 754–765. [CrossRef] [PubMed]
24. Bocsik, A.; Gróf, I.; Kiss, L.; Ötvös, F.; Zsíros, O.; Daruka, L.; Fülöp, L.; Vastag, M.; Kittel, Á.; Imre, N.; et al. Dual action of the PN159/KLAL/MAP peptide: Increase of drug penetration across Caco-2 intestinal barrier model by modulation of tight junctions and plasma membrane permeability. *Pharmaceutics* **2019**, *11*, 73. [CrossRef] [PubMed]
25. Ismail, R.; Sovány, T.; Gácsi, A.; Ambrus, R.; Katona, G.; Imre, N.; Csóka, I. Synthesis and statistical optimization of poly (lactic-*co*-glycolic acid) nanoparticles encapsulating GLP1 analog designed for oral delivery. *Pharm. Res.* **2019**, *36*, 99. [CrossRef] [PubMed]
26. Pallagi, E.; Ismail, R.; Paal, T.L.; Csoka, I. Initial risk assessment as part of the quality by design in peptide drug containing formulation development. *Eur. J. Pharm. Sci.* **2018**, *122*, 160–169. [CrossRef] [PubMed]
27. Kürti, L.; Veszelka, S.; Bocsik, A.; Dung, N.T.K.; Ózsvári, B.; Puskás, L.G.; Kittel, Á.; Szabó-Révész, P.; Deli, M.A. The effect of sucrose esters on a culture model of the nasal barrier. *Toxicol. Vitr.* **2012**, *26*, 445–454. [CrossRef] [PubMed]
28. Kiss, L.; Walter, F.R.; Bocsik, A.; Veszelka, S.; Ózsvári, B.; Puskás, L.G.; Szabó-Révész, P.; Deli, M.A. Kinetic analysis of the toxicity of pharmaceutical excipients cremophor EL and RH40 on endothelial and epithelial cells. *J. Pharm. Sci.* **2013**, *102*, 1173–1181. [CrossRef] [PubMed]
29. Hellinger, É.; Veszelka, S.; Tóth, A.E.; Walter, F.; Kittel, Á.; Bakk, M.L.; Tihanyi, K.; Háda, V.; Nakagawa, S.; Duy, T.D.H.; et al. Comparison of brain capillary endothelial cell-based and epithelial (MDCK-MDR1, Caco-2, and VB-Caco-2) cell-based surrogate blood-brain barrier penetration models. *Eur. J. Pharm. Biopharm.* **2012**, *82*, 340–351. [CrossRef] [PubMed]

30. Kiss, L.; Hellinger, É.; Pilbat, A.M.; Kittel, Á.; Török, Z.; Furedi, A.; Szakács, G.; Veszelka, S.; Sipos, P.; Ózsvári, B.É.; et al. Sucrose esters increase drug penetration, but do not inhibit p-glycoprotein in Caco-2 intestinal epithelial cells. *J. Pharm. Sci.* **2014**, *103*, 3107–3119. [CrossRef] [PubMed]
31. Fonte, P.; Araújo, F.; Silva, C.; Pereira, C.; Reis, S.; Santos, H.A.; Sarmento, B. Polymer-based nanoparticles for oral insulin delivery: Revisited approaches. *Biotechnol. Adv.* **2014**, *33*, 1342–1354. [CrossRef] [PubMed]
32. Aung, M.M.; Slade, K.; Freeman, L.A.R.; Kos, K.; Whatmore, J.L.; Shore, A.C.; Gooding, K.M. Locally delivered GLP-1 analogues liraglutide and exenatide enhance microvascular perfusion in individuals with and without type 2 diabetes. *Diabetologia* **2019**, *62*, 1701–1711. [CrossRef] [PubMed]
33. Fredenberg, S.; Wahlgren, M.; Reslow, M.; Axelsson, A. The mechanisms of drug release in poly(lactic-*co*-glycolic acid)-based drug delivery systems—A review. *Int. J. Pharm.* **2011**, *415*, 34–52. [CrossRef] [PubMed]
34. Dinarvand, R.; Sepehri, N.; Manoochehri, S.; Rouhani, H.; Atyabi, F. Polylactide-*co*-glycolide nanoparticles for controlled delivery of anticancer agents. *Int. J. Nanomed.* **2011**, *6*, 877–895. [CrossRef] [PubMed]
35. Xu, F.; Wang, K.Y.; Wang, N.; Li, G.; Liu, D. Bioactivity of a modified human glucagon-like peptide-1. *PLoS ONE* **2017**, *12*, e0171601. [CrossRef] [PubMed]
36. Silva, A.L.; Rosalia, R.A.; Sazak, A.; Carstens, M.G.; Ossendorp, F.; Oostendorp, J.; Jiskoot, W. Optimization of encapsulation of a synthetic long peptide in PLGA nanoparticles: Low-burst release is crucial for efficient CD8+ T cell activation. *Eur. J. Pharm. Biopharm.* **2013**, *83*, 338–345. [CrossRef] [PubMed]
37. Xiong, S.; George, S.; Yu, H.; Damoiseaux, R.; France, B.; Ng, K.W.; Loo, J.S.C. Size influences the cytotoxicity of poly (lactic-*co*-glycolic acid) (PLGA) and titanium dioxide (TiO_2) nanoparticles. *Arch. Toxicol.* **2013**, *78*, 1075–1086. [CrossRef] [PubMed]
38. Platel, A.; Carpentier, R.; Becart, E.; Mordacq, G.; Betbeder, D.; Nesslany, F. Influence of the surface charge of PLGA nanoparticles on their in vitro genotoxicity, cytotoxicity, ROS production and endocytosis. *J. Appl. Toxicol.* **2016**, *36*, 434–444. [CrossRef] [PubMed]
39. Silva, A.M.; Alvarado, H.L.; Abrego, G.; Martins-Gomes, C.; Garduño-Ramirez, M.L.; García, M.L.; Calpena, A.C.; Souto, E.B. In vitro cytotoxicity of oleanolic/ursolic acids-loaded in PLGA nanoparticles in different cell lines. *Pharmaceutics* **2019**, *11*, 362. [CrossRef] [PubMed]
40. Cui, K.; Chen, S.; Houston, M.; Quay, S. Tight Junction Modulator Peptide PN159 for Enhanced Mucosal Delivery of Therapeutic Compounds. US20060062758A1, 23 March 2006.
41. Trier, S.; Linderoth, L.; Bjerregaard, S.; Strauss, H.M.; Rahbek, U.L.; Andresen, T.L. Acylation of salmon calcitonin modulates in vitro intestinal peptide flux through membrane permeability enhancement. *Eur. J. Pharm. Biopharm.* **2015**, *96*, 329–337. [CrossRef] [PubMed]
42. Gupta, V.; Doshi, N.; Mitragotri, S. Permeation of insulin, calcitonin and exenatide across Caco-2 monolayers: Measurement using a rapid, 3-day system. *PLoS ONE* **2013**, *8*, e57136. [CrossRef] [PubMed]
43. Lanza, R.; Langer, R.; Vacanti, J.P. *Principles of Tissue Engineering*, 4th ed.; Academic Press: Cambridge, MA, USA, 2013. [CrossRef]
44. Cohen-Sela, E.; Chorny, M.; Koroukhov, N.; Danenberg, H.D.; Golomb, G. A new double emulsion solvent diffusion technique for encapsulating hydrophilic molecules in PLGA nanoparticles. *J. Control Release* **2009**, *133*, 90–95. [CrossRef] [PubMed]
45. Yoo, H.S.; Park, T.G. Biodegradable nanoparticles containing protein-fatty acid complexes for oral delivery of salmon calcitonin. *J. Pharm. Sci.* **2004**, *93*, 488–495. [CrossRef]
46. Panyam, J.; Dali, M.M.; Sahoo, S.K.; Ma, W.; Chakravarthi, S.S.; Amidon, G.L.; Levy, R.J.; Labhasetwar, V. Polymer degradation and in vitro release of a model protein from poly(D,L-lactide-co-glycolide) nano-and microparticles. *J. Control Release* **2003**, *92*, 173–187. [CrossRef]
47. Datrie, M.; Schumann, M.; Wieprecht, T.; Winkler, A.; Beyermann, M.; Krause, E.; Matsuzaki, K.; Murase, O.; Bienert, M. Peptide helicity and membrane surface charge modulate the balance of electrostatic and hydrophobic interactions with lipid bilayers and biological membranes. *Biochemistry* **1996**, *24*, 35–38. [CrossRef]
48. Arouri, A.; Dathe, M.; Blume, A. The helical propensity of KLA amphipathic peptides enhances their binding to gel-state lipid membranes. *Biophys. Chem.* **2013**, *180–181*, 10–21. [CrossRef] [PubMed]

pharmaceutics

MDPI

Article

Boosting Drug Discovery for Parkinson's: Enhancement of the Delivery of a Monoamine Oxidase-B Inhibitor by Brain-Targeted PEGylated Polycaprolactone-Based Nanoparticles

Miguel Pinto [1,†], Carlos Fernandes [1,†], Eva Martins [2], Renata Silva [2], Sofia Benfeito [1], Fernando Cagide [1], Ricardo F. Mendes [3], Filipe A. Almeida Paz [3], Jorge Garrido [4], Fernando Remião [2] and Fernanda Borges [1,*]

1 CIQUP, Departmento de Química e Bioquímica, Centro de Investigação em Química, Faculdade de Ciências, Universidade do Porto, 4169-007 Porto, Portugal

2 UCIBIO-REQUIMTE, Laboratório de Toxicologia, Departamento de Ciências Biológicas, Faculdade de Farmácia, Universidade do Porto, 4050-313 Porto, Portugal

3 Departamento de Química, CICECO-Instituto de Materiais de Aveiro, Universidade de Aveiro, 3810-193 Aveiro, Portugal

4 Departamento de Engenharia Química, Instituto Superior de Engenharia do Porto (ISEP), Instituto Politécnico do Porto, 4200-072 Porto, Portugal

* Correspondence: fborges@fc.up.pt

† These authors contributed equally to this work.

Received: 15 May 2019; Accepted: 9 July 2019; Published: 12 July 2019

check for updates

Abstract: The current pharmacological treatments for Parkinson's disease only offer symptomatic relief to the patients and are based on the administration of levodopa and catechol-O-methyltransferase or monoamine oxidase-B inhibitors (IMAO-B). Since the majority of drug candidates fail in pre- and clinical trials, due largely to bioavailability pitfalls, the use of polymeric nanoparticles (NPs) as drug delivery systems has been reported as an interesting tool to increase the stealth capacity of drugs or help drug candidates to surpass biological barriers, among other benefits. Thus, a novel potent, selective, and reversible IMAO-B (chromone C27, IC_{50} = 670 ± 130 *p*M) was encapsulated in poly(caprolactone) (PCL) NPs by a nanoprecipitation process. The resulting C27-loaded PEGylated PCL NPs (~213 nm) showed high stability and no cytotoxic effects in neuronal (SH-SY5Y), epithelial (Caco-2), and endothelial (*h*CMEC/D3) cells. An accumulation of PEGylated PCL NPs in the cytoplasm of SH-SY5Y and *h*CMEC/D3 cells was also observed, and their permeation across Caco-2 and *h*CMEC/D3 cell monolayers, used as in vitro models of the human intestine and blood-brain barrier, respectively, was demonstrated. PEGylated PCL NPs delivered C27 at concentrations higher than the MAO-B IC_{50} value, which provides evidence of their relevance to solving the drug discovery pitfalls.

Keywords: Parkinson disease; chromone; monoamine oxidase B inhibitor; PEGylated nanoparticles; intestinal and brain permeability

1. Introduction

Parkinson's disease (PD) is a multifactorial disorder characterized by the progressive degeneration of the structure and well-function of the central nervous system (CNS), leading to a depletion of dopaminergic neurons in the substantia nigra [1]. The decrease of the neurotransmitter dopamine in neuron cleft is associated with movement control and cognitive losses, which appear as a form of tremors, memory loss, and inconsistent speech. Nowadays, PD is the second most prevalent neurodegenerative disorder, affecting 1–2% of the world's population above 65 years of age, increasing to approximately 4% in individuals above 85 years of age [2].

Despite how clinical trials for PD have long relied on observing whether a therapy improves the symptomatology of patients, the studies performed so far were not able to reveal information about how the treatment affects the progressive neurodegeneration process. The current pharmacological treatments only offer symptomatic relief to the patients [3], and are based on the administration of levodopa and catechol-O-methyltransferase or monoamine oxidase B inhibitors (IMAO-B) [4].

Monoamine oxidase-B (MAO-B) is one of the isoforms of monoamine oxidases involved in the metabolization of dopamine in neuronal tissues, whose expression increases about 4-fold with aging. Therefore, IMAO-Bs are used to decrease the turnover rate of striatal dopamine in early PD, or as an adjunctive therapy in patients treated with levodopa that are experiencing motor complications [3]. However, until now, no IMAO-B developed so far has been able to modify or revert the progression of PD [5].

As a result, in recent years, an intensive search focused on the discovery of novel IMAO-B has been carried out and, in line, chromone has emerged as a validated scaffold (Figure 1) for the development of novel MAO inhibitors [6,7]. In fact, a chromone-based compound (Figure 1, C27) has been reported as a potent, selective, and reversible IMAO-B (IC_{50} = 670 ± 130 *p*M) [5]. This IC_{50} value is one order of magnitude lower than the reference inhibitors ((R)-(−)-deprenyl (IC_{50} = 16.73 ± 1.48 nM), rasagiline (IC_{50} = 49.66 ± 2.26 nM) and safinamide (IC_{50} = 23.07 ± 2.07 nM) [8]. Despite C27's remarkable outline, several setbacks, mainly related to poor water solubility and bioavailability, have hampered the progress toward in vivo preclinical studies.

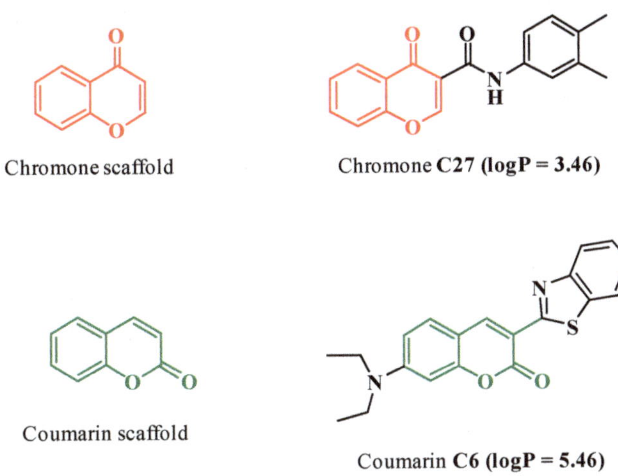

Figure 1. Chemical structures of chromone C27 and coumarin C6. The values of logP were obtained from the literature [5,9].

To overcome the attrition rates of central nervous system (CNS) drug discovery and development programs, new precision nanomedicine approaches are being developed. The main goals of these new approaches are the increment of stability, solubility, and other tunable properties and therapeutic index, namely, by improving the targeted delivery across the blood-brain barrier (BBB), of neuroactive drugs and drug candidates.

Through the years, nanoparticles (NPs) based on biodegradable polymers have been used for controlled drug delivery and to improve the therapeutic performance of drugs. Recently, they have been also used to solve shortcomings related to the solubility, stability, and bioavailability of drug candidates and other bioactive molecules [10,11]. Among the various biodegradable polymers approved by the US Food and Drug Administration, poly(lactide) (PLA), poly(D,L-lactide-*co*-glycolide) (PLGA), and poly(caprolactone) (PCL) are the most reported in the literature [12]. In particular, PCL is described as a non-toxic semi-crystalline hydrophobic polyester, miscible with a variety of polymers, with a high toughness and biocompatibility, since, in physiological conditions, it degrades slower than other biodegradable polyesters [13]. The surface of NPs can be coated with PEGylated surfactants, such as polysorbate 80 (Tween® 80, T80) or poloxamers [8,14], to increase the permeability of NPs into the brain [15]. This process is described to be related to an improvement of the blood circulation time and to the stealth nature of PEG-coated NPs [16]. The PEGylated moiety shields the surface from aggregation, opsonization, and phagocytosis, ensuring the NPs remain undetected by the reticuloendothelial system and increasing their blood circulation half-life [17]. In line with this, Wilson et al. used NPs coated with T80 to deliver rivastigmine and tacrine in rat brains [18,19]. PCL-T80 NPs were also reported to be internalized into glioma C6 cells, with a higher cellular uptake than the PCL NPs, and were able to successfully deliver a nerve growth factor into outbred C57BL/6 mice brain [20,21]. Furthermore, Wang et al. proved that PCL-T80 NPs were capable of delivering paclitaxel in the brains of male Sprague–Dawley rats after intravenous injection [22].

A handful of studies have addressed the use of PEGylated NPs to tackle drug discovery and development problems in the field of neurodegenerative diseases [23]. Even though rasagiline and selegiline (IMAO-B used in therapy) were effectively encapsulated in polymeric NPs [3,24], a gap still exists on the use of nanotechnology to solve site-targeted and absorption, distribution, metabolism, excretion, and toxicity (ADMET) problems along the pre-clinical phase. Herein, the IMAO-B chromone C27 was encapsulated into PCL NPs coated with T80 and a nanoformulation with suitable morphological and physicochemical properties was obtained after an optimization process. The release profile of C27 from PEGylated PCL NPs was evaluated, as well as the unloaded and loaded NPs cytotoxicity outline in human differentiated neuroblastoma (SH-SY5Y), epithelial colorectal adenocarcinoma (Caco-2), and endothelial brain (*h*CMEC/D3) cells. Moreover, cellular uptake (in SH-SY5Y and *h*CMEC/D3 cells) and permeability studies (in Caco-2 and *h*CMEC/D3 cells) were performed. For the cellular uptake, intracellular localization, and permeability studies, PEGylated PCL NPs containing a fluorescent coumarin-based probe (coumarin-6 dye, Figure 1) were also prepared.

2. Materials and Methods

2.1. Materials and Reagents

Polycaprolactone (PCL, $M_n \approx 10{,}000$ Da, determined by gel permeation chromatography) and polysorbate 80 (T80) were purchased from Sigma-Aldrich (Sintra, Portugal) and used without further purification. The reagents and solutions used for cell-based assays were described in [8] and are detailed in the Supplementary Materials. EndoGRO Media was acquired from Merck (Cambridge, MA, USA) and rat tail collagen type I (low viscosity) was purchased from Cultrex. Other reagents were obtained from Sigma-Aldrich (Sintra, Portugal). The water used was Milli-Q filtered (Millipore, Burlington, MA, USA).

2.2. Synthesis of N-(3′,4′-Dimethylphenyl)-4-oxo-4H-chromene-3-carboxamide (C27)

The synthesis of *N*-(3′,4′-dimethylphenyl)-4-oxo-4*H*-chromene-3-carboxamide (C27) was previously reported [5]. The structural characterization and purity were ascertained by nuclear magnetic resonance (^{1}H NMR, ^{13}C NMR and DEPT) (data available in the Supplementary Materials).

2.3. Preparation of PEGylated PCL-Based NPs

The encapsulation of chromone C27 in PEGylated PCL-based NPs was performed using the nanoprecipitation method [8]. The nanoformulation was prepared by dissolving PCL and C27 (2.5, 5 or 10% of PCL weight) in acetone and adding it dropwise to Milli-Q water containing 0.2% T80 under vigorous magnetic stirring. The resulting suspension was left under stirring for 1 h at room temperature. Acetone was then fully eliminated by evaporation under reduced pressure at room temperature. C27 loaded NPs were purified by ultrafiltration (10 min, 3000× *g*, Amicon Ultra 100 kDa MWCO, Millipore) and stored at 4 °C until use. The unloaded NPs were similarly prepared, as described above.

The encapsulation of coumarin-6 (C6) in PCL NPs was performed in a similar manner to that described above (1% *w*/*w* of PCL).

For simplicity, from this point onwards PCL nanoformulations containing C27 and C6 will be labelled PCL@C27 NPs and PCL@C6 NPs, respectively. Unloaded NPs will be referred to as PCL NPs.

2.4. Encapsulation and Drug Loading Efficiency

The quantification of chromone C27 was performed using a Shimadzu UV-Vis spectrophotometer (UV-1700 PharmaSpec, Kyoto, Japan). The C27 UV/Vis spectra were obtained using a C27 solution (50 μM) prepared in dimethyl sulfoxide (DMSO). The amount of C27 incorporated into the PCL@C27 NPs was determined directly after the complete dissolution of NPs in DMSO. The encapsulation efficiency (EE%) was calculated as the ratio between the chromone content in the freeze-dried powder and the initial chromone amount used in the NPs preparation (Equation (1)) [25]. The drug loading capacity (DLC%) was determined as the ratio between the amount of C27 encapsulated and the mass of NP powder (Equation (2)) [26].

$$\text{EE}\ \% = \frac{\text{Amount of loaded C27}}{\text{Amount of feeding C27}} * 100\%; \tag{1}$$

$$\text{DLC}\ \% = \frac{\text{Amount of loaded C27}}{\text{Amount of PCL@C27 NPs}} * 100\%. \tag{2}$$

Briefly, 1 mL of nanosuspension was centrifuged (15 min, 16,060× *g*, 4 °C) and washed three times with Milli-Q water. Then, the pellet was dried and dissolved in 1 mL of DMSO. Once completely dissolved, samples were quantified by UV-Vis spectrophotometry at 370 nm. Experiments were performed in triplicate and under sink conditions.

The quantification of C6 encapsulated inside PCL NPs was performed using a fluorescent detection method [27]. After dissolving the pellet in 1 mL of DMSO, the resulting solution was measured using fluorescence radiation in a multi-well plate reader (excitation and emission wavelengths at 485 and 528 nm, respectively). Experiments were performed in triplicate and under sink conditions.

The experiments performed in preformulation studies make it possible to conclude that PCL and T80 did not interfere in the analysis of both C27 and C6.

2.5. Spectroscopic Analysis

NMR spectra of free chromone C27, PCL@C27, and PCL NPs were acquired on a Bruker AMX 300 spectrometer operating at 400.13 MHz, at ambient temperature. Samples were dissolved in $CDCl_3$ and analyzed with a final concentration of 30 mg/mL. Chemical shifts are quoted in δ (ppm) values relative to tetramethylsilane (TMS) used as internal reference. Coupling constants (*J*) are given in Hz.

2.6. Particle Size, Zeta Potential, and Morphology Analysis

The hydrodynamic particle size (D_{DLS}), polydispersity index (PdI), and zeta potential (z-potential) of all prepared NPs were analyzed by dynamic light scattering (DLS) and electrophoretic mobility, using a Zetasizer (Litesizer™500, Anton Paar, Graz, Austria) equipped with a 4.0 mW internal laser. The resulting concentrated solutions of different NPs formulations were diluted (10% *v*/*v*) in Milli-Q water, phosphate buffered saline solution PBS (1×), and Hank's balanced salt solution, with calcium and magnesium HBSS (+/+). Each value resulted from triplicate determinations.

The morphology of the PCL and PCL@C27 NPs were analyzed by scanning electron microscopy (SEM) (JEOL JSM-6390, Tokyo, Japan). The experimental details were described in [8] and are detailed in the Supplementary Materials.

2.7. Differential Scanning Calorimetry and Powder X-ray Diffraction

Thermal analyses of chromone C27, PCL, and PCL@C27 NPs were performed in accordance with [8] and are detailed in the Supplementary Materials.

Powder X-ray diffraction (PXRD) was performed on an Empyrean PANalytical Diffractometer ($CuK_{\alpha,1,2}$ X-radiation; $\lambda_1 = 1.540598$ Å; $\lambda_2 = 1.544426$ Å), equipped with a PIXcel 1D detector and a flat-plate sample holder in a Bragg–Brentano para-focusing optics configuration (45 kV, 40 mA) (Panalytical, Almelo, The Netherlands). Data were collected at ambient temperature. Intensity data was measured by the step-counting method (step 0.01°), in continuous mode, over the scan range (2θ) between 3.5 and 50°.

2.8. In Vitro Release Studies

The in vitro release studies were carried out using PBS (1×, pH 7.4) or a sequential combination of PBS (1×, pH 7.4) followed by HCl (0.1 N, pH 1.2), to mimic the gastrointestinal track as release mediums [8]. Briefly, 1 mL of concentrated PCL@C27 nanoformulation was centrifuged (10 min, 16,060× *g*, 4 °C) to separate the supernatant from the nanoformulation. The resulting pellet was resuspended in 1 mL of release medium. The eppendorfs (1.5 mL) were kept in a bath-shaker at 37 °C and 90 rpm. At predetermined time intervals, the nanosuspension was centrifuged (10 min, 16,060× *g*, 4 °C) and the supernatant removed and replaced by fresh release medium. All collected supernatants were stored at 4 °C until quantification. Experiments were performed in triplicate and under sink conditions.

2.9. In Vitro Cellular Studies

2.9.1. Cell Lines and Culture Conditions

Human neuroblastoma differentiated cells (SH-SY5Y cell line), epithelial colorectal adenocarcinoma cells (Caco-2 cell line), and cerebral microvascular endothelial cells (hCMEC/D3 cell line) were used as in vitro models. Details on the culture conditions are described in the literature [8] and in the Supplementary Materials.

2.9.2. Cell Viability Assays

After the confluence was reached, SH-SY5Y, Caco-2 and *h*CMEC/D3 cells were exposed to different concentrations of the chromone under study (2.5, 5.0, and 10.0 μM) and PCL NPs (25–100 μg/mL of nanoformulation powder) in fresh cell culture medium. The cell viability was evaluated using the 3-(4,5-dimethylthiazol-2-yl)-2,5-diphenyltetrazolium bromide (MTT) reduction assay [28]. After 24 h of exposure (SH-SY5Y and Caco-2 cells), or 5 h in the case of *h*CMEC/D3 cells, the cell culture medium was aspirated, and fresh cell culture medium containing 0.5 mg/mL MTT was added. Afterwards, SH-SY5Y and Caco-2 cells were incubated for 1 h and *h*CMEC/D3 cells were incubated for 4 h, at 37 °C, in a humidified 5% CO_2–95% air atmosphere. Then, the cell culture medium was removed, and the formed formazan crystals dissolved in 100% DMSO. The absorbance was measured at 550 nm in a multi-well plate reader (PowerWaveX BioTek Instruments, Winooski, VT, USA). The results are expressed as a percentage of the control (nontreated) of three independent experiments (performed in triplicate).

2.9.3. Cellular Uptake and Intracellular Localization Studies

The capacity of PCL NPs to be internalized in SH-SY5Y and *h*CMEC/D3 cells was determined using a fluorescent probe (coumarin C6) encapsulated inside PCL NPs (PCL@C6 NPs) and measured by fluorescence spectroscopy. After incubation with PCL@C6 NPs at 50 and 100 μg/mL (conditions described in cell viability analysis section), cells were washed thrice with fresh HBSS (+/+) to remove the excess of PCL@C6 NPs. Then, HBSS (+/+) was added and the intracellular fluorescence assessed in a multi-well plate reader (excitation and emission wavelengths at 485 and 528 nm, respectively). For the intracellular localization, after treatment with PCL@C6 NPs (50 μg/mL), cells were washed thrice with fresh HBSS (+/+) and treated with 5 μg/mL of Hoechst 33,342 for 30 min at 37 °C in the dark. Afterwards, the cells were washed twice with fresh HBSS (+/+) and images were captured with an automated microscope (Lionheart FX, BioTek, Winooski, VT, USA). After this, the images were superimposed to determine the intracellular localization of the NPs.

2.9.4. Evaluation of MAO-B Activity by Fluorescence Kynuramine Assay

The capacity of C27 and nanoformulations to inhibit MAO-B in a cell SH-SY5Y-based model was performed, as described by Santillo et al., with minor modifications [29]. Briefly, after a prior inhibition of MAO-A with a standard MAO-A inhibitor (clorgyline), the activity of MAO-B was measured by the metabolization of kynuramine to 4-hydroxyquinoline, which presents fluorescence properties. After the confluence was reached, SH-SY5Y cells were incubated with clorgyline (100 nM) for 30 min. Then, cells were co-incubated with kynuramine (60 μM) and test compounds (C_{C27} = 1 μM) for 8 h at 37 °C. The reaction was stopped by the addition of NaOH (0.5 M) to the well and the fluorescence was measured in a multi-well plate reader (excitation and emission wavelengths at 360 and 460 nm, respectively). Selegiline (100 nM) was used as a standard inhibitor of MAO-B and cells without treatment and treated only with kynuramine were used as controls. The results are expressed as a percentage of the control (nontreated), normalized with the protein content of the three independent experiments (performed in triplicate).

2.9.5. Protein Quantification

The activity of MAO-B was normalized to the protein content of the cell lysate, which was determined by a bicinchoninic acid (BCA) assay, as described by other authors [8], using bovine serum albumin as a standard.

2.9.6. Cellular Permeability Studies

The permeability of PCL@C6 NPs in human intestinal epithelium was evaluated using Caco-2 cells monolayer as an in vitro model [30]. The experimental details were performed as described in [8] and are presented in the Supplementary Materials.

2.9.7. Rhodamine 123 Accumulation Assay

The P-glycoprotein (P-gp) inhibitory activity of both C27 and PCL@C27 NPs was determined by measuring the intracellular accumulation of rhodamine (RHO) 123 in Caco-2 and *h*CMEC/D3 cells, in the absence or presence of P-gp inhibitors, as described in the literature [31]. Briefly, after reaching the confluence, cells were pre-treated with C27 (10 μM) and PCL@C27 NPs (100 μg/mL) for 30 min. Then, 20 μM RHO 123 was added and the cells were incubated for 90 min at 37 °C. Elacridar, a potent third-generation P-gp inhibitor (10 μM), was used as a positive control. After this time, cells were washed with PBS, lysed with DMSO, and the intracellular levels of RHO 123 were quantified by fluorimetry using a multi-well plate reader (PowerWaveX BioTek Instruments, Winooski, VT, USA) (excitation and emission wavelengths were 485 and 535 nm, respectively). Data was expressed as the percentage of RHO 123 accumulation relative to control cells (untreated with P-gp inhibitors), arbitrarily set at 100%.

2.10. Statistical Analysis

Physicochemical and in vitro data are presented as the mean ± standard deviation (SD). Data analysis for all the studies are specified in the Supplementary Materials.

3. Results

3.1. Preparation and Characterization of PEGylated PCL-Based Nanoformulations

The encapsulation of chromone C27 inside PEGylated polycaprolactone nanoparticles (PCL NPs) was successfully obtained by the nanoprecipitation method using polysorbate 80 (Tween® 80, T80) as a stabilizer agent. The first step was to establish the C27 optimal concentration to be used during the encapsulation process, being PCL NPs fed with different chromone C27 amounts (2.5, 5 and 10% of PCL, *w*/*w*). It is important to note that feeding PCL NPs with amounts of C27 higher than 10% caused the destabilization of the nanoformulation, with the concomitant formation of aggregates (data not shown). After complete evaporation of the acetone, all stable nanoformulations were ultrafiltrated to remove any trace of free chromone C27 and also of T80 that were not adsorbed to the PCL NPs surface.

The resulting PCL@C27 nanoparticles were analyzed and the data of encapsulation efficiency (EE%) and drug loading capacity (DLC%) determinations are summarized in Figure 2a (detailed information in the Supplementary Materials).

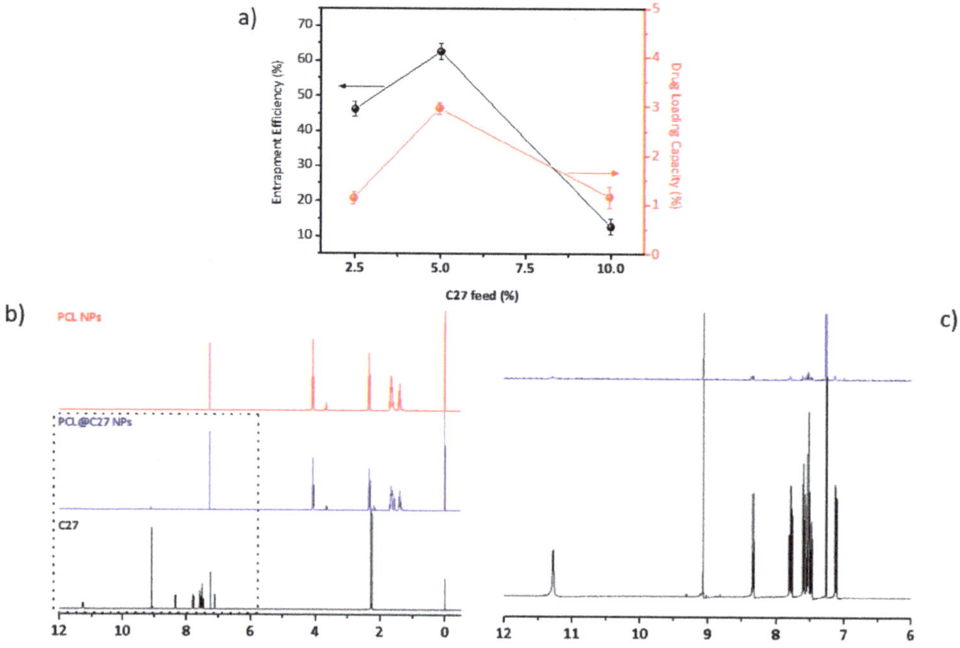

Figure 2. Physicochemical characterization of PCL@C27 nanoformulation by (**a**) determination of the encapsulation efficiency (EE%) and drug loading capacity (DLC%) of NPs fed with different amounts of chromone C27 (2.5, 5, and 10%). (**b**) Structural characterization of C27, PCL NPs, and PCL@C27 NPs using ^{1}H NMR between 0 and 12 ppm, and (**c**) respective magnification between 12 and 6 ppm of C27 ^{1}H NMR spectra. Values of entrapment efficiency (EE%, black) and drug loading capacity (DLC%, red) obtained for PCL@C27 NPs prepared with different amounts of feeding C27. Measurements of C27 quantification were performed in triplicate and results are presented as mean ± SD.

As observed in Figure 2a, the highest EE% (62.5 ± 2.3%) and DLC% (2.96 ± 0.12%) values were obtained for the NPs with a 5% C27:PCL ratio. As a result, under these conditions, the final C27 concentration (C_{C27}) value of PCL@C27 NPs was 263.3 ± 12.1 μM, presenting a concentration approximately 400,000-fold higher than the concentration required for the therapeutic effect (MAO-B IC_{50} = 670 ± 130 *p*M).

To confirm the PCL NPs structural composition ^1H NMR spectra were acquired (Figure 2b). The ^1H NMR spectra of the nanoformulations (red and blue lines) showed the PCL characteristic peak resonances at 1.38, 1.65, 2.30, and 4.06 ppm, which are related to the CH_2 protons of the PCL backbone [32]. Also, the peak located at 3.63 ppm, assigned to a CH_2, corroborates the presence of T80 in both PCL NPs [33]. In addition, in the PCL@C27 NPs spectra (blue line), the signals caused by the presence of the benzopyrone proton peaks were observed at 2.25 and 2.28 ppm, which are related to the two methyl groups located on the exocyclic aromatic ring, and between 9.1–7.1 ppm. The data confirm the presence of the C27 chromone in the NPs. The detailed C27 ^1H NMR data (black line) was included in the Supplementary Materials.

3.2. *Differential Scanning Calorimetry and Powder X-ray Diffraction Analysis*

Chromone C27, unloaded, and loaded PCL NPs were subjected to thermal analysis to obtain information regarding their crystalline morphology, as well as information on putative interactions between C27 and the PCL polymeric matrix [34,35]. The differential scanning calorimetry (DSC) curves are depicted in Figure 3a.

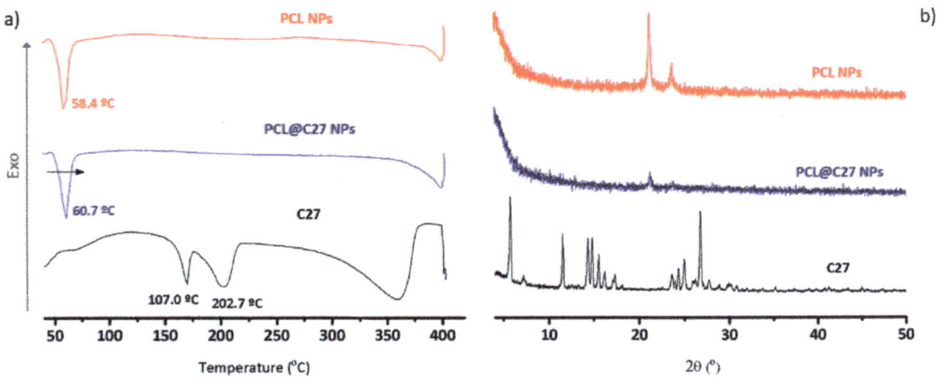

Figure 3. Physicochemical characterization of chromone C27, PCL NPs, and PCL@C27 NPs using (**a**) differential scanning calorimetry (DSC) between 40 and 400 °C, and (**b**) powder X-ray diffraction (PXRD) in the range of 4–50 °C.

For chromone C27 (black curve) a typical thermogram of a crystalline structure with two endothermic peaks at 170.0 °C and 202.7 °C was obtained. Since chromone C27 thermogravimetric analysis (data not shown) did not show any weight loss in this range of temperatures, these peaks were not due to evaporation of water. Instead, the first peak could be related to a glass transition process [36] and the second peak to the C27 melting transition, characterized by a T_{max} = 233–236 °C, which is approximately the same value of the C27 melting point obtained by capillary method (Supplementary Materials).

As none of these peaks appeared in the PCL@C27 NPs DSC curve (in blue), it can be concluded that C27 is encapsulated, probably in an amorphous or disordered phase [37]. The melting endothermic peak at 58.4 °C presented by the PCL NPs (in red) suffered a shift of 2 °C in the PCL@C27 NPs (60.7 °C, Figure 3a), a change that discloses a strong interaction between C27 and the PCL matrix [8,38].

As the crystallinity of the encapsulated C27 into PCL NPs can influence the drug release properties of the nanoparticles [39], powder X-ray diffraction (PXRD) data were acquired.

The diffractograms of C27, unloaded, and loaded NPs are depicted in Figure 3b. The diffractogram of C27 exhibited sharp peaks that are intrinsically related to the crystalline nature of chromone.

For nanoformulations, a semi-crystalline structure with two sharp peaks between 20 and 25° was detected [40]. Moreover, a decrease in polymer crystallinity was observed when comparing both PCL and PCL@C27 nanoformulations, which can be related to an interaction between the chromone and the polymer. The diffractogram of PCL@C27 NPs does not have any reflection pertaining to C27, which suggests its complete amorphization when loaded in the polymeric matrix [41]. These results are in good agreement with the literature [42]. Moreover, the data corroborate well with the results of DSC analysis.

3.3. *PEGylated PCL@C27 Particle Size, z-Potential, and Morphology*

The morphology and shape of PCL@C27 NPs were evaluated using scanning electron microscopy (SEM, Figure 4a).

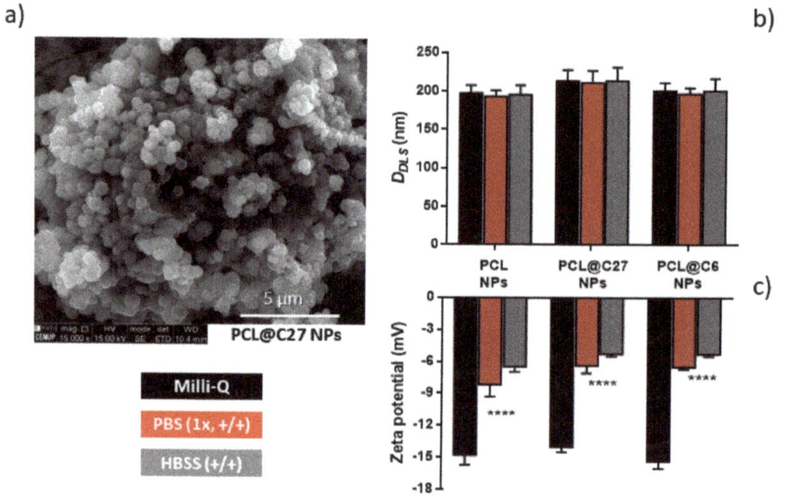

Figure 4. Morphological characterization of PCL@C27 NPs by scanning electron microscopy (SEM) (**a**) and hydrodynamic average sizes (D_{DLS}) (**b**) and zeta potential values (**c**) of PEGylated PCL NPs in Milli-Q water, phosphate buffered saline solution (PBS) (1×, +/+), and Hank's balanced salt solution (HBSS) (+/+), as measured by dynamic light scattering (DLS). All measurements were performed in triplicate and results are given as mean ± SD. Statistical comparisons were made using two-way ANOVA. In all cases, *p*-values lower than 0.05 were considered significant (**** $p < 0.0001$ versus Milli-Q water values).

The PCL@C27 nanoformulation presented NPs with a spherical shape and a uniform size distribution (Figure 4a), although some aggregation, probably due to the drying process, was observed. The hydrodynamic size (D_{DLS}) and zeta potential (z-potential) of C27-loaded and unloaded PEGylated PCL NPs were assessed in three different conditions: Milli-Q water, phosphate-buffered solution with calcium and magnesium (PBS 1×, +/+), and Hank's balanced salt solution with calcium and magnesium (HBSS +/+) (Figure 4b,c). Although particle characterization measurements are commonly conducted in Milli-Q water [43], the morphology and surface charge of NPs were also evaluated in PBS and HBSS, the mediums used in drug-controlled release and cell-based studies, respectively.

Since PEGylated PCL NPs containing a fluorescent coumarin-based probe (PCL@C6 NPs) were used for the cellular uptake, intracellular localization, and permeability studies, it was necessary to compare the morphology between both PCL@C27 and PCL@C6 nanoformulations.

For all tested media, the nanoformulations presented monodisperse profiles with D_{DLS} lower than 250 nm (Figure 4b). In fact, in physiological mediums (PBS and HBSS medium), PCL@C27 NPs had D_{DLS} values between 211 and 213 nm. As NPs sized circa 200 nm have been reported to be able to cross biological barriers, by preventing spleen filtration and reducing the opsonization by reticuloendothelial system, this is considered an encouraging result [44,45]. Despite no significant morphological differences being observed, the presence of chromone C27 seemed to influence the size of PCL NPs in both media, as they presented a slightly larger size when compared to unloaded NPs (~3–7% higher size values). This data is in good agreement with the literature [25,46].

The stability of NPs in aqueous medium is often assured by the presence of a surface charge, as it avoids the aggregation process. Without surfactant, PCL NPs usually present a z-potential between −35 and −30 mV in Milli-Q water, due to the negatively charged ionized carboxylic acid groups of the polymer [37]. In our case, the presence of T80 in NPs surface led to a reduction of the z-potential value to −14.0 and −15.3 mV in Milli-Q water for PCL NPs and PCL@C27 (Figure 4c), respectively. In physiological medium, the z-potential values (between −5.3 and −8.2 mV) were significantly different ($p < 0.0001$) from those obtained in Milli-Q water. This data is in accordance with what has been previously reported [8], and can be ascribed to the presence of interactions of opposite charged ions with the NPs surface [47,48]. The presence of T80 and negative charge in NPs surface could justify the high storage stability at 4 °C over three months, since both D_{DLS} and z-potential NPs remained unchanged and no aggregates were observed (data not shown).

The data showed non-significant differences in terms of NPs size and surface charge density, when comparing PCL@C6 to PCL@C27 NPs (Figure 4b,c). These results allow us to use the nanoformulation PCL@C6 as a model of C27 delivery carrier in cellular studies.

3.4. *In Vitro C27 Release Kinetics*

The evaluation of C27 sustainable release from PCL@C27 NPs was performed in PBS (pH 7.4) at 37 °C for seven days, and at pH 1.2 for 2 h, followed by pH 7.4 for 5 h, to simulate the passage through the upper human gastrointestinal tract [49]. In both conditions, the in vitro release profile from PCL@C27 NPs was obtained by graphing the cumulative percentage of the released C27 with respect to the amount of chromone encapsulated as a function of the time (Figure 5).

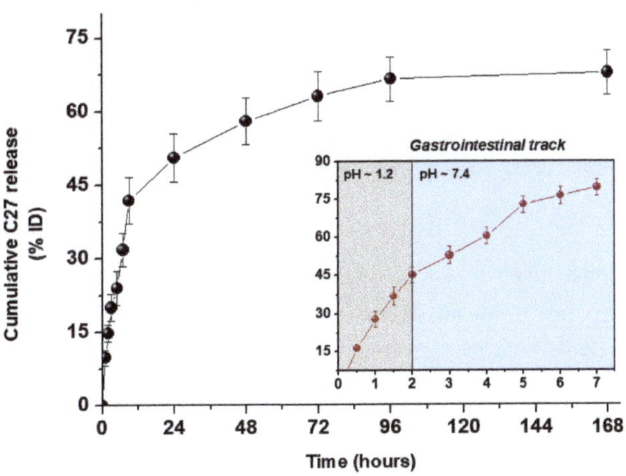

Figure 5. In vitro release profile of chromone C27 from PCL@C27 NPs in PBS (pH 7.4) conducted for seven days (black dot). Inset: In vitro release profile in 0.1 N HCl, pH 1.2, for 2 h followed by PBS, pH 7.4, for 5 h (red data). Results are presented as means ± SD of three independent experiments.

The in vitro release profile showed a typical biphasic pattern, with an initial burst release in the first 9 h, followed by a slow and continuous release up to 168 h. The initial burst estimated for PCL@C27 NPs was 41.8 ± 4.8% and can be related to the dissolution in the release media of C27 chromone adsorbed in the NPs surface [19]. After the initial burst, C27 was slowly released from NPs until the end of the experiment (seven days), reaching a plateau with a total C27 cumulative release amount of 67.7 ± 4.4%. The final C_{C27} in release medium was 174.2 ± 11.5 μM (approximately 275,000-fold higher than C27 MAO-B IC_{50} value). The second phase of the biphasic release pattern is assumed to be related to: (1) a slow C27 diffusion from the PCL matrix, (2) a progressive degradation/erosion of the polymeric matrix in physiological medium, or (3) a mixed course involving both processes [50].

The in vitro release studies in gastrointestinal simulated fluids (inset of Figure 5) make it possible to conclude that the C27 release is accelerated in acidic medium, in a process probably related to a faster degradation of PCL. Comparing the data obtained in both media in the first 2 and 7 h, a circa 3- and 2.5-fold increase of C27 release was observed in the acidic medium, respectively. This data is in good agreement with data previously reported by our group related to the release of a coumarin-based compound from PEGylated PLGA NPs [8].

The data obtained from in vitro drug release studies was fitted to the Korsmeyer–Peppas model [12] and the results are summarized in Table 1.

Table 1. Correlation values (R^2) and release exponent (n) of kinetic data analysis of chromone C27 release from PCL@C27 NPs in different medium.

Medium pH	**Korsmeyer–Peppas**		
	R^2	***n***	**K (h^{-1})**
7.4	0.972	0.464	11.2
1.2–7.4	0.985	0.594	26.9

The regression coefficient (R^2) of the plot of log $Mt/M\infty$ versus log t for NPs in PBS (1×) was found to be 0.972, with values of the release exponent (n) and release constant (K) of 0.464 and 11.2, respectively. In the case of the use of a combined acidic and PBS medium (R^2 = 0.985), the release exponent and release constants values were 0.594 and 26.9, respectively. To sum up, the type of medium affected the way and rate of C27 release: (a) when PBS was used as medium, the n value was <0.5, showing that the release of C27 from NPs can be explained by Fickian diffusion [12]; (b) when an acidic medium was employed, the n value was higher than 0.5, suggesting a release process controlled by polymer erosion [51]. This assumption was corroborated by the higher release constant (26.9 h^{-1}) in the first 2 h of the experiment.

In conclusion, C27 was successfully released from PCL NPs, with good yields and in a sustained way, regardless of its low solubility profile and strong interaction with the PCL matrix.

3.5. *In Vitro Cellular Studies*

3.5.1. In Vitro Cytotoxicity in Neuronal, Intestinal, and Endothelial Cells

Since IMAO-B operate by metabolic inactivation, and thus increase dopamine activity in the neuron synaptic cleft and at respective postsynaptic receptor sites [52], a human model of neuronal cells was chosen (differentiated neuroblastoma SH-SY5Y cells) to evaluate the cytotoxic profile of free C27 and PCL@C27 NPs. The neuroblastoma SH-SY5Y cell line is widely used in studies requiring neuronal-like cells, as they express a number of dopaminergic neuronal markers [53,54]. Furthermore, the cytotoxic effects of chromone C27 and PCL@C27 NPs were also evaluated in epithelial colorectal adenocarcinoma (Caco-2) and in human brain microvascular endothelial (*h*CMEC/D3) cells, as they are widely used in vitro models to determine the intestinal and blood-brain barrier (BBB) permeability, respectively [25,30].

The cellular viability of C27 (2.5, 5.0, and 10.0 μM) and PCL@C27 NPs (25–100 μg/mL of nanoformulation powder) was indirectly determined by the measurement of the cellular metabolic activity using the MTT reduction method, 24 h after exposure (Figure 6). The PCL@C27 NPs concentration range was chosen in order to test nanoformulation at same C27 concentration as free C27 and taking into account the final C_{C27} (263.3 ± 12.1 μM). Furthermore, an end-point of 24 h was performed to avoid possible interferences of cytotoxic effects of free C27, since, as observed in controlled release data (Figure 5), after this period of time, around 50% of C27 is released from NPs.

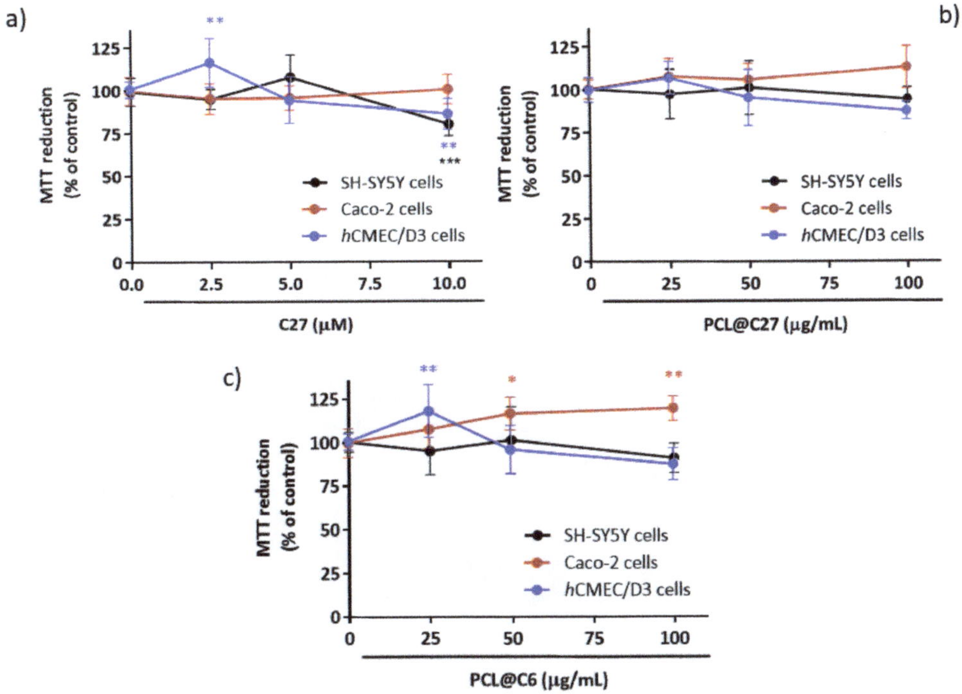

Figure 6. Cytotoxicity profile of C27 at 2.5, 5, and 10 µM (**a**) as well as nanoformulations of PCL@C27 (**b**) and PCL@C6 NPs (**c**) at 25, 50, and 100 µg/mL in different cells lines, evaluated by the MTT reduction assay 24 h after exposure. Results are expressed as mean percentage of MTT reduction ± SD of three independent experiments. In all cases, p-values lower than 0.05 were considered significant (* $p < 0.05$, ** $p < 0.01$, *** $p < 0.001$ versus control cells).

No significant cytotoxic effects were observed in Caco-2 cells when exposed to C27, when compared to control cells, and for the conditions tested (Figure 6a). On the contrary, a slight but significant decrease in the metabolic activity was observed when SH-SY5Y (80.3 ± 2.4%, $p < 0.01$) and *h*CMEC/D3 cells (86.5 ± 3.2%, $p < 0.05$) were exposed to C27 at the highest concentration (10 µM) for 24 h.

In the case of PCL@C27 NPs, in the present experimental conditions, no significant cytotoxic effects were noticed (Figure 6b), even for the highest concentration tested (100 µg/mL), which corresponded to a final encapsulated C27 concentration of around 10.6 µM.

Although efforts were made to evaluate the uptake and permeability of chromone C27 and PCL@C27 NPs across SH-SY5Y, Caco-2, and *h*CMEC/D3 cells, these studies were hampered by the chromone spectral features and the sensitivity of the analytical method. Therefore, a model probe was used to validate the carrier properties of the PCL nanoformulation. As previously mentioned, a probe based on coumarin scaffold (coumarin C6, Figure 1) [55], a benzopyran analogue, was chosen to attain the goal, due to its structural similarity and unique fluorescent properties. After the preparation and morphologic characterization of PCL@C6 NPs, the cytotoxic profile of the PCL@C6 nanoformulation was evaluated using the same conditions described for the PCL@C27 NPs studies. For all tested cell lines, no significant reduction in the metabolic activity was detected, for concentrations up to 100 µg/mL, and after 24 h of incubation.

Summing up the data herein obtained reinforced the benefit of using PEGylated PCL NPs as carriers of bioactive compounds that present bioavailability drawbacks.

3.5.2. In Vitro Cellular Uptake and Intracellular Localization in Neuronal and Endothelial Cells

The intracellular localization of PEGylated PCL NPs was verified by fluorescence microscopy in neuronal and endothelial (SH-SY5Y and *h*CMEC/D3) cells that express MAO-B [56,57]. For this, cells were exposed to PCL@C6 NPs (50 µg/mL) for 24 and 5 h, respectively. After that, they were washed with HBSS (+/+) to remove the excess PEGylated PCL NPs and the cell's nucleus was marked with Hoechst 33,342 (blue stain) and the nucleic acid (either DNA or RNA) with acridine orange (orange stain). From the data (Figure 7a), it was observed that PCL@C6 NPs (green stain) were dispersed in the cytoplasm of the cells.

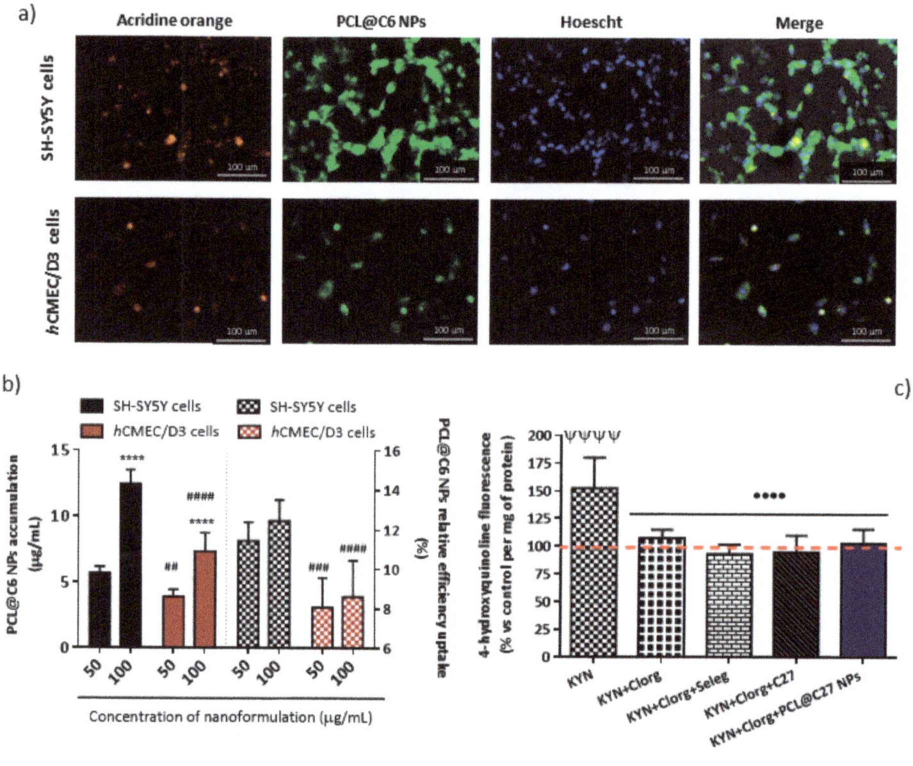

Figure 7. Representative images of acridine-orange and Hoescht stained SH-SY5Y and *h*CMEC/D3 cells treated with PCL@C6 NPs at 50 µg/mL for 24 and 5 h, respectively (**a**); accumulation (solid bars) and relative efficiency uptake (patterned bars) of PCL@C6 NPs (50 and 100 µg/mL) in SH-SY5Y and *h*CMEC/D3 cells after 24 and 5 h of exposure, respectively (**b**); evaluation of the MAO-B inhibitory capacity in SH-SY5Y cells after 8 h of exposure, using kynuramine (KYN) as the substrate and clorgyline (Clorg) and selegiline (Seleg) as standard inhibitors of MAO-A and MAO-B, respectively (**c**). The results are expressed as mean ± SD ($n = 3$) and statistical comparisons were made using two-way ANOVA and one-way ANOVA, in cases (**b**) and (**c**), respectively. In all cases, p-values lower than 0.05 were considered significant (**** $p < 0.0001$ by comparison of different concentrations tested concentration; ## $p < 0.01$, ### $p < 0.001$, and #### $p < 0.0001$ by comparison the results obtained between the two cell lines tested; ΨΨΨΨ $p < 0.0001$ by comparison with untreated cells; •••• $p < 0.0001$ by comparison with cells treated only with kynuramine). The dashed red line represents the control data.

As described above, after the exposure to PCL@C6 NPs at 50 and 100 μg/mL, cells were washed with HBSS (+/+) and the remaining fluorescence was measured by exciting the media with a radiation of 485 nm. The results are presented as the intracellular accumulation and relative efficiency uptake of PCL@C6 NPs (50 and 100 μg/mL, Figure 7b).

The cellular uptake of PCL@C6 NPs is supposed to occur via endocytosis [58], and in both cell lines is significantly different and influenced by the initial concentration (IC) used in the treatment of the cells (Figure 7a). For the same concentration, PCL@C6 NPs uptake is higher in neuronal cells than in endothelial cells (Figure 7a, solid bars). Actually, a proportionality between the concentration and the uptake was established: the exposure to PCL@C6 NPs at 100 μg/mL resulted in a 2.2- and 1.9-fold higher cellular uptake (**** $p < 0.0001$, Figure 7a, solid bars) in neuronal and endothelial cells, respectively, when compared to the 50 μg/mL concentration. However, after comparing the efficiency of cellular uptake for the concentrations tested (50 and 100 μg/mL), no significant difference was observed (Figure 7a, patterned bars).

Since the structure and morphology of PCL@C27 and PCL@C6 NPs are similar, we suggested that both nanoformulations presented identical cellular uptake profiles, which allowed us to correlate the concentrations of nanoformulations PCL@C6 and PCL@C27 inside the cells. After the end-points, minimum nanoformulation concentrations of 5.7 ± 0.2 and 3.9 ± 0.2 μg/mL were determined for SH-SY5Y and *h*CMEC/D3 cells (Figure 7a, solid bars), respectively, which corresponded to a final C27 concentration between 410 and 606 nM. This range of concentration is 651- to 963-fold higher than the C27 MAO-B IC_{50} value.

To demonstrate the MAO-B inhibitory activity of the systems in a cell-based model (SH-SY5Y), additional studies were performed in the presence and absence of C27 and nanoformulation PCL@C27 [29]. The activity of MAO-A was inhibited by pre-treatment with clorgyline (100 nM) and the fluorescence intensity was normalized to the protein content. The same procedure was followed with the MAO-B inhibitor selegiline (100 nM).

After 8 h of experiments, and in the absence of MAO inhibitors, an increment of fluorescence intensity caused by the metabolization of kynuramine to 4-hydroxyquinoline (fluorescent metabolite) was observed, when compared with untreated cells (^{ΨΨΨΨ} $p < 0.0001$, Figure 7c). Meanwhile, when both MAO-A and MAO-B inhibitors were used, a reduction in the intensity of fluorescence was observed, when compared with cells treated only with kynuramine (•••• $p < 0.0001$, Figure 7c). The same protocol was applied with selegiline. In this case, a slight decrease in fluorescence, when compared with the treatment done only with clorgyline, was detected. This effect is related to the inhibition of both MAO isoforms. Since the treatment with C27 and PCL@C27 NPs showed the same tendency as the data obtained with selegiline, it was concluded that C27 works as a MAO-B inhibitor in neuronal cells.

The combined data of microscopy analysis, cellular uptake, and MAO-B activity in cell-based models suggested that PEGylated PCL NPs can deliver C27 into the cellular cytoplasmatic matrix. Even further, it was shown that C27 and PCL@C27 NPs are capable of inhibiting MAO-B in neuronal cells.

3.5.3. In Vitro Permeability Studies in Epithelial and Endothelial Cells

Several in vitro BBB models, also used by the (bio)pharma industry, have been developed to assist the selection and preclinical evaluation of CNS drug candidates [59]. Among those based on primary cultures of cerebral endothelial cells or immortalized cell lines, the human brain endothelial cell line (*h*CMEC/D3) is one of the most studied in drug transport and uptake experiments [60]. However, other models based on epithelial cell lines have been also used to predict the permeability of CNS drug candidates, due to their advantages in terms of costs [59,60]. In fact, the Caco-2 human intestinal epithelial cell line is one of the most widely in vitro models used to predict human drug absorption, mainly for small intestine absorption [61].

Accordingly, the ability of PEGylated PCL NPs to cross biological barriers was evaluated in two different cell lines (Caco-2 and *h*CMEC/D3 cells) [8]. These cell lines retain the expression of most transporters and receptors expressed in vivo in the human intestine and BBBs and have been widely used in several permeability studies related to Alzheimer's and Parkinson's diseases [8,62,63]. Both cell lines were cultured in a cell culture insert filter (Transwell) and used after obtaining cellular monolayers with high integrity, a parameter that was verified by the measurement of transendothelial electrical resistance values. The number of PCL@C6 NPs accumulated in basal medium was normalized with the weight of protein presented in the respective Transwell.

The data presented in Figure 8 show that PCL@C6 NPs (100 μg/mL) exhibited two permeability profiles with a biphasic tendency, which can be ascribed to a mixed passive and transport-mediated mechanism [8,64]. Although, in terms of the particle size, the transport mechanism established for particles up to 200 nm is the receptor-mediate endocytosis [65], the use of T80 as a coating agent in PCL NPs could enhance the paracellular permeation in epithelial and endothelial cell monolayers [66,67]. In fact, it was demonstrated in vivo that the accumulation of donepezil in the brain was significantly higher when the drug was encapsulated in PLGA NPs, and that Tween 80 can improve the opening of the tight junctions at the BBB and inhibit the P-glycoprotein efflux system [68].

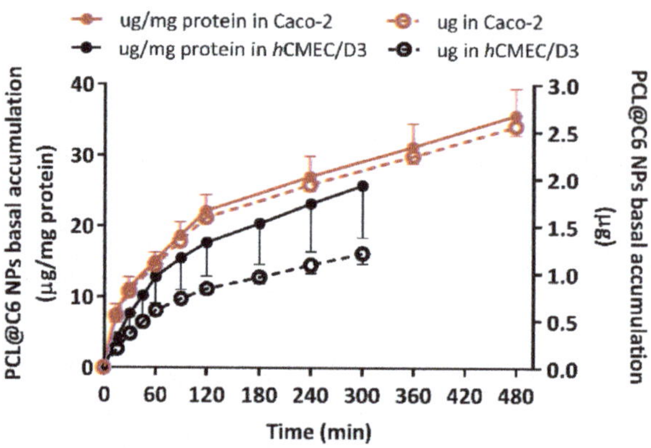

Figure 8. Amount of PCL@C6 NPs, with (μg/mg of protein, solid circles and line) or without (μg, empty circles points and dash line) protein mass normalization, accumulated in basal medium after 5 h and 8 h of permeability experiment in *h*CMEC/D3 (black data) and Caco-2 (red data) cells with a 100 μg/mL initial concentration. Results are expressed as mean ± SD of three independent experiments.

In Caco-2 cell monolayers, a final accumulation of 35.7 ± 3.8 μg/mg protein (solid red circles), were obtained with an apparent permeability coefficient (P_{app}) value calculated of 4.7×10^{-7} cm/s [69]. Meanwhile, after 5 h of experiment, a final PEGylated PCL NPs concentration of 25.7 ± 7.4 μg/mg protein (solid black circles) was found in basal medium, with a concomitant P_{app} value of 1.2×10^{-7} cm/s [69]. The combined data showed that PEGylated PCL NPs were capable to cross both intestinal and BBB barriers in a time-dependent manner.

In terms of nanoformulation mass (empty circles, Figure 8), 2.56 ± 0.09 (red empty circles, Figure 8) and 1.21 ± 0.11 μg (black empty circles, Figure 8) were detected at the end of experiment, in the basal medium of Caco-2 and *h*CMEC/D3 cells, respectively. The slight differences between the values obtained in permeability assays could be associated with the type of cells and intrinsic mechanisms of passive transport [70].

In line with the results of cellular uptake, we hypothesize that PCL@C27 nanoformulation is also able to cross intestine and BBB in vitro monolayers in the same extension as PCL@C6 NPs. Thus, extrapolating the results obtained for PCL@C6 NPs, a final C27 concentration of 117.7 and 55.8 nM (at least one order of magnitude higher than C27 MAO-B IC_{50} value) could be expected.

PEGylated PCL NPs showed the ability to circumvent in vitro BBB cell monolayer hindrances and deliver an IMAO-B at concentrations higher enough to induce a therapeutic effect.

3.5.4. Rhodamine 123 Accumulation in Epithelial and Endothelial Cells

The inhibition of efflux pumps is described to be a parameter of utmost importance in drug delivery [71]. In fact, after the discovery that polymeric pharmaceutical excipients can inhibit efflux pumps, various other polymers have been studied regarding their potential efflux pump inhibitory activity [72]. Excipients containing polyethylene glycol and PEGylated surfactants, such T80 or poloxamers, have been described to efficiently inhibit efflux pumps, namely P-glycoprotein (P-gp) [73–75].

In this work, Caco-2 and *h*CMEC/D3 cells were used as in vitro models and the P-gp activity assessed using RHO 123 as a P-gp fluorescent substrate. Therefore, a decrease in P-gp activity results in a decreased amount of RHO 123 effluxed by this pump, which is followed by an increase in the intracellular fluorescence intensity (increased Rho 123 intracellular content). For that purpose, cells were pre-treated for 30 min with C27 (10 μM), PCL@C27 NPs (100 μg/mL) or with elacridar (10 μM), a well-known third-generation P-gp inhibitor, used as a positive control for this experiment. After that, 20 μM RHO 123 was added and cells were incubated for 90 min. The obtained results are presented in Figure 9.

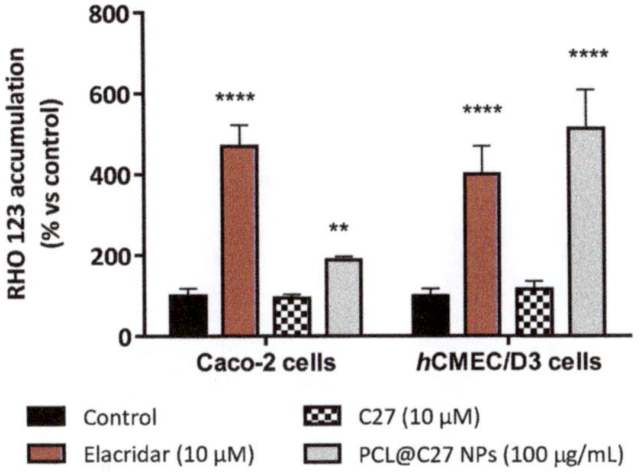

Figure 9. Accumulation of Rhodamine (RHO) 123 in Caco-2 and hCMEC/D3 cells. Data are expressed as percentage of fluorescence dye accumulation in control cells exposed only to RHO 123, arbitrarily set as 100%, and are the means ± SD of three independent assays. In all cases, *p*-values lower than 0.05 were considered significant (** $p < 0.01$ and **** $p < 0.0001$).

As observed in Figure 9, no significant difference in RHO123 intracellular fluorescence was verified when C27 was tested, when compared to control cells. On the contrary, elacridar (10 μM) acted as a P-gp inhibitor, avoiding the RHO 123 efflux from both cellular models, which resulted in a 5-fold increase in RHO123 intracellular fluorescence, when compared to control cells ($p < 0.0001$). The same behavior was obtained with PCL@C27 NPs. In this case, an increase of 1.9- and 5-fold of RHO 123 intracellular levels was found for the Caco-2 and *h*CMEC/D3 cells, respectively. This data is in accordance with other works, which showed that NPs coated with Tween 80 have a good capacity to inhibit P-gp [76,77].

4. Conclusions

In this work, a potent, selective, and reversible IMAO-B was successfully encapsulated in PCL NPs coated with T80. The nanoprecipitation method was employed to obtain stable particles in physiological conditions with D_{DLS} lower than 213 nm and a z-potential of around −5 mV, with a high stability in physiological medium. The optimization process of C27 encapsulation gave rise to NPs with a final C27 concentration of 263.3 ± 12.1 μM, 400,000-fold higher than its IC_{50} value (670 ± 130 *p*M).

In physiological conditions, the nanoformulation sustainably released C27 over seven days, with a final release amount of 67.7 ± 4.4%. Furthermore, no cytotoxic effects of C27 and PCL@C27 NPs were observed in Caco-2 cells. In SH-SY5Y and *h*CMEC/D3 cells, the encapsulation of C27 was essential to decrease its cytotoxicity in the conditions tested.

The uptake and permeability studies, performed with PCL@C6 NPs, showed that PEGylated PCL NPs were capable of accumulating in the cytoplasm of both neuronal and endothelial cells and surpassing the in vitro epithelial and endothelial cell monolayers. By extrapolation, one can infer that PCL@C27 NPs are able to deliver C27 across the intestine and BBB monolayer at concentrations higher than its MAO-B IC_{50} value.

Overall, our results provide evidence of the effectiveness of PCL nanocarriers to deliver neuroactive compounds that have bioavailability and physicochemical drawbacks. Further improvement and research, focused on in vivo long-term safety and efficacy studies, on polymeric nanomedicines will be, in the future, a promising tool to boost the success of clinical trials. Given the significant costs associated with drug discovery and development, it is becoming increasingly important to engineer targeted nanomaterial carriers allowing a sustained release of new neuroactive drug candidates.

Supplementary Materials: The following are available online at http://www.mdpi.com/1999-4923/11/7/331/s1. Table S1 is related to the EE% and DLC% values obtained for different formulations prepared with different amounts of feeding C27. Table S2 is related to D_{DLS} and z-potential values obtained in Milli-Q, PBS (1×, +/+), and HBSS media (+/+). Figure S1 is related to the UV-Vis spectra of chromone C27.

Author Contributions: C.F., F.R. and F.B. conceived the overall study design; M.P. and F.C. prepared, purified and fully characterized the chromone C27; C.F. and M.P. produced and optimized the encapsulation of C27 in PCL NPs; M.P. measured the hydrodynamic size of NPs by DLS; J.G. performed the DSC studies; R.F.M. and F.A.A.P. performed the P-XRD characterization; C.F., E.M. and R.S. evaluated the cytotoxicity in SH-SY5Y and Caco-2 cells as well as the cellular uptake and P-gp inhibition in both Caco-2 and *h*CMEC/D3 cell lines; S.B. tested the cytotoxicity and permeability of NPs in hCMEC/D3 cells; S.B. conducted the quantification of coumarin C27; C.F. performed all rest of experiments presented in this manuscript; C.F. and F.B. wrote the draft manuscript; and all authors contributed to manuscript revision.

Funding: Pinto, M., Fernandes, C., Benfeito, S., Cagide, F. (NORTE-01-0145-FEDER-000028), Silva, R. (SFRH/BPD/110201/2015) and Mendes, R.F. (SFRH/BD/84231/2012) grants are supported by FCT, POPH and FEDER/COMPETE. The authors also thank the COST action CA15135 for support. This work was funded by FEDER funds through the Operational Programme Competitiveness Factors—FEDER/COMPETE-by national funds (FCT—Foundation for Science and Technology) and Norte Portugal Regional Operational Programme (Grants POCI-01-0145-FEDER-006980-UID/QUI/00081/2013, POCI-01-0145-FEDER-007679-UID/CTM/50011/2013, EXPL/CTM-NAN/0013/2013, POCI-01-0145-FEDER-029164 and NORTE-01-0145-FEDER-000028).

Conflicts of Interest: The authors declare no conflict of interest.

References

1. Taylor, J.P.; Hardy, J.; Fischbeck, K.H. Toxic Proteins in Neurodegenerative Disease. *Science* **2002**, *296*, 1991–1995. [CrossRef] [PubMed]
2. Mattson, M.P. Apoptosis in neurodegenerative disorders. *Nat. Rev. Mol. Cell Biol.* **2000**, *1*, 120. [CrossRef] [PubMed]
3. Fernández, M.; Barcia, E.; Fernández-Carballido, A.; Garcia, L.; Slowing, K.; Negro, S. Controlled release of rasagiline mesylate promotes neuroprotection in a rotenone-induced advanced model of Parkinson's disease. *Int. J. Pharm.* **2012**, *438*, 266–278. [CrossRef] [PubMed]
4. Massano, J.; Bhatia, K.P. Clinical Approach to Parkinson's disease: Features, Diagnosis, and Principles of Management. *Cold Spring Harb. Perspect. Med.* **2012**, *2*, a008870. [CrossRef] [PubMed]
5. Reis, J.; Cagide, F.; Chavarria, D.; Silva, T.; Fernandes, C.; Gaspar, A.; Uriarte, E.; Remião, F.; Alcaro, S.; Ortuso, F.; et al. Discovery of New Chemical Entities for Old Targets: Insights on the Lead Optimization of Chromone-Based Monoamine Oxidase B (MAO-B) Inhibitors. *J. Med. Chem.* **2016**, *59*, 5879–5893. [CrossRef] [PubMed]
6. Gaspar, A.; Matos, M.J.; Garrido, J.; Uriarte, E.; Borges, F. Chromone: A Valid Scaffold in Medicinal Chemistry. *Chem. Rev.* **2014**, *114*, 4960–4992. [CrossRef] [PubMed]

7. Reis, J.; Gaspar, A.; Milhazes, N.; Borges, F. Chromone as a Privileged Scaffold in Drug Discovery: Recent Advances. *J. Med. Chem.* **2017**, *60*, 7941–7957. [CrossRef] [PubMed]
8. Fernandes, C.; Martins, C.; Fonseca, A.; Nunes, R.; Matos, M.J.; Silva, R.; Garrido, J.; Sarmento, B.; Remião, F.; Otero-Espinar, F.J.; et al. PEGylated PLGA Nanoparticles As a Smart Carrier to Increase the Cellular Uptake of a Coumarin-Based Monoamine Oxidase B Inhibitor. *ACS Appl. Mater. Interfaces* **2018**, *10*, 39557–39569. [CrossRef] [PubMed]
9. Molinspiration. Molinspiration Cheminformatics. Available online: http://www.molinspiration.com (accessed on 2 March 2017).
10. Singh, R.; Lillard, J.W. Nanoparticle-based targeted drug delivery. *Exp. Mol. Pathol.* **2009**, *86*, 215–223. [CrossRef]
11. Ekladious, I.; Colson, Y.L.; Grinstaff, M.W. Polymer–drug conjugate therapeutics: Advances, insights and prospects. *Nat. Rev. Drug Discov.* **2018**. [CrossRef]
12. Seju, U.; Kumar, A.; Sawant, K.K. Development and evaluation of olanzapine-loaded PLGA nanoparticles for nose-to-brain delivery: In vitro and in vivo studies. *Acta Biomater.* **2011**, *7*, 4169–4176. [CrossRef] [PubMed]
13. Mei, L.; Zhang, Y.; Zheng, Y.; Tian, G.; Song, C.; Yang, D.; Chen, H.; Sun, H.; Tian, Y.; Liu, K.; et al. A Novel Docetaxel-Loaded Poly (ε-Caprolactone)/Pluronic F68 Nanoparticle Overcoming Multidrug Resistance for Breast Cancer Treatment. *Nanoscale Res. Lett.* **2009**, *4*, 1530–1539. [CrossRef] [PubMed]
14. Nance, E.A.; Woodworth, G.F.; Sailor, K.A.; Shih, T.-Y.; Xu, Q.; Swaminathan, G.; Xiang, D.; Eberhart, C.; Hanes, J. A Dense Poly (Ethylene Glycol) Coating Improves Penetration of Large Polymeric Nanoparticles Within Brain Tissue. *Sci. Transl. Med.* **2012**, *4*, 119–149. [CrossRef] [PubMed]
15. Calvo, P.; Gouritin, B.; Chacun, H.; Desmaële, D.; D'Angelo, J.; Noel, J.-P.; Georgin, D.; Fattal, E.; Andreux, J.P.; Couvreur, P. Long-Circulating PEGylated Polycyanoacrylate Nanoparticles as New Drug Carrier for Brain Delivery. *Pharm. Res.* **2001**, *18*, 1157–1166. [CrossRef] [PubMed]
16. Rizvi, S.A.A.; Saleh, A.M. Applications of nanoparticle systems in drug delivery technology. *Saudi Pharm. J.* **2018**, *26*, 64–70. [CrossRef] [PubMed]
17. Suk, J.S.; Xu, Q.; Kim, N.; Hanes, J.; Ensign, L.M. PEGylation as a strategy for improving nanoparticle-based drug and gene delivery. *Adv. Drug Deliv. Rev.* **2016**, *99*, 28–51. [CrossRef]
18. Wilson, B.; Samanta, M.K.; Santhi, K.; Kumar, K.P.S.; Paramakrishnan, N.; Suresh, B. Poly(n-butylcyanoacrylate) nanoparticles coated with polysorbate 80 for the targeted delivery of rivastigmine into the brain to treat Alzheimer's disease. *Brain Res.* **2008**, *1200*, 159–168. [CrossRef]
19. Wilson, B.; Samanta, M.K.; Santhi, K.; Kumar, K.P.S.; Paramakrishnan, N.; Suresh, B. Targeted delivery of tacrine into the brain with polysorbate 80-coated poly(n-butylcyanoacrylate) nanoparticles. *Eur. J. Pharm. Biopharm.* **2008**, *70*, 75–84. [CrossRef]
20. Ma, Y.; Zheng, Y.; Zeng, X.; Jiang, L.; Chen, H.; Liu, R.; Huang, L.; Mei, L. Novel docetaxel-loaded nanoparticles based on PCL-Tween 80 copolymer for cancer treatment. *Int. J. Nanomed.* **2011**, *6*, 2679–2688. [CrossRef]
21. Kurakhmaeva, K.B.; Djindjikhashvili, I.A.; Petrov, V.E.; Balabanyan, V.U.; Voronina, T.A.; Trofimov, S.S.; Kreuter, J.; Gelperina, S.; Begley, D.; Alyautdin, R.N. Brain targeting of nerve growth factor using poly(butyl cyanoacrylate) nanoparticles. *J. Drug Target.* **2009**, *17*, 564–574. [CrossRef]
22. Wang, Y.; Wang, C.; Gong, C.; Wang, Y.; Guo, G.; Luo, F.; Qian, Z. Polysorbate 80 coated poly (ε-caprolactone)–poly (ethylene glycol)–poly (ε-caprolactone) micelles for paclitaxel delivery. *Int. J. Pharm.* **2012**, *434*, 1–8. [CrossRef] [PubMed]
23. Saraiva, C.; Praça, C.; Ferreira, R.; Santos, T.; Ferreira, L.; Bernardino, L. Nanoparticle-mediated brain drug delivery: Overcoming blood–brain barrier to treat neurodegenerative diseases. *J. Control. Release* **2016**, *235*, 34–47. [CrossRef] [PubMed]
24. Le Droumaguet, B.; Souguir, H.; Brambilla, D.; Verpillot, R.; Nicolas, J.; Taverna, M.; Couvreur, P.; Andrieux, K. Selegiline-functionalized, PEGylated poly(alkyl cyanoacrylate) nanoparticles: Investigation of interaction with amyloid-β peptide and surface reorganization. *Int. J. Pharm.* **2011**, *416*, 453–460. [CrossRef] [PubMed]
25. Gomes, M.J.; Fernandes, C.; Martins, S.; Borges, F.; Sarmento, B. Tailoring Lipid and Polymeric Nanoparticles as siRNA Carriers towards the Blood-Brain Barrier—From Targeting to Safe Administration. *J. Neuroimmune Pharmacol.* **2017**, *12*, 107–119. [CrossRef] [PubMed]

26. Gou, P.-F.; Zhu, W.-P.; Shen, Z.-Q. Synthesis, Self-Assembly, and Drug-Loading Capacity of Well-Defined Cyclodextrin-Centered Drug-Conjugated Amphiphilic A14B7 Miktoarm Star Copolymers Based on Poly(ε-caprolactone) and Poly(ethylene glycol). *Biomacromolecules* **2010**, *11*, 934–943. [CrossRef] [PubMed]
27. Gu, C.; Huang, N.; Xu, F.; Gao, J.; Jiang, D. Cascade exciton-pumping engines with manipulated speed and efficiency in light-harvesting porous π-network films. *Sci. Rep.* **2015**, *5*, 8867. Available online: https://www.nature.com/articles/srep08867#supplementary-information (accessed on 11 July 2019). [CrossRef] [PubMed]
28. Teixeira, J.; Oliveira, C.; Amorim, R.; Cagide, F.; Garrido, J.; Ribeiro, J.A.; Pereira, C.M.; Silva, A.F.; Andrade, P.B.; Oliveira, P.J.; et al. Development of hydroxybenzoic-based platforms as a solution to deliver dietary antioxidants to mitochondria. *Sci. Rep.* **2017**, *7*, 6842. [CrossRef]
29. Santillo, M.F.; Liu, Y.; Ferguson, M.; Vohra, S.N.; Wiesenfeld, P.L. Inhibition of monoamine oxidase (MAO) by β-carbolines and their interactions in live neuronal (PC12) and liver (HuH-7 and MH1C1) cells. *Toxicol. In Vitro* **2014**, *28*, 403–410. [CrossRef]
30. Van Breemen, R.B.; Li, Y. Caco-2 cell permeability assays to measure drug absorption. *Expert Opin. Drug Metab. Toxicol.* **2005**, *1*, 175–185. [CrossRef]
31. Jouan, E.; Le Vée, M.; Mayati, A.; Denizot, C.; Parmentier, Y.; Fardel, O. Evaluation of P-Glycoprotein Inhibitory Potential Using a Rhodamine 123 Accumulation Assay. *Pharmaceutics* **2016**, *8*, 12. [CrossRef]
32. Gao, X.; Wang, B.; Wei, X.; Rao, W.; Ai, F.; Zhao, F.; Men, K.; Yang, B.; Liu, X.; Huang, M.; et al. Preparation, characterization and application of star-shaped PCL/PEG micelles for the delivery of doxorubicin in the treatment of colon cancer. *Int. J. Nanomed.* **2013**, *8*, 971–982. [CrossRef] [PubMed]
33. Bensaid, F.; du Boullay, O.T.; Amgoune, A.; Pradel, C.; Reddy, L.H.; Didier, E.; Sable, S.; Louit, G.; Bazile, D.; Bourissou, D. Y-Shaped mPEG-PLA Cabazitaxel Conjugates: Well-Controlled Synthesis by Organocatalytic Approach and Self-Assembly into Interface Drug-Loaded Core-Corona Nanoparticles. *Biomacromolecules* **2013**, *14*, 1189–1198. [CrossRef] [PubMed]
34. Gamisans, F.; Lacoulonche, F.; Chauvet, A.; Espina, M.; Garcia, M.L.; Egea, M.A. Flurbiprofen-loaded nanospheres: Analysis of the matrix structure by thermal methods. *Int. J. Pharm.* **1999**, *179*, 37–48. [CrossRef]
35. Calvo, P.; Vila-Jato, J.L.; Alonso, M.J. Comparative in vitro evaluation of several colloidal systems, nanoparticles, nanocapsules, and nanoemulsions, as ocular drug carriers. *J. Pharm. Sci.* **1996**, *85*, 530–536. [CrossRef] [PubMed]
36. Guo, F.; Wu, F.; Mu, Y.; Hu, Y.; Zhao, X.; Meng, W.; Giesy, J.P.; Lin, Y. Characterization of organic matter of plants from lakes by thermal analysis in a N2 atmosphere. *Sci. Rep.* **2016**, *6*, 22877. [CrossRef] [PubMed]
37. Zhang, Y.; Tang, L.; Sun, L.; Bao, J.; Song, C.; Huang, L.; Liu, K.; Tian, Y.; Tian, G.; Li, Z.; et al. A novel paclitaxel-loaded poly (ε-caprolactone)/Poloxamer 188 blend nanoparticle overcoming multidrug resistance for cancer treatment. *Acta Biomater.* **2010**, *6*, 2045–2052. [CrossRef] [PubMed]
38. Sanchez-Lopez, E.; Egea, M.A.; Cano, A.; Espina, M.; Calpena, A.C.; Ettcheto, M.; Camins, A.; Souto, E.B.; Silva, A.M.; Garcia, M.L. PEGylated PLGA nanospheres optimized by design of experiments for ocular administration of dexibuprofen-in vitro, ex vivo and in vivo characterization. *Colloids Surf. B Biointerfaces* **2016**, *145*, 241–250. [CrossRef]
39. Panyam, J.; Williams, D.; Dash, A.; Leslie-Pelecky, D.; Labhasetwar, V. Solid-state solubility influences encapsulation and release of hydrophobic drugs from PLGA/PLA nanoparticles. *J. Pharm. Sci.* **2004**, *93*, 1804–1814. [CrossRef]
40. Hombreiro Pérez, M.; Zinutti, C.; Lamprecht, A.; Ubrich, N.; Astier, A.; Hoffman, M.; Bodmeier, R.; Maincent, P. The preparation and evaluation of poly (ϵ-caprolactone) microparticles containing both a lipophilic and a hydrophilic drug. *J. Control. Release* **2000**, *65*, 429–438. [CrossRef]
41. Silveira, N.; Longuinho, M.M.; Leitão, S.G.; Silva, R.S.; Lourenço, M.C.; Silva, P.E.; Maria do Carmo, F.R.; Abraçado, L.G.; Finotelli, P.V. Synthesis and characterization of the antitubercular phenazine lapazine and development of PLGA and PCL nanoparticles for its entrapment. *Mater. Sci. Eng. C* **2016**, *58*, 458–466. [CrossRef]
42. Deshmukh, R.K.; Naik, J.B. Aceclofenac microspheres: Quality by design approach. *Mater. Sci. Eng. C* **2014**, *36*, 320–328. [CrossRef] [PubMed]
43. Pinkerton, N.M.; Grandeury, A.; Fisch, A.; Brozio, J.; Riebesehl, B.U.; Prud'homme, R.K. Formation of Stable Nanocarriers by in Situ Ion Pairing during Block-Copolymer-Directed Rapid Precipitation. *Mol. Pharm.* **2013**, *10*, 319–328. [CrossRef] [PubMed]

44. Li, J.W.; Zhang, C.; Li, J.; Fan, L.; Jiang, X.G.; Chen, J.; Pang, Z.Q.; Zhang, Q.Z. Brain Delivery of NAP with PEG-PLGA Nanoparticles Modified with Phage Display Peptides. *Pharm. Res.* **2013**, *30*, 1813–1823. [CrossRef] [PubMed]
45. Shilo, M.; Motiei, M.; Hana, P.; Popovtzer, R. Transport of nanoparticles through the blood-brain barrier for imaging and therapeutic applications. *Nanoscale* **2014**, *6*, 2146–2152. [CrossRef] [PubMed]
46. Massella, D.; Celasco, E.; Salaün, F.; Ferri, A.; Barresi, A. Overcoming the Limits of Flash Nanoprecipitation: Effective Loading of Hydrophilic Drug into Polymeric Nanoparticles with Controlled Structure. *Polymers* **2018**, *10*, 1092. [CrossRef] [PubMed]
47. Fornaguera, C.; Dols-Perez, A.; Calderó, G.; García-Celma, M.J.; Camarasa, J.; Solans, C. PLGA nanoparticles prepared by nano-emulsion templating using low-energy methods as efficient nanocarriers for drug delivery across the blood–brain barrier. *J. Control. Release* **2015**, *211*, 134–143. [CrossRef] [PubMed]
48. Pfeiffer, C.; Rehbock, C.; Hühn, D.; Carrillo-Carrion, C.; de Aberasturi, D.J.; Merk, V.; Barcikowski, S.; Parak, W.J. Interaction of colloidal nanoparticles with their local environment: The (ionic) nanoenvironment around nanoparticles is different from bulk and determines the physico-chemical properties of the nanoparticles. *J. R. Soc. Interface* **2014**, *11*. [CrossRef] [PubMed]
49. Sanna, V.; Siddiqui, I.A.; Sechi, M.; Mukhtar, H. Resveratrol-Loaded Nanoparticles Based on Poly(epsilon-caprolactone) and Poly(D,L-lactic-co-glycolic acid)–Poly(ethylene glycol) Blend for Prostate Cancer Treatment. *Mol. Pharm.* **2013**, *10*, 3871–3881. [CrossRef] [PubMed]
50. Kamaly, N.; Yameen, B.; Wu, J.; Farokhzad, O.C. Degradable Controlled-Release Polymers and Polymeric Nanoparticles: Mechanisms of Controlling Drug Release. *Chem. Rev.* **2016**, *116*, 2602–2663. [CrossRef] [PubMed]
51. Sriamornsak, P.; Thirawong, N.; Weerapol, Y.; Nunthanid, J.; Sungthongjeen, S. Swelling and erosion of pectin matrix tablets and their impact on drug release behavior. *Eur. J. Pharm. Biopharm.* **2007**, *67*, 211–219. [CrossRef]
52. Riederer, P.; Laux, G. MAO-inhibitors in Parkinson's disease. *Exp. Neurobiol.* **2011**, *20*, 1–17. [CrossRef] [PubMed]
53. Schneider, L.; Giordano, S.; Zelickson, B.R.; Johnson, M.; Benavides, G.; Ouyang, X.; Fineberg, N.; Darley-Usmar, V.M.; Zhang, J. Differentiation of SH-SY5Y cells to a neuronal phenotype changes cellular bioenergetics and the response to oxidative stress. *Free Radic. Biol. Med.* **2011**, *51*, 2007–2017. [CrossRef] [PubMed]
54. Lopes, F.M.; Schröder, R.; da Frota Júnior, M.L.C.; Zanotto-Filho, A.; Müller, C.B.; Pires, A.S.; Meurer, R.T.; Colpo, G.D.; Gelain, D.P.; Kapczinski, F.; et al. Comparison between proliferative and neuron-like SH-SY5Y cells as an in vitro model for Parkinson disease studies. *Brain Res.* **2010**, *1337*, 85–94. [CrossRef] [PubMed]
55. Je, H.J.; Kim, E.S.; Lee, J.-S.; Lee, H.G. Release Properties and Cellular Uptake in Caco-2 Cells of Size-Controlled Chitosan Nanoparticles. *J. Agric. Food Chem.* **2017**, *65*, 10899–10906. [CrossRef] [PubMed]
56. Lu, H.; Chen, J.; Huang, H.; Zhou, M.; Zhu, Q.; Yao, S.Q.; Chai, Z.; Hu, Y. Iron modulates the activity of monoamine oxidase B in SH-SY5Y cells. *BioMetals* **2017**, *30*, 599–607. [CrossRef] [PubMed]
57. Shawahna, R.; Uchida, Y.; Declèves, X.; Ohtsuki, S.; Yousif, S.; Dauchy, S.; Jacob, A.; Chassoux, F.; Daumas-Duport, C.; Couraud, P.-O.; et al. Transcriptomic and Quantitative Proteomic Analysis of Transporters and Drug Metabolizing Enzymes in Freshly Isolated Human Brain Microvessels. *Mol. Pharm.* **2011**, *8*, 1332–1341. [CrossRef] [PubMed]
58. Geldenhuys, W.; Mbimba, T.; Bui, T.; Harrison, K.; Sutariya, V. Brain-targeted delivery of paclitaxel using glutathione-coated nanoparticles for brain cancers. *J. Drug Target.* **2011**, *19*, 837–845. [CrossRef]
59. Stanimirovic, D.B.; Bani-Yaghoub, M.; Perkins, M.; Haqqani, A.S. Blood–brain barrier models: In vitro to in vivo translation in preclinical development of CNS-targeting biotherapeutics. *Expert Opin. Drug Discov.* **2015**, *10*, 141–155. [CrossRef]
60. Veszelka, S.; Tóth, A.; Walter, F.R.; Tóth, A.E.; Gróf, I.; Mészáros, M.; Bocsik, A.; Hellinger, É.; Vastag, M.; Rákhely, G.; et al. Comparison of a Rat Primary Cell-Based Blood-Brain Barrier Model with Epithelial and Brain Endothelial Cell Lines: Gene Expression and Drug Transport. *Front. Mol. Neurosci.* **2018**, *11*, 166. [CrossRef]
61. Artursson, P.; Palm, K.; Luthman, K. Caco-2 monolayers in experimental and theoretical predictions of drug transport1PII of original article: S0169-409X(96)00415-2. The article was originally published in Advanced Drug Delivery Reviews 22 (1996) 67–84.1. *Adv. Drug Deliv. Rev.* **2001**, *46*, 27–43. [CrossRef]

62. Weksler, B.; Romero, I.A.; Couraud, P.-O. The hCMEC/D3 cell line as a model of the human blood brain barrier. *Fluids Barriers CNS* **2013**, *10*, 16. [CrossRef] [PubMed]
63. Fernandes, C.; Pinto, M.; Martins, C.; Gomes, M.J.; Sarmento, B.; Oliveira, P.J.; Remião, F.; Borges, F. Development of a PEGylated-Based Platform for Efficient Delivery of Dietary Antioxidants Across the Blood–Brain Barrier. *Bioconj. Chem.* **2018**, *29*, 1677–1689. [CrossRef] [PubMed]
64. Sugano, K.; Kansy, M.; Artursson, P.; Avdeef, A.; Bendels, S.; Di, L.; Ecker, G.F.; Faller, B.; Fischer, H.; Gerebtzoff, G.; et al. Coexistence of passive and carrier-mediated processes in drug transport. *Nat. Rev. Drug Discov.* **2010**, *9*, 597–614. [CrossRef] [PubMed]
65. Win, K.Y.; Feng, S.S. Effects of particle size and surface coating on cellular uptake of polymeric nanoparticles for oral delivery of anticancer drugs. *Biomaterials* **2005**, *26*, 2713–2722. [CrossRef] [PubMed]
66. Murugan, K.; Choonara, Y.E.; Kumar, P.; Bijukumar, D.; du Toit, L.C.; Pillay, V. Parameters and characteristics governing cellular internalization and trans-barrier trafficking of nanostructures. *Int. J. Nanomed.* **2015**, *10*, 2191–2206. [CrossRef]
67. Deli, M.A. Potential use of tight junction modulators to reversibly open membranous barriers and improve drug delivery. *Biochim. Biophys. Acta Biomembr.* **2009**, *1788*, 892–910. [CrossRef] [PubMed]
68. Bhavna Md, S.; Ali, M.; Baboota, S.; Sahni, J.K.; Bhatnagar, A.; Ali, J. Preparation, characterization, in vivo biodistribution and pharmacokinetic studies of donepezil-loaded PLGA nanoparticles for brain targeting. *Drug Dev. Ind. Pharm.* **2014**, *40*, 278–287. [CrossRef]
69. Zhu, Q.; Song, W.; Xia, D.; Fan, W.; Yu, M.; Guo, S.; Zhu, C.; Gan, Y. A poly-l-glutamic acid functionalized nanocomplex for improved oral drug absorption. *J. Mater. Chem. B* **2015**, *3*, 8508–8517. [CrossRef]
70. Poller, B.; Gutmann, H.; Krähenbühl, S.; Weksler, B.; Romero, I.; Couraud, P.-O.; Tuffin, G.; Drewe, J.; Huwyler, J. The human brain endothelial cell line hCMEC/D3 as a human blood-brain barrier model for drug transport studies. *J. Neurochem.* **2008**, *107*, 1358–1368. [CrossRef]
71. Amin, M.L. P-glycoprotein Inhibition for Optimal Drug Delivery. *Drug Target Insights* **2013**, *7*, 27–34. [CrossRef]
72. Werle, M. Natural and Synthetic Polymers as Inhibitors of Drug Efflux Pumps. *Pharm. Res.* **2008**, *25*, 500–511. [CrossRef] [PubMed]
73. Shen, Q.; Lin, Y.; Handa, T.; Doi, M.; Sugie, M.; Wakayama, K.; Okada, N.; Fujita, T.; Yamamoto, A. Modulation of intestinal P-glycoprotein function by polyethylene glycols and their derivatives by in vitro transport and in situ absorption studies. *Int. J. Pharm.* **2006**, *313*, 49–56. [CrossRef] [PubMed]
74. Friche, E.; Jensen, P.; Sehested, M.; Demant, E.J.; Nissen, N.N. The solvents Cremophor EL and Tween 80 modulate daunorubicin resistance in the multidrug resistant Ehrlich ascites tumor. *Cancer Commun.* **1990**, *2*, 297–303. [PubMed]
75. Hugger, E.D.; Audus, K.L.; Borchardt, R.T. Effects of Poly(ethylene glycol) on Efflux Transporter Activity in Caco-2 Cell Monolayers. *J. Pharm. Sci.* **2002**, *91*, 1980–1990. [CrossRef] [PubMed]
76. Netsomboon, K.; Laffleur, F.; Suchaoin, W.; Bernkop-Schnürch, A. Novel in vitro transport method for screening the reversibility of P-glycoprotein inhibitors. *Eur. J. Pharm. Biopharm.* **2016**, *100*, 9–14. [CrossRef] [PubMed]
77. Tian, X.-H.; Lin, X.-N.; Wei, F.; Feng, W.; Huang, Z.-C.; Wang, P.; Ren, L.; Diao, Y. Enhanced brain targeting of temozolomide in polysorbate-80 coated polybutylcyanoacrylate nanoparticles. *Int. J. Nanomed.* **2011**, *6*, 445–452. [CrossRef]

MDPI
St. Alban-Anlage 66
4052 Basel
Switzerland
www.mdpi.com

Pharmaceutics Editorial Office
E-mail: pharmaceutics@mdpi.com
www.mdpi.com/journal/pharmaceutics